THE STABILITY OF THE SOLAR SYSTEM AND OF SMALL STELLAR SYSTEMS

INTERNATIONAL ASTRONOMICAL UNION
UNION ASTRONOMIQUE INTERNATIONALE

SYMPOSIUM No. 62

(COPERNICUS SYMPOSIUM I)

HELD AT WARSAW, POLAND, SEPTEMBER 5–8, 1973

THE STABILITY OF THE SOLAR SYSTEM AND OF SMALL STELLAR SYSTEMS

EDITED BY

Y. KOZAI

Tokyo Astronomical Observatory, Mitaka, Tokyo, Japan

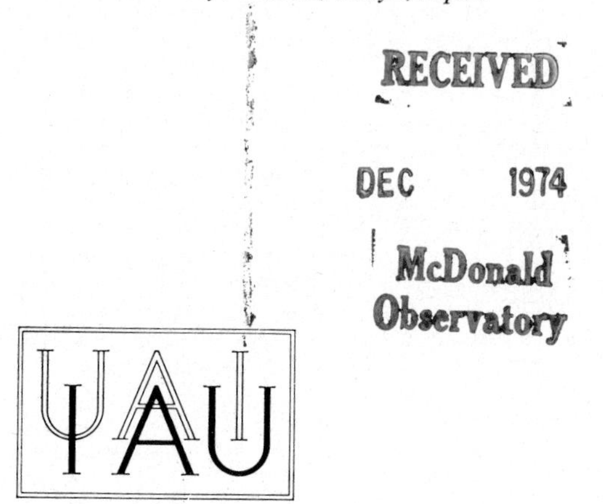

D. REIDEL PUBLISHING COMPANY

DORDRECHT-HOLLAND / BOSTON-U.S.A.

1974

Published on behalf of
the International Astronomical Union and
the International Union of Theoretical and Applied Mechanics
by
D. Reidel Publishing Company, P.O. Box 17, Dordrecht, Holland

Sold and distributed in the U.S.A., Canada, and Mexico
by D. Reidel Publishing Company, Inc.
306 Dartmouth Street, Boston,
Mass. 02116, U.S.A.

Library of Congress Catalog Card Number 74–76475

Cloth edition: ISBN 90 277 0458 9
Paperback edition: ISBN 90 277 0459 7

Printed in The Netherlands by D. Reidel, Dordrecht

TABLE OF CONTENTS

PREFACE

The IAU Symposium No. 62, 'The Stability of the Solar System and of Small Stellar Systems' was held in Warsaw in Poland during the Extraordinary General Assembly of the IAU in commemoration of the 500th anniversary of the birth of Nicolaus Copernicus. The Symposium was sponsored by Commission 7 (Celestial Mechanics) and cosponsored by Commissions 4 (Ephemerides) and 37 (Star Clusters and Associations) of the IAU and by IUTAM. The Organizing Committee included Y. Kozai (Chairman), J. A. Agekjan, A. Deprit, G. N. Duboshin, S. Gąska (Local representative), M. Hénon, B. Morando and C. Parkes (IUTAM representative).

The Symposium was supported financially by the IAU, the IUTAM and the Polish Academy of Sciences.

Y. KOZAI
Chairman of the Organizing Committee

STABILITY THEORY IN CELESTIAL MECHANICS

J. MOSER

Courant Institute of Mathematical Sciences, New York University, New York, N.Y. 10012, U.S.A.

Abstract. This expository lecture surveys recent progress of the stability theory in Celestial Mechanics with emphasis on the analytical problems. In particular, the old question of convergence of perturbation series are discussed and positive results obtained, in the light of the work by Kolmogorov Arnold and Moser. For the three body problem, classes of quasi-periodic solutions and doubly asymptotic (or homoclinic) orbits are discussed.

1. Introduction

Over the years celestial mechanics has presented scientists with a variety of difficult problems challenging to astronomers, numerical analysts and mathematicians. One of the most puzzling questions is related to the stability of the solar system and the long time behavior of the solutions of the n-body problem which has been attacked again and again over the last hundred years. The analytical difficulties of these problems has to do with the presence of the notorious 'small divisors'. In the last decade definite progress has been made in this direction and the main goal of this lecture is to report on some of these advances and their significance for celestial mechanics. Speaking as a mathematician the emphasis will naturally be on the theoretical aspects and much remains to be done in the applications of these results to the realistic problems.

We want to point out that the recent work has led to a new concept of stability. It is the conventional view that one can delineate some 'blobs' or open regions in phase space in which the solutions remain bounded or stable, while outside of such regions the solutions escape or behave unboundedly. A sharp mathematical formulation and related results were developed by Liapounov. His work refers to dissipative systems, however, and is not applicable to the conservative systems of celestial mechanics. The recent mathematical work in this area has shown that for Hamiltonian systems this crude picture has to be replaced by another model: One finds complicated Cantor sets, which we may compare with a sponge, in which the solutions are stable and bounded for all time while the solutions lying in the many fine holes of the sponge may gradually seep out and become unstable. The filament of these holes is connected and give rise to a slow diffusion while the majority of the solutions belong to the solid part of the sponge consisting of stable solutions.

From a physical point of view this model is obviously hard to accept, but one cannot escape these conclusions if one idealizes the problem mathematically and studies the motion for *all* time and not only for a reasonable finite time interval. In fact, the idealization goes further: We are not talking about the motion of the planets under realistic forces but of rigorous solutions of the n-body problem taking into account only Newton's force law and referring to mass points with some smallness

Y. Kozai (ed.), The Stability of the Solar System and of Small Stellar Systems, 1–9.

restrictions on the masses. However, all these restrictions are less severe since the theory to be discussed allows for small additional forces as long as they are conservative and time independent, as well as for spherical homogeneous masses.

The complicated structure of the stability region is due to resonance phenomena which occur between the various frequencies $\omega_1, \omega_2, \ldots, \omega_n$. We speak of resonance, whenever one has a relation $\sum_{k=1}^{n} j_k \omega_k = 0$ with integer $j_1, j_2, \ldots, j_n$ which do not all vanish. While for practical purposes only resonances of small order matter, i.e. when $0 < |j_1| + |j_2| + \cdots + |j_n|$ is reasonably small, for the mathematical problem of long time behavior *all* these resonances affect the motion over long time. But even the definition of frequency requires the knowledge of the solution over all time, and a frequency is really defined by a limit for $t \to \infty$. A solution, oscillating for a very long time, but finally escaping, can at most approximately, but never rigorously, be assigned frequencies. On the other hand the above definition of resonances requires the exact knowledge of frequencies and we find ourselves in a vicious circle.

The way out of this dilemma is to give up the *initial value problem* which asks for the long time behavior of a solution for given initial data and instead to ask for solutions which are almost periodic with prescribed frequencies. Geometrically this requires that the solutions lie on a torus in phase space and gives rise to a *boundary value problem.* Historically, the expansion methods described by H. Poincaré are precisely based on such a boundary value problem, the boundary condition expressing the periodicity of the unknown function.

Therefore we begin with the classical approach to approximate the orbits by perturbation series in terms of trigonometric expansions. Poincaré's work put these techniques on a firm foundation, but it came as a disappointment to many that his investigations indicated that these series expansions were divergent. In fact, this was the starting point of 'asymptotic expansions' which have proved so useful in fluid dynamics and other fields of mathematical physics. It is perhaps the most remarkable recent discovery that some of these classical series expansions in celestial mechanics are actually convergent and give rise to a rigorous description of solutions of the n-body problem valid for all time. The convergence proof is based on new techniques in functional analysis which are due to Kolmogorov, Arnold and the author (Moser, 1973*). We will describe more precisely the type of expansions whose convergence can be established and explain why this result is not in contradiction to Poincaré's assertion about divergence.

2. Convergence of Perturbation Series

A. VARIABLE FREQUENCY EXPANSION

The main idea of the classical expansion techniques is *not* to solve in a straightforward way the initial value problem of integrating the equations with prescribed initial data. Instead we seek such initial conditions for which the solutions are pe-

* We refer to this booklet for further information.

riodic or quasi-periodic, i.e. remain on a torus in phase space. This leads to a boundary value problem – mathematically speaking – which has very different features from the initial value problem. From the practical point of view this amounts to the construction of a reference orbit.

Consider a Hamiltonian,

$$H(x, y, \varepsilon) = \sum_{\nu=0}^{\infty} \varepsilon^{\nu} H^{(\nu)}(x, y),$$

where $H^0(x, y) = F(x)$ independent of y, $H(x, y, \varepsilon)$ of period 2π in $y_1 \dots y_m$, and real analytic in x, y and ε, for $x \in D$, all real y and $|\varepsilon| < \varepsilon_0$. Roughly, the goal of Lindstedt's method is to find

$$S(\xi, y, \varepsilon) = \sum_{\nu=0}^{\infty} \varepsilon^{\nu} S^{(\nu)}(\xi, y),$$

such that

$$H(S_y, y, \varepsilon) = \Phi(\xi, \varepsilon)$$

is independent of η and where S_ξ and S_y must be real analytic and of period 2π in $y_1 \dots y_m$ and

$$\det S_{\xi y} \neq 0.$$

A solution to this problem would mean that the canonical transformation $(x, y) \to \to (\xi, \eta)$,

$$\eta = S_\xi, \qquad x = S_y,$$

leads to a Hamiltonian independent of η, showing that the system is integrable for all small $|\varepsilon|$. It is well known that this is not possible, in general, and we require only that

$$H(S_y, y, \varepsilon) = \Phi(\xi, \varepsilon) + O(|\xi|^2); \qquad \Phi(\xi, \varepsilon) = \Phi_0(\varepsilon) + \sum_{k=1}^{n} \Phi_k \xi_k, \tag{1}$$

so that the new Hamiltonian, say $\tilde{\Phi}(\xi, \eta, \varepsilon)$, satisfies

$$\tilde{\Phi}_\xi = \Phi_\xi, \qquad \tilde{\Phi}_\eta = 0 \quad \text{for} \quad \xi = 0,$$

and, thus

$$\xi = 0, \qquad \eta = \Phi_\xi(0, \varepsilon)\, t + \eta(0)$$

gives rise to particular solutions of the system. These solutions are called quasi-periodic, with frequencies

$$\omega_k(\varepsilon) = \Phi_{\xi_k}(0, \varepsilon), \quad (k = 1, \dots, n),$$

which are, in general, dependent on ε.

We describe three types of standard expansion methods: (i) a variable frequency expansion, (ii) a fixed frequency expansion, and (iii) a fixed frequency ratio expansion.

For (i) assume that x^0 can be chosen so that the quantities,

$$\omega_k = F_{x_k}(x^0),$$

are rationally independent, and set

$$\begin{aligned} &S^{(0)}(\xi, y) = (x^0 + \xi, y), \\ &S^{(\nu)}(\xi, y), \quad \text{period } 2\pi \text{ in } y \text{ and linear in } \xi \text{ for } \nu \geqslant 1. \end{aligned} \tag{2}$$

Inserting the series of S into Equation (1) we get by comparison of coefficients of ε^ν,

$$(F_x(x^0 + \xi), S_y^{(\nu)}) = \Phi^{(\nu)} + g(\xi, y) + O(|\xi|^2),$$

where $g(\xi, y)$ is a known function, determined by $S^{(0)}, \ldots, S^{(\nu-1)}$, $\Phi^{(0)}, \ldots, \Phi^{(\nu-1)}$. Separating the coefficients of ξ_k by setting

$$g = g_0(y) + \sum_{k=1}^{n} g_k(y)\,\xi_k + O(|\xi|^2), \qquad S^{(\nu)} = S_0^{(\nu)}(y) + \sum_{k=1}^{n} S_k^{(\nu)}(y)\,\xi_k,$$

we get

$$\begin{aligned} &(\omega, S_{0y}^{(\nu)}) = \Phi_0^{(\nu)} + g_0(y), \\ &(\omega, S_{ky}^{(\nu)}) + (F_{x_k x}(x^0),\ S_{0y}^{(\nu)}) = \Phi_k^{(\nu)} + g_k(y). \end{aligned} \tag{3}$$

It is well known how to solve such equations for the functions $S_k^{(\nu)}$ and constants $\Phi_k^{(\nu)}$ $(k=0, 1, \ldots, n)$ with the help of Fourier series, provided the frequencies $\omega = (\omega_1, \ldots, \omega_n)$ are not only rationally independent but also satisfy a condition of the type,

$$|(j, \omega)| \geqslant c\,|j|^{-\mu}, \tag{4}$$

for all integer vectors $j = (j_1, \ldots, j_m) \neq 0$ with some positive constants c and μ. We have to check that the mean value of the right-hand side vanishes, which is achieved by choice of $\Phi_0^{(\nu)}$ and $\Phi_k^{(\nu)}$, thus determining $\Phi(\xi, \varepsilon) = \Phi_0 + \sum_k \Phi_k \xi_k$ in (1). The resulting expansion approximates a motion with frequencies,

$$\omega_k(\varepsilon) = \Phi_k(\varepsilon), \quad (k = 1, \ldots, n),$$

which generally depends on ε. Therefore we refer to it as variable frequency expansion. Incidentally, S is determined up to an additive function $\sigma(\xi, \varepsilon)$ independent of y, and S can be normalized uniquely by the requirement that

$$\int_{T^n} (S - S^0)\,\mathrm{d}y = 0.$$

This type of expansion* was described in Poincaré's *Méthodes nouvelles de la mécanique céleste*, Vol. 2, and attributed to Lindstedt.

* An article by G. A. Krasinsky contained in the Russian book *Minor Planets* (edited by N. S. Samoylova-Yakhontova, Nauka, Moscow, 1973), in Chap. VI, Section 1, contains a similar expansion. The author is grateful to Professor Krasinsky for pointing out this reference.

B. FIXED FREQUENCY EXPANSION

The above expansion has the shortcoming that the irrationality conditions (4) get destroyed for $\varepsilon \neq 0$. But it is easy to modify the above procedure so as to have the frequencies ω_k independent of ε; in fact, such a procedure was also described in Poincaré's *Méthodes nouvelles de la mécanique céleste*, Vol. 2. One simply replaces S by

$$S + \sum_{k=1}^{n} \alpha_k(\varepsilon)\, y_k = S + (\alpha, y); \qquad \alpha(0) = 0.$$

Then the above procedure is the same except that $S^{(\nu)}$ has to be replaced by $S^{(\nu)} + (\alpha^{(\nu)}, y)$, so that Equations (3) take the form

$$\begin{aligned} &(\omega, S_{0y}^{(\nu)}) + (\omega, \alpha^{(\nu)}) = \Phi_0^{(\nu)} + g_0(y), \\ &(\omega, S_{ky}^{(\nu)}) + (F_{x_k x}(x^0),\ S_{0y}^{(\nu)} + \alpha^{(\nu)}) = g_k(y), \end{aligned} \tag{5}$$

where we set $\Phi_k^{(\nu)} = 0$ to assure that the frequencies are independent of ε. To make the mean values equal on both sides we use the constants $\alpha^{(\nu)} = (\alpha_1^{(\nu)}, \ldots, \alpha_n^{(\nu)})$ in the last n equations, assuming that

$$\det(F_{xx}(x^0)) \neq 0,$$

and $\Phi_0^{(\nu)}$ in the first equation. Then the above equation can again be solved uniquely if we require

$$\int_{T^n} S^{(\nu)}\, dy = 0, \quad \text{for} \quad \nu = 1, 2, \ldots, \tag{6}$$

yielding the fixed frequency expansion.

C. FIXED FREQUENCY RATIO EXPANSION

Actually, in order to preserve the irrationality conditions for small ε it suffices to keep $\omega_k(\varepsilon) = \lambda(\varepsilon)\, \omega_k(0)$ proportional to themselves, so that the frequency ratio is independent of ε. We use the extra freedom to make

$$\Phi(0, \varepsilon) = F(x^0),$$

independent of ε, ensuring that the resulting solutions lie on a fixed energy surface $H(x, y, \varepsilon) = F(x^0)$. To obtain such an expansion we set

$$\Phi_0^{(\nu)} = 0; \qquad \Phi_k^{(\nu)} = \lambda^{(\nu)} \omega_k, \quad \text{for} \quad \nu = 1, 2, \ldots.$$

The resulting equations for $S^{(\nu)}$, $\alpha^{(\nu)}$ and $\lambda^{(\nu)}$ are, instead of (5), of the form

$$\begin{aligned} &(\omega, S_{0y}^{(\nu)}) + (\omega, \alpha^{(\nu)}) = g_0, \\ &(\omega, S_{ky}^{(\nu)}) + (F_{x_k x}(x^0), S_{0y}^{(\nu)} + \alpha^{(\nu)}) = \lambda^{(\nu)} \omega_k + g_k. \end{aligned}$$

To balance the $n+1$ mean values of these equations we use the $n+1$ constants

$\alpha_1^{(\nu)}, \ldots, \alpha_n^{(\nu)}$ and $\lambda^{(\nu)}$, and need the assumption that

$$\det \begin{pmatrix} 0 & \omega \\ \omega^T & F_{xx}(x^0) \end{pmatrix} \neq 0.$$

Under this assumption again $S^{(\nu)}$, $\alpha^{(\nu)}$ and $\lambda^{(\nu)}$ are uniquely determined if normalized by (6).

These expansion procedures have been known since the last century. Also the solvability of Equations (3) or (5) at each step has been known for a long time under an irrationality condition (4) showing that all coefficients of ε^ν are well defined by convergent Fourier series. However, the convergence question of such series for small positive $|\varepsilon|$ is a much more difficult problem as it relates to a nonlinear system of differential equations. The results of the work by Kolmogorov, Arnold and the author imply, in fact, that the fixed frequency expansion and the fixed frequency ratio expansions both converge for $|\varepsilon| < \varepsilon_0$ where ε_0 is a positive number depending on the given Hamiltonian and the constants c and μ in (4), but not on ω itself. Thus these expansions do not only yield an approximation for solutions useful for long time intervals but the existence of solutions valid for all real time which can be used as reference orbits. On the other hand the variable frequency expansion can, in general, not be expected to converge because of the violation of the nonresonance conditions (4) for arbitrary small $|\varepsilon|$. Thus the crucial feature for the convergence is that the nonresonance conditions are maintained under perturbation. At first such conditions can be imposed only on the formal expansion; the convergence of the resulting series ensures that these formal expansions actually belong to existing solutions. Incidentally, the convergence proof does not succeed with the conventional majorant method but a new rapidly convergent iteration process. A posteriori, one can identify the resulting solutions with those obtained by formal expansion (Moser, 1967). Presently the relevant existence theorems are being derived from an abstract implicit function theorem in Banach spaces (Zehnder, 1974).

3. Applications

The above results, and their extensions, have a great many applications in celestial mechanics. We list a few:

(a) The plane restricted three-body problem. It is possible to give bounds for the eccentricity valid for all time, i.e. given $\delta > 0$ one can establish for the eccentricity ε,

$$\varepsilon(0) - \delta \leqslant \varepsilon(t) \leqslant \varepsilon(0) + \delta \quad \text{for all} \quad t,$$

if the mass ratio μ is small enough. The periodic solutions of the first kind can be shown to be stable for small μ if resonances of order $\leqslant 3$ are avoided. The stability of the equilibria L_4 and L_5 can be established, provided one has linear stability and excludes 3 more exceptional values for μ.

More generally, one finds stability criteria for equilibria and periodic solutions for systems of two degrees of freedom.

(b) For systems of higher degrees of freedom such stability criteria have not been found except in the trivial case where the Hamiltonian is positive or negative definite. However, another weaker concept of 'stability in measure' can be verified: An equilibrium solution is called stable in measure if for any $\varepsilon>0$ any neighborhood (in phase space) contains an open set V, containing the equilibrium, and a measurable set $S\subset V$ invariant under the flow such that

$$m(V-S)<\varepsilon m(V).$$

Here m denotes the Lebesgue or Liouville measure in phase space. In other words, most solutions starting in V will remain in V for all time, while no assertion is made about the exceptional set $V-S$. This is decidedly a weaker concept of stability than the usually accepted one due to Liapounov where it is required that $S=V$. But it has the advantage that one can give a stability criterion in terms of inequalities on finitely many coefficients of the Hamiltonian. Moreover, one can choose the set S in such a way that all solutions in S are given by quasi-periodic solutions obtained by the series expansions described above.

Of course, there are a great number of different stability concepts but it is a basic requirement of stability that the entire orbit changes little if the initial data are changed by a small amount. In addition, it is of importance that the orbit changes little under small perturbation of the system of the differential equation, at least if one restricts oneself to the class of Hamiltonian systems of differential equations. This latter requirement reflects simply that the differential equations, or some parameters in them, are not known with absolute precision but only to a limited degree of accuracy. To put this into more mathematical terms, we should like to aim at a stability concept in which one can predict the orbit within narrow bounds for initial data in an open set V in phase space and all Hamiltonian systems in an open set $\mathscr{W}$ of Hamiltonian functions in an appropriate topology. The latter condition would be satisfied, for example, if our desired stability criterion imposes only finitely many inequalities on finitely many derivatives of the Hamiltonian and is independent of the 'tail' of the expansion.

To summarize the result presently known: There are such stability criteria for Hamiltonian systems of $n\leqslant 2$ degrees of freedom, but for $n\geqslant 3$ degrees of freedom it is at present not possible to give a stability criterion of the above type except in the trivial case of an equilibrium with definite Hamiltonian. But one does have such criteria for stability in measure.

(c) The results are applicable to the three-body problem and yield Cantor sets of positive measure containing quasi-periodic motions, provided two masses are much smaller than the third mass. The solution constructed are close to circular orbits with zero inclination and a large ratio of the major axes. Thus the solutions in the set S are bounded for all time and never experience collisions. Since S has positive measure one concludes that the flow of the three-body problem is not ergodic. On the other hand the complement of S is a connected set and solutions arbitrarily close to one in S may become unbounded. This result is due to Arnold (1963) who also considered generalizations to the n-body problem for $n>3$.

4. Random Motions, open Problems

The weakened type of stability concept discussed above brings up the question about the behavior of the solution in the exceptional sets which are frequently called regions of instability. Although little is known about these solutions there are results showing the presence of orbits of very erratic behavior. In fact, their long time behavior is topologically related to that of the shift of doubly infinite sequences. Such orbits for the restricted three-body problem were discussed by Sitnikov (1960) and Alekseev (1970). This topic is discussed at length by Moser (1973) and we forego a further discussion here. Instead we list a few problems showing how little is known about the n-body problem, in spite of the progress stated.

(1) Although the existence theorems mentioned above give a finite range for the perturbation, say, $|\varepsilon| < \varepsilon_0$, in which quasi-periodic solutions can be found this range seems to be much too pessimistic. Numerical calculations (especially by Hénon and his collaborators) indicate that a realistic range is several magnitudes larger. So far this phenomenon is not understood. In particular, what is the dependence of ε_0 on n, the number of degrees of freedom?

(2) Can one find stability criteria for generic Hamiltonian systems of $n \geqslant 3$ degrees of freedom in the sense of Liapounov, i.e. without exceptional regions? For example, is there a criterion for stability of an equilibrium solution in terms of finitely many coefficients of the Taylor expansion of the Hamiltonian? At present the expectation is that this question has a negative answer which would justify the use of the weaker concept of stability in measure.

(3) Are there solutions of the n-body problem which become unbounded as t approaches a finite time t^*? Sundman's study showed that this is impossible for $n \leqslant 3$ but it has not been excluded that such a solution exists for $n \geqslant 4$.

Acknowledgement

This work was partially supported by the U.S. Air Force, Grant AFOSR-71-2055.

References

Alekseev, V. M.: 1970, *Actes Congrès Int. Math.* **2**, 893–907, Gauthier-Villars, Paris.
Arnold, V. I.: 1963, *Sup. Math. Nauk* **18**, No. 6 (114), 91–192.
Moser, J.: 1967, *Math. Ann.* **169**, 136–76.
Moser, J.: 1973, *Stable and Random Motions in Dynamical Systems*, Princeton University Press, Annals of Mathematics Studies 77.
Sitnikov, K.: 1960, *Dokl. Akad. Nauk U.S.S.R.* **133**, No. 2, 303–06.
Zehnder, E.: 1974, *Bull. Am. Math. Soc.* **80**, No. 1, 174–179.

DISCUSSION

D. Saari: I would like to comment on your last problem. As you know, if noncollision singularities do exist in the n-body problem, then n is greater than 3.

J. Moser: Yes, indeed.

J. Palous: Since an infinite number of rational numbers exist we may conclude that an infinite number of resonant periodic orbits exist. Does this mean the existence of ergodic orbits?

J. Moser: Yes. There exist resonance regions for each rational number in some interval, and for each of these regions the complicated motion modelled by the sequence shift will, in general, take place. Unfortunately, the set of these orbits form a set of measure zero. Whether there exist generic systems having a region of positive measure of ergodicity and also quasi-periodic motion is not known.

L. Perek: With a student of mine, we got interested in studying in some detail the boundary of ergodic orbits in a problem of stellar dynamics. Instead of finding a simple boundary we found a very complicated outline of many lobes.

J. Moser: Thank you.

MODERN DYNAMICAL SYSTEMS THEORY

L. MARKUS

University of Minnesota, Minneapolis, Minn. 55455, U.S.A., and University of Warwick, Coventry, Warwickshire CV 4 7AU, U.K.

Abstract. In order to analyse generic or typical properties of dynamical systems we consider the space $\mathscr{V}$ of all C^1-vector fields on a fixed differentiable manifold M. In the C^1-metric, assuming M is compact, $\mathscr{V}$ is a complete metric space and a generic subset is an open dense subset or an intersection of a countable collection of such open dense subsets of $\mathscr{V}$. Some generic properties (i.e. specifying generic subsets) in $\mathscr{V}$ are described. For instance, generic dynamic systems have isolated critical points and periodic orbits each of which is hyperbolic. If M is a symplectic manifold we can introduce the space $\mathscr{H}$ of all Hamiltonian systems and study corresponding generic properties.

1. Copernican Astronomy as a Natural System

Nicolaus Copernicus designed his model of the Solar system to achieve the greatest simplicity, within his physical and philosophical axiomatic framework. He sought to eliminate the unnecessary hypotheses of the Ancients concerning coincidences and correlations of planetary orbits by explaining these orbits in terms of his heliocentric kinematics. He wrote, "I correlate all the movements of the other planets and their spheres or orbital circles with the mobility of the Earth."

Much current research in the mathematical theory of dynamical systems proceeds in this same spirit – to eliminate the special situations depending on coincidences and correlations and to concentrate on the mathematically typical or generic cases. Of course, what is considered typical depends on the range of possibilities permitted. For instance, Copernicus allowed only uniform circular motions; but within the framework of Kepler–Newton geometry all circular orbits would be discarded as ungeneric ellipses.

The Copernican model of the Solar system consists on the central star Sol surrounded by a family of planets in concentric circular orbits. Of course, Copernicus perturbed the basic circular orbits by epicycles, and Kepler and Newton later introduced more intricate elliptical perturbations, but the fundamental conceptual picture remains that of the Sun encircled by its family of planets.

A different conceptual framework for the Solar system views this astronomical system as basically a double star, with components of Sol and Jupiter moving in almost circular orbits around their common center of mass. More detailed structure includes Saturn as a third star (or proto-star gas ball), and further the gas balls Uranus and Neptune, each moving in ever large circular orbits about the system mass center. Moreover each of these stars carries its own planetary system – namely Mercury, Venus, Earth and Mars about Sol, the Jovian satellites about Jupiter, and the other corresponding satellite systems accordingly placed.

Whether we accept the Copernican model or the multi-star model of the Solar system is a philosophical choice rather than a scientific decision in our current state

Y. Kozai (ed.), The Stability of the Solar System and of Small Stellar Systems, 11–18.

of astronomical knowledge. In fact, the two physical models are kinematically, even dynamically equivalent – they differ only in the psychological emphasis arising from the renaming of Jupiter as a star rather than a planet. In some future era when there exists a comprehensive theory of planetary and stellar evolution and structure, it may be possible to select one of these two physical models as the prefered and established description. In terms of such an evolutionary stellar theory one of the above models might appear as 'more natural, simpler, and more typical' within the framework of the theory. If the Copernican model is vindicated then the evolutionary theory would have explained why there should be two types of solar planets, inner terrestrial planets and outer gaseous giants. If the multi-star model becomes accepted, then the evolutionary theory would be expected to predict that most stars develop planetary systems in a natural way. But until a coherent astronomical theory of planetary systems is developed, there is no way of selecting one of these models as simpler or more natural than the other.

The purpose of this philosophical discussion on the methodology of mathematical astronomy has been to underline the importance of the concept of natural or generic properties of dynamical systems. Accordingly much of the research of the past decade in the mathematical theory of ordinary differential equations has concentrated on the discovery and examination of such generic properties, with the attempt to discard ungeneric systems displaying any special 'coincidences and correlations'.

2. Fundamentals of Global Differentiable Dynamics

In order to present an exposition of the modern theory of generic dynamical systems, as conducted by many pure and applied mathematicians during the past decade, we shall briefly review some of the fundamental concepts of global differential geometry and analysis.

We consider dynamical systems described by vector differential equations of first order, say

$$\mathrm{d}x^i/\mathrm{d}t = v^i(x^1, x^2, \ldots, x^n), \quad i = 1, 2, \ldots, n,$$

since any higher order differential equations can be reduced to this form. For simplicity of exposition we take time-independent or autonomous differential systems. But we emphasize that there is no supposition that these dynamical systems are conservative, or even arise from any Newtonian mechanical problem. At the end of this paper we shall mention some quite recent results that pertain to conservative Hamiltonian dynamics, but at present we make no such assumptions.

In the theory of local dynamical systems we study differential equations in an open subset of the real number space R^n; whereas in global dynamics the space is a general differentiable manifold M^n. We specify a global dynamical system as a tangent vector field v on M^n; and in any local chart $(x^1, \ldots, x^n)$ on M^n we denote the dynamical system v by its components $v^i(x^1, \ldots, x^n)$, say

$$(v) \qquad \mathrm{d}x^i/\mathrm{d}t = v^i(x^1, \ldots, x^n), \quad i = 1, 2, \ldots, n,$$

or

(v) $\dot{x} = v(x)$.

There are two basic motivations for studying differential systems on general differentiable manifolds.

(i) *Mathematical motivation.* It is of interest to study differential systems within the most general context for which the concepts of differentiation and mathematical analysis are meaningful. Thus we take the ambiant space M^n to be of arbitrary finite dimension n, and locally differentiably equivalent to R^n. That is, M^n is a differentiable n-manifold (separable, metrizable C^∞-manifold without boundary), for instance R^n, or the n-sphere S^n, or the n-torus T^n.

(ii) *Physical motivation.* Physical dynamical systems are often described by first-order vector differential equations involving the displacements, velocities, angles, and other generalized coordinates. If the generalized coordinates are unrestricted real variables, then the space in which the system evolves in time is some real vector space R^n. However, frequently the generalized coordinates are restricted by constraint or energetic equalities, or account for some angular periodicities of the physical configuration, and in these cases the space of the system is some differentiable manifold M^m.

For instance, an ordinary planar pendulum has a configuration space of a circle S^1, and a velocity-phase space of a product cylinder $S^1 \times R^1$. A spherical pendulum has a configuration space of a sphere S^2, and a velocity-phase space that is the tangent bundle TS^2. The configuration space of a rigid rotor is the rotation matrix group $SO(3)$, which is diffeomorphic to the real projective space P^3, and the velocity-phase space is the 6-manifold TP^3 (which incidently is the product $P^3 \times R^3$).

Besides the greater generality and applicability, the main advantages of the global viewpoint for dynamics are:

(i) The notation and methodology of global differential geometry (involving manifolds, vector fields, trajectory curves, etc.) are highly suited to the requirements of the problems of dynamics. Old problems can be carefully phrased and solved, and new problems and concepts are suggested. For example, a careful discussion of the spherical pendulum requires knowledge that the tangent bundle TS^2 is not the product $S^2 \times R^2$. New concepts of structural stability and genericity arise naturally when various dynamical systems are compared globally.

(ii) The global viewpoint emphasizes the unified family of all the trajectories as a portrait of a given dynamical system, rather than singling out special trajectories by their initial data. This is particularly important in physical systems where we wish to classify and compare the diverse modes of asymptotic behavior of the trajectories of the system.

Let M^n be a differentiable manifold and let v be a tangent C^r-vector field on M^n. Then in overlapping charts $(x^1, \ldots, x^n)$ and $(\bar{x}^1, \ldots, \bar{x}^n)$ on an open set of M^n, the components of v are,

(v) $\dot{x}^i = v^i(x)$ and $\dot{\bar{x}}^i = \bar{v}^i(\bar{x})$, where the contravariant vector transformation law holds for the C^r-functions $(r=1, 2, \dots, \infty)$ v^i and $\bar{v}^i$,

$$\bar{v}^i = (\partial \bar{x}^i / \partial x^j)\, v^j, \quad (\text{sum on } j), \quad i=1, \dots, n.$$

A solution or trajectory of this vector field or differential system v is a C^1-curve,

$$I \to M^n : t \to P_t, \quad (I \text{ open interval in } R),$$

whose tangent vector at each point coincides with the vector of the field v.

The usual local existence, uniqueness, regularity results are valid. That is, for each initial point $P_0 \in M^n$ there exists a unique trajectory P_t of v through P_0 at $t=0$ (and defined on some maximal time duration I). If M^n is compact (and we shall assume this henceforth for simplicity of exposition), the maximal interval I is all R. In this case the solutions of v define a C^r-flow or action of R on M^n. That is, there is a C^r-map,

$$\Phi : R \times M^n \to M^n : (t, P_0) \to P_t,$$

and for fixed $t \in R$,

$$\Phi_t : P_0 \to P_t,$$

is a C^r-diffeomorphism of M^n onto itself, and the group property holds for all times t_1, $t_2 \in R$,

$$\Phi_{t_1} \circ \Phi_{t_2} = \Phi_{t_1 + t_2}, \qquad \Phi_0 = \text{Identity}.$$

Every trajectory of v on M^n is either
(i) a point P_0,
(ii) a C^r-diffeomorphic image of a circle S^1,
(iii) a bijective regular C^r-differentiable image of a line R^1.

The case (i) is a critical point where v vanishes at P_0, case (ii) corresponds to a periodic solution or closed orbit, but case (iii) can lead to curves that are not topological lines, say for an irrational flow on the torus T^n.

An invariant set Σ of v on M^n is a subset that is the union of whole trajectories of v. That is, a subset $\Sigma \subset M^n$ is invariant for the flow of v in case each trajectory initiating in Σ remains forever in Σ. Of course, a critical point or a closed orbit is necessarily an invariant set. Also, for each initial point P_0 in the compact manifold M^n, the past (negative) and future (positive) limit set of the trajectory P_t

$$\alpha(P_0) = \bigcap_{\tau < 0} \overline{\bigcup_{t < \tau} P_t} \quad \text{and} \quad \omega(P_0) = \bigcap_{\tau > 0} \overline{\bigcup_{t > \tau} P_t},$$

are each compact connected invariant sets. Clearly $\alpha(P_0)$ and $\omega(P_0)$ depend only on the trajectory P_t and not on the initial point P_0. If $P_0 \in \omega(P_0)$ then P_t is called future recurrent or Poisson stable, and P_0 is recurrent if it is both past and future recurrent. A rather weaker property is regional recurrence of P_0 – namely each neighborhood U of P_0 in M^n has a trajectory U_t that meets U for some arbitrarily large past and future times. The set Ω of all regionally recurrent points, usually called the nonwandering set, is a compact invariant set containing all critical points, periodic orbits, and recurrent trajectories of the dynamical system v.

An invariant set Σ for the flow v in M^n is called future stable in case: for each neighborhood W of Σ in M^n there exists a subneighborhood $W_1 \subset W$ such that $P_0 \in W_1$ implies that $P_t \in W$ for all future times $t > 0$. If in addition $\omega(P_0) \subset \Sigma$ then the set Σ is future asymptotically stable (and analogues statements hold for past times).

We illustrate these concepts of recurrence and stability by some examples of invariant sets for flows in vector spaces and cylinders.

Example 1. $\dot{x} = Ax$ for $x \in R^n$ and A real constant matrix. If the matrix A is nonsingular, with complex eigenvalues $\lambda_1, \lambda_2, \ldots, \lambda_n$ not zero, then the origin $x=0$ is the unique critical point.

If no eigenvalue of A is pure imaginary, that is $\operatorname{Re} \lambda_j \neq 0$, then the origin $x=0$ is called a hyperbolic critical point. Define the attractor (stability) set of all points $P \in R^n$ for which $\omega(P)=0$, and similarly the repellor (instability) set by $\alpha(P)=0$. Then the attractor set is a linear subspace whose dimension equals the number of eigenvalues whose real parts are negative. Also $x=0$ is asymptotically stable just in case the attractor space is all R^n.

Now consider a C^1-differential system v on a differentiable manifold M^n. Let P_0 be a critical point for v and take a local chart (x) centered at P_0 to write the differential system

$$(v) \qquad \dot{x} = Ax + \cdots.$$

We define P_0 to be a hyperbolic critical point of v in case no eigenvalue of A is pure imaginary. In this case we define the attractor and repellor sets and prove that these are each C^1-differentiable submanifolds of M^n, in fact they are each regular bijective images of vector spaces of the appropriate dimensions.

Example 2. $\dot{x} = Ax$ and $\dot{\theta} = 1$, where $x \in R^{n-1}$ and $\theta \in S^1$. Here the manifold $M^n = R^{n-1} \times S^1$ is a cylinder. There are no critical points but the circle, $x=0$, $0 \leqslant \theta < 1$, is a periodic orbit σ. The Poincaré map around this periodic orbit σ is given by $x_0 \to e^A x_0$, and its eigenvalues $\mu_1, \ldots, \mu_{n-1}$ are the (nontrivial) characteristic multipliers, and $\mu_n = 1$.

If none of the characteristic multipliers has a modulus of unity, $|\mu_j| \neq 1$ for all $1 \leqslant j < n-1$, then the periodic orbit is defined to be hyperbolic. Again we define the attractor set by $\omega(P) \subset \sigma$ and the repellor set by $\alpha(P) \subset \sigma$, and it is clear that these are each cylinders of dimensions specified by the moduli of the characteristic multipliers.

Now consider a C^1-differential system v on a differentiable manifold M^n. Let σ be a periodic orbit of v with Poincaré map of a transversal section yielding the characteristic multipliers $\mu_1, \ldots, \mu_{n-1}$. If all $|\mu_j| \neq 1$ then σ is a hyperbolic periodic orbit. Again the attractor and repellor sets are C^1-submanifolds which are regular bijective images of either cylinders or generalized nonorientable Mobius bands, depending on the moduli and arguments of the complex characteristic multipliers.

3. Generic Properties of Dynamical Systems

Let M^n be a compact differentiable manifold. Denote by $\mathscr{V}$ the set of all dynamical systems, that is, C^1-vector fields on M^n. In order to make precise our ideas of perturbation and approximation of dynamical systems in $\mathscr{V}$ we specify the C^1-topology in $\mathscr{V}$. Namely, two vector fields u and v in $\mathscr{V}$ are nearby one another in case their vector components, and also the corresponding first partial derivatives, are nearly equal,

$$|u^i(x) - v^i(x)| < \varepsilon,$$

and

$$\left|\frac{\partial u^i}{\partial x^j} - \frac{\partial v^i}{\partial x^j}\right| < \varepsilon, \quad i, j = 1, \ldots, n.$$

Here the components are expressed in some fixed finite collection of local charts covering M^n. Define the metric distance $\|u - v\|_1$ between the vector fields u and v in $\mathscr{V}$ as the infimum of all such bounds $\varepsilon > 0$ for which the above inequalities hold globally on M^n. Then it is known that $\mathscr{V}$ is a complete metric space, in fact a Banach space as defined by that distinguished Polish mathematician of the past generation.

For the study of perturbations of differential systems the open sets of $\mathscr{V}$ are important. Recall that a set $\mathscr{O}$ is open in $\mathscr{V}$ in case any small perturbation of a member of $\mathscr{O}$ always yields a member of $\mathscr{O}$. For approximation theory of differential systems the dense sets of $\mathscr{V}$ are important. Namely, a set $\mathscr{D}$ is dense in $\mathscr{V}$ in case each member of $\mathscr{V}$ can be approximated by members of $\mathscr{D}$.

We define a generic set $\mathscr{G}$ of $\mathscr{V}$, or refer to a generic property specifying $\mathscr{G}$, when $\mathscr{G}$ is a countable intersection of open and dense subsets of $\mathscr{V}$. It is a standard theorem of Baire that a generic set of a complete metric space $\mathscr{V}$ is dense in $\mathscr{V}$. We think of a generic subset $\mathscr{G}$ of $\mathscr{V}$ as comprising almost all the members of $\mathscr{V}$, with the complement $\mathscr{V} - \mathscr{G}$ being a negligible collection of differential systems. With this terminology we can now state some results concerning the generic properties of dynamical systems.

DEFINITION. Let $\mathscr{V}$ be the metric space of all C^1-vector fields on a compact differentiable manifold M^n.

Define the subset $\mathscr{G}_1$ of $\mathscr{V}$ to consist of all differential systems v in $\mathscr{V}$ for which: $v \in \mathscr{G}_1$ has only a finite number of critical points and each of these is hyperbolic.

Define the subset $\mathscr{G}_2$ of $\mathscr{V}$ to consist of all differential systems v in $\mathscr{V}$ for which: $v \in \mathscr{G}_2$ has, for each integer $N \geqslant 1$, only a finite number of periodic orbits with period less than N, and each of these orbits is hyperbolic.

THEOREM. $\mathscr{G}_1$ is open and dense in $\mathscr{V}$. $\mathscr{G}_2$ is generic in $\mathscr{V}$.

We remark that it is then immediate that $\mathscr{G}_1$, as well as the intersection $\mathscr{G}_1 \cap \mathscr{G}_2$, is generic in $\mathscr{V}$. In this sense almost all differential systems in $\mathscr{V}$ have isolated and hyperbolic critical points and periodic orbits.

4. Generic Properties of Hamiltonian Dynamical Systems

Locally we specify a Hamiltonian differential system,

$$\frac{dx^i}{dt}=\frac{\partial H}{\partial y_i}, \qquad \frac{dy_i}{dt}=-\frac{\partial H}{\partial x^i}, \quad i=1,\ldots,n,$$

by a real differentiable Hamiltonian function $H(x^1,\ldots,x^n,y_1,\ldots,y_n)$ in a real number space R^{2n}. A differentiable coordinate transformation $(x,y)\to(\bar{x},\bar{y})$ allows us to write the differential system in the Hamiltonian format,

$$\frac{d\bar{x}^i}{dt}=\frac{\partial H}{\partial \bar{y}_i}, \qquad \frac{d\bar{y}_i}{dt}=-\frac{\partial H}{\partial \bar{x}^i}, \quad i=1,\ldots,n,$$

where $H(\bar{x},\bar{y})=H(x(\bar{x},\bar{y}),y(\bar{x},\bar{y}))$, in case the transformation is canonical (or symplectic). That is, the Jacobian matrix $T=\partial(\bar{x},\bar{y})/\partial(x,y)$ satisfies everywhere the symplectic condition,

$$TJT^t=J, \quad \text{with} \quad J=\begin{pmatrix} 0 & I \\ -I & 0\end{pmatrix}.$$

This discussion shows that a global theory of nonlinear autonomous Hamiltonian systems can be formulated on a differentiable manifold M^{2n} which is covered by a family of canonical charts interrelated by canonical coordinate transformations. Such a manifold M^{2n}, with the family of canonical charts, is called a symplectic manifold. Each real differentiable function H on a symplectic manifold M^{2n} specifies a C^1-vector field on M^{2n} which has the required Hamiltonian format in each canonical chart (x,y) of M^{2n}.

Let M^{2n} be a compact symplectic manifold and denote by $\mathscr{H}$ the set of all real differentiable functions on M^{2n}. Use the C^∞-topology on $\mathscr{H}$, wherein two Hamiltonian functions H_1 and H_2 are nearby in case the values of a finite set of partial derivatives are approximately equal everywhere on M^{2n} (normalize each H in $\mathscr{H}$ to have a minimum value of zero on M^{2n}, since augmenting H by a constant does not modify the corresponding Hamiltonian differential system). In this case $\mathscr{H}$ is a complete metric space and we can seek generic Hamiltonian systems in $\mathscr{H}$. Note that $\mathscr{H}$ itself is negligible in the space $\mathscr{V}$ of all vector fields on M^{2n}, but we agree to compare vector fields only within the fixed space $\mathscr{H}$ in order to find generic sets of $\mathscr{H}$.

The mathematical theory of generic Hamiltonian systems of $\mathscr{H}$ on a compact symplectic manifold M^{2n} is only in the preliminary stages of study and research. Some of the major mathematical discoveries of the past few decades can be formulated in these terms, but many problems remain unsolved. We close this short review by listing several properties which are valid for a generic set of Hamiltonian in $\mathscr{H}$.

(1) Only finitely many critical points, each with nonzero eigenvalues. But there exists a critical point with all eigenvalues pure imaginary, and rationally independent in the usual sense.

(2) Noncountably many periodic orbits. But only a countable number of degenerate periodic orbits (having more than two characteristic multipliers of unity).

(3) Noncountably many almost periodic orbits each dense in some toral submanifold.

References

Markus, L.: 1971, *Lectures in Differentiable Dynamics*, Regional Conference Series in Mathematics No. 3.

Markus, L. and Meyer, K.: 1974, *Mem. Am. Math. Soc.* (in press).

Robinson, R. C.: 1971, *Lectures on Hamiltonian Systems*, Instituto de Matematica Pura e Aplicado, Rio de Janeiro, Brazil.

Siegel, C. L. and Moser, J.: 1971, *Lectures on Celestial Mechanics*, Springer Verlag, New York.

Smale, S.: 1967, *Bull. Am. Math. Soc.* **73**, 747.

THE PRESENT STATE OF THE *n*-BODY PROBLEM

H. POLLARD
Purdue University, Ind., U.S.A.

The development of the n-body problem has occurred along four lines:

Determination of collision singularities. The main results here are due to von Zeipel–Sperling, and Pollard–Saari who have found necessary and sufficient conditions that singularities be due to collisions.

Behavior of the system as $t \to \infty$. The main results are due to Pollard and more especially Saari, and are concerned with the long behavior of a system unimpaired by singularities.

The topology of the solution surfaces. Smale has examined the differential equations by modern topological methods. The results which are satisfactory in the case of planar motion, still need development in the general case.

Numerical results. The results, due to a number of investigators, are far from definitive.

Y. Kozai (ed.), The Stability of the Solar System and of Small Stellar Systems, 19.

BODE'S LAW

M. LECAR

Center for Astrophysics, Cambridge, Mass. 02138, U.S.A.

M. W. Ovenden (*Nature* **239** (1972) 508) has outlined a theory intended to provide a dynamical explanation for Bode's law. The author suggests instead that the approximately constant spacing ratio expressed in Bode's mnemonic can be generated by a sequence of random numbers subject to the constraint that adjacent planets cannot be too close to each other.

The full text of the paper appeared in *Nature* **242** (1973) 318.

DISCUSSION

R. H. Miller: You should be able to get almost any distribution you want by choosing the radial distribution of gas mass (the initial angular momentum distribution) of the right form.

S. F. Dermott: I have recently analyzed the statistical significance of Bode's law (*Nature* **244** (1973) 17) and found that if it is allowed that ratios of adjacent orbital periods are not too close to unity then it is not too difficult to fit random periods into a Bode-type law.

D. C. Heggie: If it were true that Bode's law, obtained by plotting log a, were accidental, surely one should obtain equally successful Bode's laws by plotting a itself.

S. F. Dermott: It is not disputed that the planetary orbits are nonrandom on a linear scale. Bode's law effectively states that the orbits are nonrandom on a logarithmic scale. This is a far more interesting postulate, but I consider it to be untrue.

Y. Kozai (ed.), The Stability of the Solar System and of Small Stellar Systems, 21.

DISTRIBUTION OF THE MEAN MOTIONS OF PLANETS AND SATELLITES AND THE DEVELOPMENT OF THE SOLAR SYSTEM

H. JEHLE

Center for Theoretical Physics, Department of Physics and Astronomy, University of Maryland, College Park, Md. 20742, U.S.A., and Physics Department, George Washington University, Washington, D.C. 20006, U.S.A.

Abstract. In this paper the point of view is taken that the distribution of orbital elements in the solar system should be discussed first on a purely gravitational basis, i.e. on the basis of a set of particles entirely under gravitational interaction, before hydromagnetic and other effects are taken in consideration too. One might indeed assume that there has been a time in the history of the solar system from when on hydromagnetic and gas laws ceased to play an important role in comparison to gravity. In the epoch since that time the solar system might have developed from a set of a large number of smaller particles into the present solar system by way of transitions which these particles made to preferential orbital elements, and by accretion. Means had been found to handle the development of this set of particles under gravitational interaction, by defining the set appropriately in terms of a statistical distribution. In considering the problem of the evolution of the solar system, such a gravitational approach, which was encouraged by Einstein, seems the reasonable first step.

Y. Kozai (ed.), The Stability of the Solar System and of Small Stellar Systems, 23.

CAN THE SOLAR SYSTEM BE QUANTIZED?

J. M. BARNOTHY

833 Lincoln Street, Evanston, Ill., U.S.A.

Abstract. It is suggested that the general form of the constant of quantization, K in Schrödinger's equation, is not $h/2\pi$, but $K=2s\alpha^{-k}$, with s being the spin of the orbiting object, α the fine structure constant (1/137.0361), and k a small positive integer, or zero. For atoms $k=0$; for planets and satellites $k=2$, 3 or 4; for the solar system as a whole, revolving around the center of the Galaxy, $k=6$. The probability that 16 objects of the solar system would follow this quantum rule by chance alone is 1 in 10^{16}, suggestive that quantum mechanics, as we know it today, can be seen as a special case of a more general quantum mechanics of the future; it also supports the view expressed by Dirac, that h is probably not a fundamental constant.

Section 1 contains the basic idea which induced me to undertake an investigation of a relationship between rotational and orbital angular momenta of planets; Sections 2–7 contain the experimental data, the application of the new quantum rule and the statistical evaluation whether the relationship proposed in Section 1, could have occurred by chance alone. The results obtained in Sections 2–7 are noteworthy in themselves, independently whether the basic idea is accepted or not.

February 1973 is the quinquecentennial anniversary of the birth of Copernicus, who recognized that the planets revolve around the Sun, and in doing so revolutionized the premises of all subsequent astronomy and cosmology. The year 1973 also marks the 60 th anniversary of Bohr's discovery of his atomic model: electrons revolve on Keplerian orbits around a nucleus, like planets around the Sun. Is there a further physical law common to both?

1. A New Quantum Rule

Let us ask the question what would have happened to quantum theory if the electron spin (s_e) had been known in 1900? Let us take a ride in a time machine capable of rearranging the sequence of events.

We start with Goudsmith and Uhlenbeck's discovery of the electron spin. Millikan and Einstein show that in the photoelectric effect light is absorbed in quanta, each quantum carrying an energy $4\pi s_e \nu$. At the next stop of our journey we find Planck saying that oscillators do not emit light continuously but in quanta, with energies as given by Einstein's equation. Obviously a light quantum sometimes behaves as a wave, and other times as a corpuscle.

Then the Broglie recognizes that since nature seems to have a preference for symmetries, if light behaves in a dualistic manner, then particles should also behave at times undulatorily; whereby, in analogy with light quanta, particle waves should also exist with wavelength equal to the ratio of 4π times the spin of the particle to its momentum. Next Bohr explains the discrete set of Balmer lines, by postulating that electrons revolving around the nucleus do not emit radiation as long as they move on orbits, along which the de Broglie waves are standing waves. This means, of course, that the

Y. Kozai (ed.), The Stability of the Solar System and of Small Stellar Systems, 25–35.

orbital angular momentum of the electron must always be an integer multiple of the electron spin. Unfortunately Heisenberg shows that the simultaneous observation of conjugate quantities always involves an error equal to the electron spin, hence it is impossible to observe Bohr's orbits. Schrödinger saves the situation by setting up a partial differential equation, leading for negative energies to a discrete set of Eigenvalues. He achieves this by substituting in the Hamiltonian a new variable $S = K \ln \psi$. To remain in agreement with observations, he sets the numerical value of K equal to twice the spin of the electron.

We have now covered the events leading up to Schrödinger without the need to mention h, that is, Planck's constant. This curious circumstance raised in the '40 s the heretical thought (Barnothy 1946) in me that h is probably not a fundamental constant of nature, but a constant, the value of which changes in accordance with the spin of the orbiting object. I found that the general formula for the constant K in Schrödinger's equation would then be:

$$K = 2s_x \alpha^{-k}, \tag{1}$$

where s_x is the spin of the oscillating or revolving body, k is a small positive integer or zero. To remain in agreement with observations, the constant α has to be set equal to the fine structure constant, $\frac{1}{137}$.

In the special case of the atom, $k=0$; and since $s_x = s_e$, $K = h/2\pi$ as assumed by Schrödinger. But in the case of the Earth $k=3$, and s_x is the spin of the Earth, consequently 'Planck's constant' according to which the orbit of the Earth should be quantized is 1.7×10^{45} cgs. Since the Earth is in its lowest quantum state, its de Broglie wavelength equals the circumference of the ecliptic. It is conceivable that when the Earth was formed, it occupied a higher quantum state, and by releasing energy in the form of gravitational waves it 'jumped' to its present 'ground state'. I do not wish as yet to speculate further on a possible gravitational wave spectrum emitted from newly formed planetary systems in the universe.

2. Application to the Planetary System

The formula to compute n, the quantum number of the planets, is:

$$n = \frac{\text{orbital angular momentum}}{2 \text{ rotational spin}} \alpha^k = (t/T)(R/r)^2 \times 137.04^{-k} (2\tau \cos\delta)^{-1}, \tag{2}$$

where r and t are radius and period of rotation (in tropical days) of the planet; T and R the orbital period and distance from the Sun, respectively; $\tau = I/Mr^2$ is the moment of inertia factor; and δ the inclination of the equator to the orbital plane.

No actual reason can be given for the occurrence of the factor α in Equations (1) and (2). But we should remember that the fine structure constant appears unexplanably in many places in physics. Some theoretical physicists believe that the fine structure constant may have something to do with geometrical calculations. Such an approach was suggested by Barnothy (1947) and recently by Wyler (1969). Barnothy's approach to the problem is based on an application of Friedmann's field equations to nuclear forces, leading to a geometrical structure of elementary particles, which in turn permits the exact calculation of dimensionless numbers such as α, gyromagnetic ratios, and the mass ratio of proton and electron. Wyler's approach is purely geometrical and compares the volume elements of a seven-dimensional group having five real and two imaginary timelike dimensions, with those of the subgroup of the five real dimensions.

Table I shows the quantum levels of the nine planets. Three different k values were

TABLE I

Quantum levels of major planets

	k	n	$\Delta[n]$
Mercury	4	4.15	+0.15
Venus [a]	4	0.69	−0.31
Earth	3	0.97	−0.03
Mars	3	3.83	−0.17
Jupiter	2	1.30	−0.30
Saturn	2	3.14	+0.14
Uranus [a]	3	0.99	−0.01
Neptune	3	0.29	–
Pluto	4	0.91	−0.09

[a] Retrograde rotation.

needed. Among the nine planets, only Neptune does not fit into the scheme. When two planets occupy the same n and k quantum numbers – e.g. Pluto and Venus, and again Earth and Uranus – one of them happens to have retrograde rotation, suggestive that Pauli's exclusion principle might be applicable.

The probability to find among nine planets eight with integer n values [1], [2], [3] and [4] by chance alone is 1 in 10^8. (For the computations see Section 5.) Such a small probability suggests that Equation (1) may in fact be a generally valid formula to compute the factor of quantization (K) in Schrödinger's equation.

This in turn leads to four further inferences:

(1) It is possible to quantize macroscopic rotators.

(2) Planck's constant is not a fundamental physical constant.

(3) The uncertainty relation depends on what kind of signal carriers we use. Photons give very blurred pictures of Bohr's orbits, but very sharp pictures of the orbits of the planets. Gravitational waves, on the other hand, would give very hazy pictures of the planetary system, and what one could say would merely be that the probability to find the Earth somewhere around the Sun is proportional to ψ^2. This means that model-like descriptions should not be dismissed as worthless: they are complementary to statistical mathematical descriptions. Nature appears dualistic even in its recognition patterns (Barnothy, 1947).

(4) Should the above considerations be correct, their effect on present quantum mechanics would be similar to Einstein's special relativity on Newtonian mechanics, current quantum mechanics being a special case of a more general quantum mechanics of the future.

3. Observational Data

The orbital parameters of the planets and satellites are rather precisely known, while radii, periods of rotation, and moment of inertia factors (τ) are occasionally uncertain. For the latter the best values as reported in Dollfus (1970), *Surfaces and Interiors of Planets and Satellites*, were used. In the following I refer to this book as *SIPS*, indicating authors and page number. The used values are listed in Table III.

According to Plagemeann's (1965) thermal calculations (see also Majeva, 1969) the interior of Mercury was never molten, hence a coreless model, with uniform distribution of metallic iron must be preferred (Levin in *SIPS*, p. 481), with a moment of inertia factor $\tau = 0.388$, at a radius of $r = 2432 \pm 7$ km (Dollfus in *SIPS*, p. 136). The sidereal period of rotation of Mercury is 58.65 days (Dyce in *SIPS*, p. 160), $\frac{2}{3}$ times its orbital period. One solar day on Mercury (which we shall call the tropical period of rotation of the planet or satellite) is $t = (t_s^{-1} - T^{-1})^{-1} = 176.01$ days.

The three-zone model of Venus, consisting of a rock crust, a metallized silicon layer and a metallic iron core is probably correct (Levin in *SIPS*, *p.* 493 and 495). We adopt $\tau = 0.341$ and $r = 6052 \pm 6$ km (Dollfus in *SIPS*, p. 136). The recent radar observations of Shapiro (1967) give a sidereal period of rotation in astounding agreement with the Earth–Venus resonance value of 243.16 days. Considering the retrograde rotation of Venus, this corresponds to a tropical period of rotation of $t = (t_s^{-1} + T^{-1})^{-1} = 116.78$ days.

The moment of inertia of the Earth is well established from mass distribution models and from its hydrostatic flattening, yielding $\tau = 0.3309$.

Several models have been developed for Mars, but each of them was fitted to the observed flattening, hence they cannot yield additional information regarding the moment of inertia factor of this planet. The observations of Phobos and Deimos yield $\tau = 0.375$ (Levin in *SIPS*, p. 487).

No reliable models are available for Jupiter, Saturn, Uranus and Neptune. I have computed their moment of inertia factor (τ) as for Mars from the observed hydrostatic flattening (f) and the mass (M) of the planet, using the equation:

$$\tau = \tfrac{2}{3}\left[1 - \tfrac{2}{5}(5m/2f - 1)^{1/2}\right], \tag{3}$$

where $m = 4\pi^2 r^3 / G t_s^2 M$ is the ratio of centrifugal to gravitational accelerations on the equator of the planet. Such a computation of the moment of inertia factor around the polar axis is based on the assumption that the surface of the planet is an equipotential surface. Table II lists the used equatorial and polar radii, the f and M and the resulting τ values.

The flattening of Uranus computed from the optically observed equatorial and polar diameters leads to an impossibly small τ value. Cook (1972) believes that a flattening between $f=4.5\times10^{-2}$ and 5.5×10^{-2} would be reasonable. Kovalevsky (in *SIPS*, p.32) suggests a flattening of $\frac{1}{18}$. I have accepted this latter value in my computations.

The difference between the equatorial and the polar radii of Neptune is not known. The coefficient of the second harmonic in the gravitational potential of Neptune has been derived from the motion of the satellite Triton, with the result $J_2=5\times10^{-3}$ (Kovalevsky in *SIPS*, p. 39). If the surface of the planet is assumed to be equipotential,

TABLE II

Moment of inertia factor

	r_e (km)	r_p (km)	f	M (10^{24} g)	τ
Jupiter	70850 [a]	66550 [a]	0.0607	1900 [b]	0.233
Saturn	60000 [a]	53450 [a]	0.1092	569.3 [b]	0.213
Uranus	25400 [a]	24700 [a]	0.0555	87.7 [c]	0.263
Neptune	25225 [d]		0.0217 [b]	102.9 [e]	0.264

[a] Dollfus in *SIPS*, p. 136.
[b] Kovalevsky in *SIPS*, pp. 31, 39.
[c] Klepcynski *et al.*, 1970.
[d] Kovalevsky and Link, 1969.
[e] Seidelmann *et al.*, 1969.

the polar flattening is given, to first order, by the formula $f=\frac{3}{2}J_2+\frac{1}{2}m$, whence the flattening is 2.17×10^{-2}.

Among all the planets Pluto has the most eccentric orbit, which moreover intercepts the orbit of Neptune. This is one of the reasons to regard Pluto as an escaped satellite. Its size and density, however, place it in the class of terrestrial planets. Recently all available transit circle observations of Neptune, from its discovery in 1846 to the present, have been reexamined by Seidelmann *et al.* (1971). Pluto's mass was found to be $(0.107\pm0.001)\,M_\oplus$ very close to that of Mars. We adopt, therefore, for the moment of inertia factor of Pluto the same value $\tau=0.375$ as for Mars, and take its radius to 3200 km, the value used by Seidelmann *et al.*

We may assume that the very small asteroids behave as rigid bodies, whence τ reaches for spherically shaped objects its maximal value of $\frac{2}{5}=0.40$. Only two asteroids, Ceres and Vesta, could be included in Table III. The period of rotation of Pallas and the diameter of Juno and Eros are too uncertain; for the others only orbital data are known.

The moment of inertia factor of the Moon was very exactly determined to $0.3906\pm\pm0.0003$ from the orbits of the many artificial satellites which during the last years orbited the Moon (Tolson *et al.*, 1967). The same τ value was adopted for all the other satellites.

The resulting quantum numbers k and n of 9 planets, 2 asteroids and 8 satellites are listed in the last columns of Table III.

TABLE III

Quantum levels of major and minor planets and satellites

	t (days)	T (days)	r_e (km)	R (10^8 km)	τ	δ (degree)	k	n	$\Delta[n]$
Mercury	176.01	87.96	2432 ± 7	0.579	0.388	0	4	4.15	+0.15
Venus [a]	116.78	224.69	6052 ± 6	1.082	0.341	1.2	4	0.69	−0.31
Earth	1	365.24	6370 ±	1.496	0.3309	23.45	3	0.97	−0.03
Mars	1.027	686.95	3402 ± 8	2.279	0.375	24.86	3	3.83	−0.17
Jupiter	0.4097	4332	70850 ± 100	7.782	0.233	3.083	2	1.30	+0.30
Saturn	0.4264	10796	60000 ± 240	14.30	0.213	26.81	2	3.14	+0.14
Uranus [a]	0.450	30662	25402 ± 280	28.68	0.263	98	3	0.99	−0.01
Neptune	0.658	59863	25225 ± 450	44.81	0.264	27	3	0.29	–
Pluto	6.375	90324	3200 ?	58.93	0.375	?	4	0.91	−0.09
Ceres	0.378	1682	385 ± 20	4.141	0.40	?	4	0.92	−0.08
Vesta	0.223	1326	210 ± 17	3.532	0.40	?	4	1.69	−0.31
E. Moon	29.53	365	1739 ± 5	1.496	0.3906	~0	4	2.17	+0.17
J. Io	1.77	4332	1750 ± 75	7.782	0.39	3	4	0.29	–
Europe	3.55	4332	1550 ± 75	7.782	0.39	3	4	0.75	−0.25
Ganymede	7.14	4332	2775 ± 65	7.782	0.39	3	4	0.47	–
Callisto	16.75	4332	2500 ± 75	7.782	0.39	3	4	1.36	+0.36
S. Titan	15.96	10796	2425 ± 150	14.30	0.39	27	4	2.10	+0.10
Iapetus	53.15	10796	575 ± 50	14.30	0.39	27	5	0.92	−0.08
N. Triton [b]	5.88	59863	1885 ± 750	44.81	0.39	~0	4	2.03	+0.03

[a] Retrograde rotation.
[b] Retrograde orbit.

4. Specific Questions

One way to quantize the satellites would be to take their respective planets as central body. As far as we know, all satellites always turn the same side toward their mother planet, hence, their period of rotation is equal to their orbital period and Equation (2) is simplified to:

$$n=(R/r)^2\,\alpha^k(2\tau)^{-1}. \tag{4}$$

However, the results one obtains from Equation (4) is a completely random dispersion of n values, scattered between 2 and 50, indicating that satellites cannot be quantized in such a manner.

It seems to me that the explanation of this behavior is that the spin of the orbiting object has a physical meaning for quantization purposes only as related to the radius vector of its orbit. Satellites, which always turn the same side toward their planet, do not rotate relative to their radius vector, whence their spin is zero in the coordinate system of their orbit around the planet!

Satellites, however, rotate relative to the radius vector of their orbit around the Sun, and can be quantized in the same manner as planets, orbiting around the Sun, as this is illustrated in Table III.

Artificial satellites (Apollo, Mariner, etc.) turn always the same side toward the Moon, or the planet they orbit, thus their spin is zero relative to the radius vector of their orbit. Similarly as celestial satellites, they cannot be quantized when orbiting a planet. During their passage from the Earth to their target, artificial satellites are stabilized relative to the direction of the Sun. Consequently, they cannot be quantized as being in orbit around the Sun either.

I could only include the Moon, four satellites of Jupiter and two satellites of Saturn in Table III, because the diameters of the other larger satellites are not yet known with sufficient accuracy. Although the diameter of Triton is uncertain to $\pm 40\%$, nevertheless I included it because it is the second largest satellite, and because it has a retrograde orbit relative to its mother planet. There are three satellites in the $n=2$, $k=4$ quantum state, and only one has retrograde rotation; two satellites are in the $n=1$, $k=4$ quantum state, neither has retrograde rotation. This does not seem to support the applicability of Pauli's exclusion principle.

The circumstance that the rotational spin of the orbiting objects has to be computed relative to the radius vector of the orbit requires use of the tropical period of rotation, instead of the sidereal period of rotation. This difference becomes significant only for the planets Mercury and Venus, where the tropical periods of rotation are 176.01 and 116.78 days, instead of 58.65 and 243.16 sidereal days, respectively.

In 1946 when the periods of rotation of Mercury and Venus were not yet known, it was generally assumed that the period of rotation of these two innermost planets were the same as those of the two other terrestrial planets Earth and Mars that is, one day. Nevertheless, the quantum number n which I computed at that time for Mercury and Venus was the same (Barnothy, 1946), as found now, that is 4 and 1, respectively. The quantum number k was 3 instead of 4.

This suggests that all four terrestrial planets were formed with the same $k=3$ value, Venus and Earth being then in the same quantum state $n=1$ and $k=3$, and Pauli's exclusion principle required that one member of the pair should have retrograde rotation. This was the reason why I concluded 30 years ago that Venus must have retrograde rotation, a prediction born out by later observations.

In the course of time the direction of rotation has not changed, but Mercury and Venus have increased their quantum number k from 3 to 4 by lengthening their sidereal period of rotation by factors 56 and 244, respectively, changing thereby the original ratio of rotational to orbital angular momenta by a factor of about 137, without thereby changing their quantum level n. Mercury and Venus are much closer to the Sun than the Earth, and the solar tides are therefore much more powerful. It is generally assumed that tidal friction was the reason which increased the periods of rotation of the two innermost planets to their present rates. The astonishing agreement of the n values, whether computed with one day, or with the present tropical periods of rotation, supports the assumption that at the start the period of rotation was the same for all four terrestrial planets.

The large deviation of the n value of Venus from an integer is rather surprising, because its radius and moment of inertia factor are well known. We may seek an explanation in the phenomenon that Venus turns the same point of its surface toward the Earth at each inferior conjunction, indicating the existence of an Earth lock. It seems, therefore, possible that during the course of the slowing down of the rotation of Venus, this process was halted through the Earth lock.

It is generally assumed that spiral galaxies are systems in gravitation equilibrium, their star systems revolving on Keplerian orbits around the galactic center. The total angular momentum of the solar system is $s = 3.21 \times 10^{50}$ cgs, while its mass is 2×10^{33} g. The distance to the Galactic center is $R = 9.4$ kpc (Van den Bergh, 1972), and its rotational velocity $v = 250 \pm 20$ km s^{-1} (de Vaucouleurs and Peters, 1968). The equation to quantize the solar system revolving around the center of the Galaxy is:

$$n = \frac{vRM}{2s} \alpha^k (\cos \delta)^{-1}, \tag{5}$$

where $\delta = 63.6°$ is the inclination of the orbital plane of the major planets to the galactic plane. With $k = 6$, we obtain $n = 0.76$. The deviation from the integer number [1] could be due to an error of 30% in the distance from the Galactic center, but could also arise from the presence of a heavy trans-Plutonian planet with retrograde revolution. From the irregularities observed in the orbit of Neptune, Brady (1972) concluded upon the existence of a trans-Plutonian planet of 300 $M_\oplus$ mass, with its orbital plane 120° inclined to the ecliptic.

5. Probability that *n* is an Integer Number by Chance Alone

Altogether there are 19 objects (9 planets, 8 satellites and 2 asteroids) for which the radius r, period of rotation t, and moment of inertia factor τ are known with an accuracy better than 10% (see Table III). With the exception of Neptune, and two of the inner satellites of Jupiter, all the other 16 objects have n values which deviate less than $+0.36$ and -0.31 from the integer numbers 1, 2, 3 or 4. If we compute n from Equation (2) without the factor α^k, then n is of the order of 10^4 to 10^9. Applying the factor α^k divides the total range of n in subintervals, each containing the same range of n values from 1 to 137. Let us assume that we accept an n value as being an integer quantum number if its deviation from [1], [2], [3] or [4] is less than ± 0.25. Then, the probability that a random number falls within ± 0.25 from [1], [2], [3] or [4] would be $\frac{2}{137} = 1.5\%$. One has, however, to take into account that whenever n is found to be a large integer, say between 103 and 137, this number could be changed into an n value between 0.75 and 1, simply by changing α^k to α^{k+1}. We have thus to divide the $16n$ values into two groups: one where n is less than 1, and another where n is greater than 1, the groups being populated by 7 and 9 objects, respectively. In the first group the chance of a success in one trial is about $\frac{137}{7} \sim 20$ times higher than in the second group.

The probability to find 7 objects in the first group and 9 in the second follows a biniminal distribution:

$$P = \frac{N!}{x!(N-x)!} p^x (1-p)^{N-x} \tag{6}$$

where N is the number of trials, x the number of successes (integer), and p the probability for one trial. Here p is determined by the average deviation of n from integer values. Because the average deviation from [1] in the first group is -0.11, $p=0.11$ and with $x=7$ and $N=19$ we have

$$P_1=\frac{19!}{7!\,12!}\,0.11^7\times 0.89^{12}=2.4\times 10^{-3}. \tag{7}$$

In the second group we have an average positive deviation of $+0.33$ for [1], but ± 0.15 for [2], [3], or [4]; thus $p=(0.33+6\times 0.15)\,137^{-1}=9.0\times 10^{-3}$. With $x=9$ and $N=19$,

$$P_2=\frac{19!}{9!\,10!}\,(9.0\times 10^{-3})^9=3.6\times 10^{-14}. \tag{8}$$

The total probability is thus:

$$P=P_1P_2=8.6\times 10^{-17}. \tag{9}$$

The chance of 1 in 10^{16} cannot be called anymore a chance, but a proof that the solar system, hence macroscopic objects, can be quantized. Unless, of course, we want to believe that our solar system was specially selected by the Divine Creator from 10^{16} similar planetary systems to satisfy this curious agreement. This would be, however, equivalent to return to the old, pre-Copernician geocentric cosmological philosophy.

6. Objections

The possibility to launch artificial satellites in whatever orbits, contradicts the hypothesis of quantization of macroscopic objects. This objection is answered in Section 4. Even for a satellite rotating around an axis perpendicular to its orbit, it would take millions of years before its energy loss through gravitational radiation would suffice to settle it in a stable quantized orbit.

The angular velocity of the Earth is slowing down at a rate of $(2.65\pm 0.58)\times 10^{-10}$ yr^{-1} (Newton, 1972). During the past life time of the Earth, this would have amounted to a considerable change in the ratio of the Earth's spin to its orbital angular momentum, contradicting quantization. A decrease of $0.000136''\ \mathrm{yr}^{-1}$ from the present $23°45'$ inclination of the rotational axis to the normal of the ecliptic, would compensate the effect of the slowing down of the rotation of the Earth on the quantum number n, as this can be seen from Equation (2). On account of the wobbling of the polar axis by $\pm 0.3''$, the detection of such a quantum mechanical 'readjustement' of the rotational axis would be difficult to establish even within a century, although the six stations of the international network for latitude determinations are capable to determine the momentary direction of the polar axis to an accuracy of $0.01''$.

7. Further Consequences

A quantization of binary systems consisting of two neutron stars reinstates the possibility that pulsars could be binary systems, because there is no emission of gravitational radiation while in stable orbits. To date some 20 different pulsar models were proposed. Some predict that the radiation is emitted from the first few centimeters thick surface layers of a neutron star, while in others the emission occurs at a height of 10000 km. This indicates that in spite of the great sophistication of some models, the question of what pulsars are, is not finally settled. A few years ago I have proposed at the 127th meeting of American Astron. Soc. that should one be able to disregard for one reason or another, the slowing down process through the emission of gravitational radiation, a binary system of two neutron stars could explain without further assumptions many peculiarities of pulsars (Barnothy, 1968).

Quantization of a binary pulsar raises the hope that the 'glitches' in the pulse frequency of the Crab and Vela pulsars may be explanable through quantum jumps from a higher to a lower quantum state.

If the planetary system can be indeed quantized, it would entail a re-evaluation of the uncertainty principle, and with it, of quantum mechanics, certainly in its Copenhagen interpretation, which renounces reality as a metaphysical unscrutable. True, if an observer would look on our planetary system using gravitational waves with wavelength equal to those the system would emit when the planets would change from one quantum state to another, he could describe the position of the planets and their motion merely with the help of ψ-functions, and may thus say, that the existence of the Earth, for instance, is only a statistical probability. But, if he would use light, and would have a sufficiently powerful telescope to see people walking on the streets, I seriously doubt whether he would still question the physical reality of the Earth.

The solar system seems thus to be a quantized system, which depending on the signal carrier we use, obeys or rejects the uncertainty principle. The main importance of what I have here shown and hopefully proven is: the repudiation of John von Neumann's theorem, namely that no parameter previously 'hidden' from quantum physics could later be discovered and permit a precise measurement that violates the uncertainty law – thereby forever exiling cause-and-effect from the scene of physics.

Tycho Brahe's and Copernicus' astronomy could develop into modern astrophysics only with the help of physics. It seems that astronomy is now in the position to repay this debt by helping to understand the meaning of the discarded concepts of causality and dualism, concepts, though used for a long time, but understood only vaguely by the physicists themselves.

References

Barnothy, J.: 1946, *Nature*, June 15 and August 31.
Barnothy, J.: 1947, *Papers of Terrestrial Magnetism*, Hung. Inst. for Meteor. Terres. Magn., No. 2.
Barnothy, J.: 1968, *Astron. J.* **73**, 164.
Brady, J.: 1972, *Publ. Astron. Soc. Pacific* **84**, 314.
Cook, A. H.: 1972, *Observatory* **92**, 84.
de Vaucouleurs, G. and Peters, L.: 1968, *Nature* **220**, 368.

Dirac, P. A. M.: 1963, *Sci. Am.* **208**, No. 5, 45.
Dollfus, A. (ed.): 1970, *Surfaces and Interiors of Planets and Satellites*, Academic Press, New York.
Klepcynski, W. J., Seidelmann, P. K., and Duncombe, R. L.: 1970, *Astron. J.* **75**, 739.
Kovalevsky, J. and Link, F.: 1969, *Astron. Astrophys.* **2**, 398.
Majeva, S. S. V.: 1969, *Astrophys. Letters* **7**, 11.
Newton, R. R.: 1972, *Astrophys. Space Sci.* **16**, 179.
Plagemann, S.: 1965, *J. Geophys. Res.* **70**, 985.
Seidelmann, P. K., Duncombe, R. L., and Klepcynski, W. J.: 1969, *Astron. J.* **74**, 776.
Seidelmann, P. K., Klepcynski, W. J., Duncombe, R. L., and Jackson, E. S.: 1971, *Bull. Am. Astron. Soc.* **3**, 270.
Shapiro, I. I.: 1967, *Science* **157**, 806.
Tolson, H. and Gapcinsky, J. P.: 1967, in A. Dollfus (ed.), *Moon and Planets*, II, Academic Press, New York, p. 178.
van den Bergh, S.: 1972, *Dearborn Obs. Colloquium.*
Wyler, A.: 1969, *C.R. Acad. Sci. Paris* **269A**, 743.

ON THE ORIGINAL DISTRIBUTION OF THE ASTEROIDS

M. LECAR and F. A. FRANKLIN
Smithsonian Astrophysical Observatory and Harvard College Observatory, Cambridge, Mass. 02138, U.S.A.

Abstract. The depletion of an initially uniform distribution of asteroids extending from Mars to Saturn, caused by the gravitational perturbations of Jupiter and Saturn, is calculated by numerical integration of the asteroid orbits. Almost all (about 85%) the asteroids between Jupiter and Saturn are ejected in the first 6000 years. Most of the asteroids between the 2/3 Jupiter resonance (4.0 AU) and Jupiter are ejected in the first 2400 years with the exception of the stable librators (e.g., the Hilda group). Interior to the 2/3 resonance the depletion was small, and interior to the 1/2 resonance (3.3 AU), no asteroids were ejected in the first 2400 years.

1. Introduction

The recently reported Palomar-Leiden Survey (PLS) of asteroids (Van Houten *et al.*, 1970) was complete, in a selected area (12° × 18°) of the sky, to apparent magnitude 20. The PLS, apparently free from selection effects, confirmed in some detail the features of the space distribution of the asteroids that were established in previous studies (see, for example, Brown *et al.*, 1967, for a pictorial representation of the qualitative aspects of asteroid statistics).

In Figure 1, we present – in the form of a histogram – the surface number density (i.e., number of asteroids with $a_0 \leqslant a \leqslant a_0 + \delta a_0$ divided by $2\pi a_0 \delta a_0$) for the 981 PLS asteroids with well-determined orbits.

The 'Kirkwood gaps' at positions where the asteroid periods are 1/3, 2/5, and 3/7 of Jupiter's period are apparent in this presentation, and have – in the past – received the most attention from celestial mechanicians. We refer the reader to the classic paper by Brouwer (1963) for an elegant treatment of the gaps.

We were struck by the precipitous fall-off of the surface density at the 1/2 resonance. We find it difficult to imagine any formation mechanism (such as accretion or gravitational instability) that would vary so drastically over so small a scale. Furthermore, the existence of stable librators at the 2/3 Jupiter resonance (the 'Hilda' group) and at the 1/1 Jupiter resonance (the 'Trojans') suggests that whatever mechanisms formed the asteroids operated also at those larger distances. Likewise, we are uncomfortable with the notion that a catastrophic event (e.g., an exploded planet) created the asteroids, because of the low probability of capturing so many fragments in the 2/3 and 1/1 libration regions. By the same token, we reject the idea that a catastrophic event removed asteroids exterior to the 1/2 resonance, leaving the librators intact.

Thus, we are led to suggest that the asteroids had initially a more or less uniform surface density in the region between Mars and Jupiter and that the present asteroids, tightly confined between 2.0 and 3.3 AU, are the remnants that survived ejection by the gravitational perturbations of the planets. However, we hasten to point out that the present surface mass density in the asteroid belt is well below what one would infer from the 'heavy element' content of the planets. Figure 2 presents the surface density

Y. Kozai (ed.), The Stability of the Solar System and of Small Stellar Systems, 37–56.

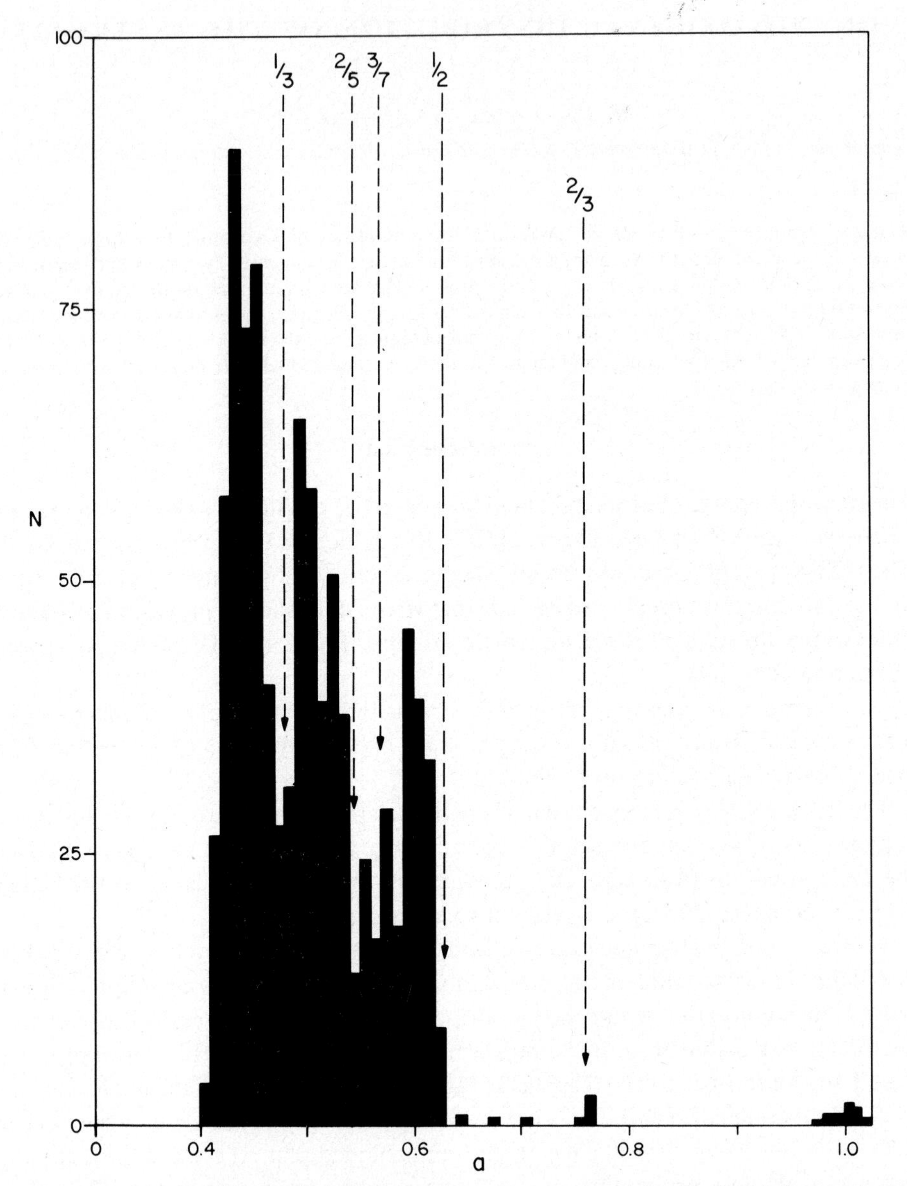

Fig. 1. Histogram of the surface number density vs semimajor axis ($a_J = 1$) of 981 first-class orbits from the Palomar-Leiden Survey of Faint Minor Planets.

heavy elements versus the distance from the Sun on the assumption that the planets collected all the heavy elements inside their present Lagrangian (L_1 to L_2) points (upper curve) or all the heavy elements between the successive planets (lower curve). The heavy-element content of the giant planets is taken as 1/200 of their present mass. In either case, the mass density in the asteroid belt is anomalously low. Our work sheds no light on this anomaly. We confine ourselves, in this study, to examining the gravita-

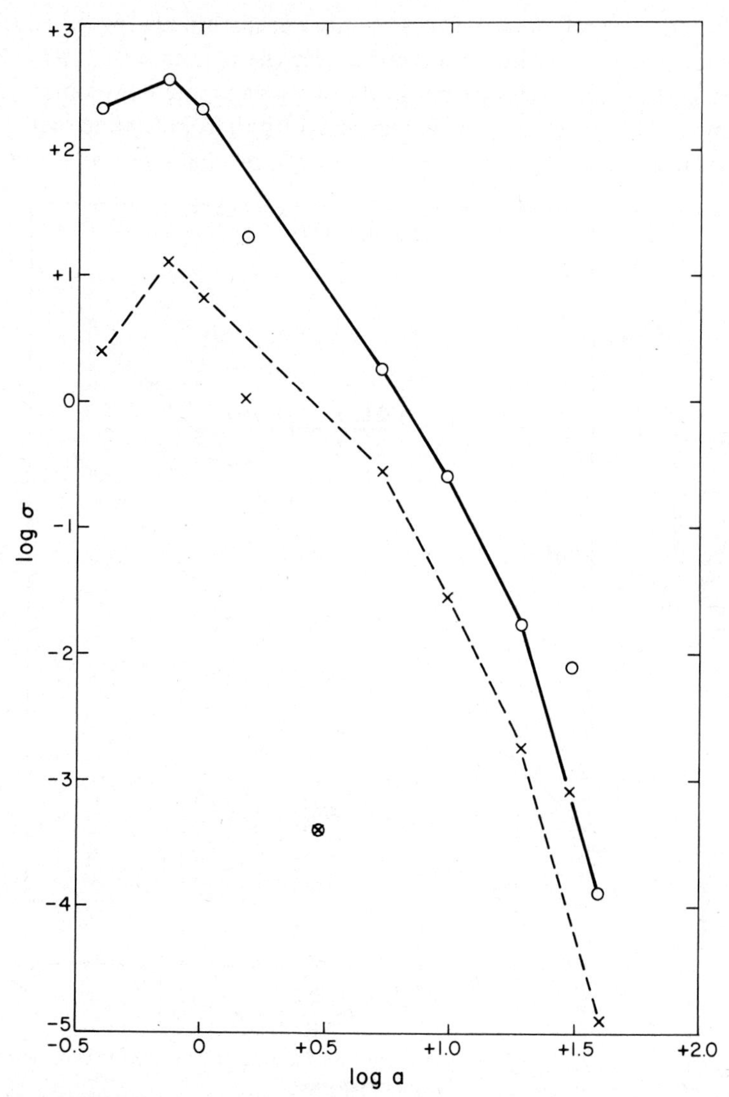

Fig. 2. Surface density in g cm^{-2} vs log of the semimajor axis in AU of the 'heavy element' material, now collected into planets, when spread uniformly between the present values of the L_1 and L_2 points of each planet (upper curve – ○). Lower curve (×) assumes the material to be spread uniformly to a distance proportional to L_1 or L_2 but extended to fill the entire area between successive planets. The asteroids are at $\log a \cong +0.5$.

tional perturbations of the planets on an initially uniform (albeit low) distribution of asteroids.

Having supposed that the initial distribution of asteroids extended well past its present boundaries, we are led to ask whether initially there were also asteroids beyond Mars and Jupiter. In this preliminary survey, we look also at the region between Jupiter and Saturn.

We conducted this preliminary survey by numerically integrating the equations of motion (the integration algorithm is described elsewhere, Lecar *et al.*, 1973). Figure 3 gives the integration accuracies (errors in the energy and angular momentum) for a typical example. The errors are growing linearly with time, indicating that round-off errors are still negligible. The integration times are somewhat longer than the shortest

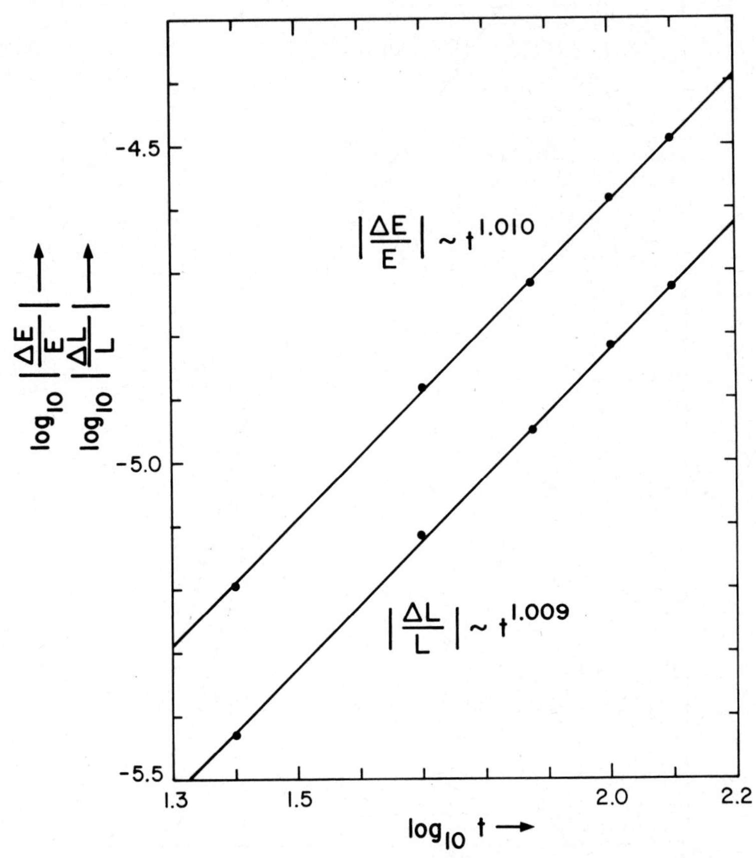

Fig. 3. Errors in energy and angular momentum as a function of time in periods of Jupiter for a typical calculation.

time scale ($\mu^{-1/2}$ Jupiter periods = 375 yr) and less than μ^{-1} Jupiter periods ≅ 12000 yr. We used numerical integration because, although we trust perturbation theory to give sufficiently accurate trajectories for stable asteroids, we do not trust perturbation theory to single out the unstable orbits of interest in this study. The combined effects of close approaches to Jupiter and overlapping resonances contribute to the invalidity of perturbation theory. We will return to this problem in a subsequent communication.

The results of a typical integration are shown in Figure 4, where the semimajor axes (in units of Jupiter's semimajor axis = 5.203 AU), are shown as a function of time (in units of Jupiter's period = 11.86 yr). The 1/2 resonance at 0.630 and the 2/3 resonance at 0.763 show up clearly in this presentation. Figure 5 is essentially a blowup of

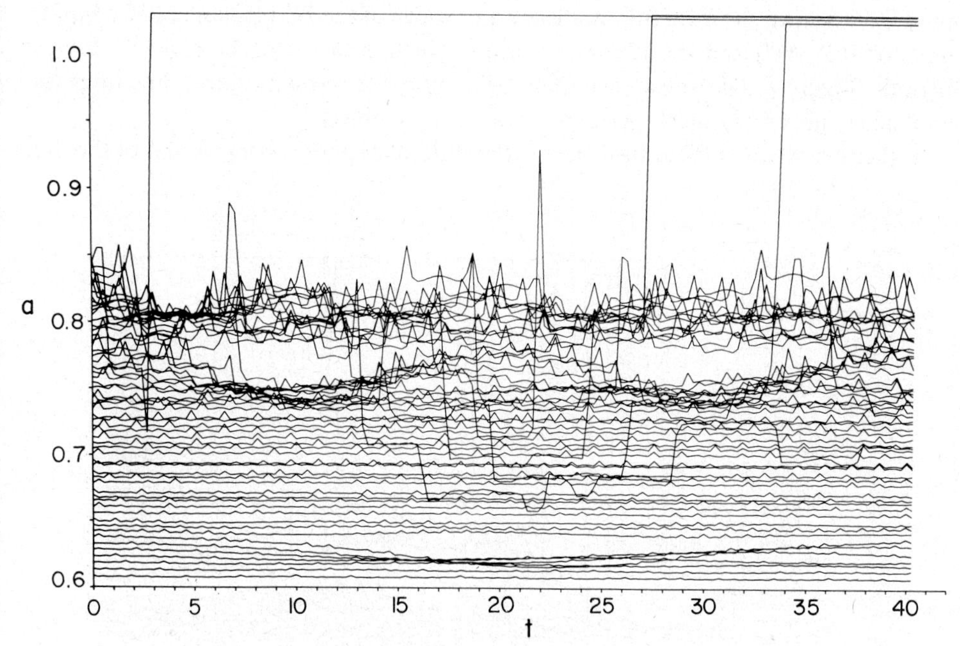

Fig. 4. Semimajor axis vs time in periods of Jupiter, where Jupiter's eccentricity = 0.0, for a set of asteroids whose initial eccentricities are all zero.

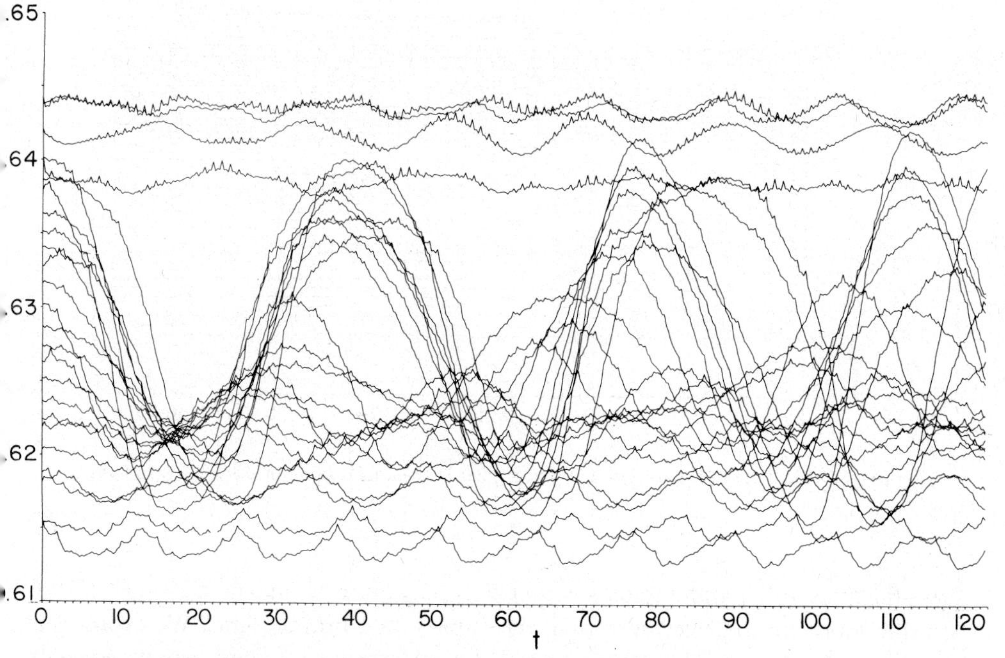

Fig. 5. Semimajor axis vs time in periods of Jupiter, where Jupiter's eccentricity = 0.06, for 25 bodies started near the 1/2 resonance, all with initial eccentricities of zero. Plot is slightly smoothed.

the 1/2 resonance showing the amplitude and width of the 1/2 resonance (0.01 Jupiter units or 0.05 AU) and the libration period, which is of the order of $\mu^{-1/2}$ Jupiter periods. Figure 6 also shows asteroids below the 1/2 resonance and illustrates the weakened effect of Jupiter's perturbations on these objects.

In the surveys to be described below, the ratio of Jupiter's mass to that of the Sun

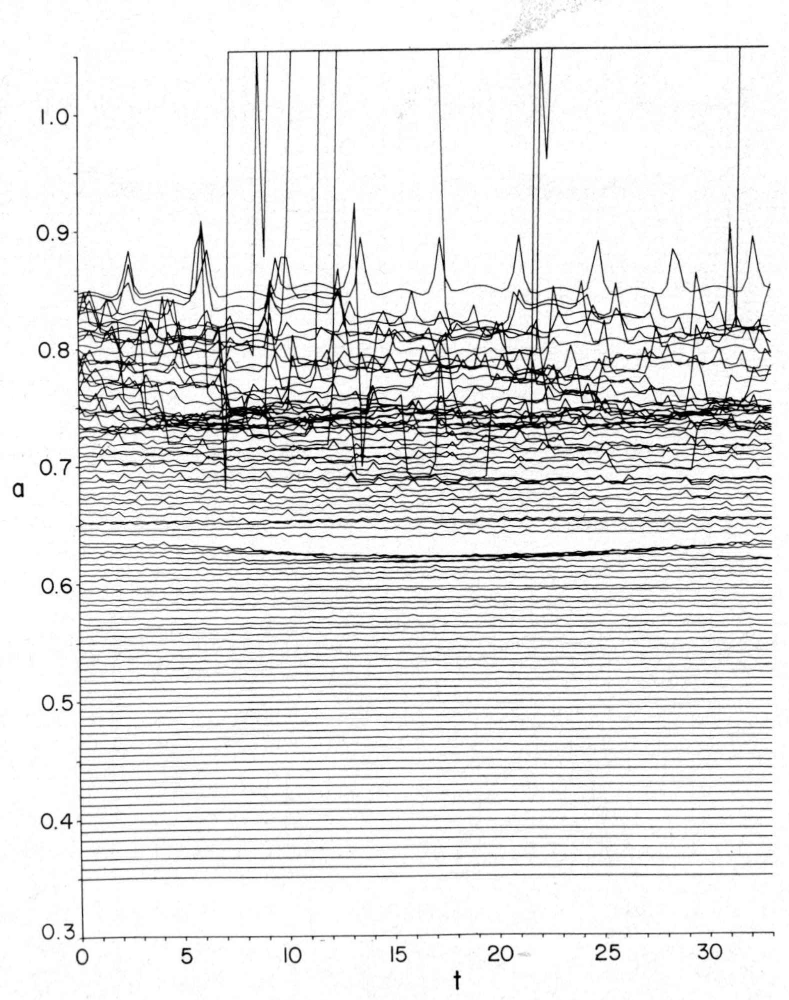

Fig. 6. Identical to Figure 4 except that Jupiter is given its maximum eccentricity of 0.06.

was taken as 10^{-3}, and Jupiter's eccentricity was given its maximum value of 0.06. Furthermore, all the asteroids are in the Jupiter–Sun orbital plane. We expect that inclined asteroids would behave in a qualitatively similar way, but would evolve on a longer time scale.

2. The Region Between Mars and Jupiter

Two-hundred and sixty asteroids were started with semimajor axes distributed with uniform surface density between 0.55 and 0.85 Jupiter units. All other orbital elements were distributed randomly, with the eccentricities lying between 0.0 and 0.3. From preliminary studies (e.g., Figures 4 and 6), we expected that asteroids interior to 0.55 would not be perturbed much by Jupiter and that asteroids exterior to 0.85 would be ejected immediately. Figure 7 shows the semimajor axis vs time for the first 50 Jupiter periods. The almost vertical lines indicate a very rapid change of semimajor axis (i.e., energy) following a close encounter with Jupiter. Figures 8 and 9 continue the plot of semimajor axis vs time to 200 Jupiter periods. In Figures 8 and 9, bodies with initial semimajor axes below 0.633 have been deleted as they seemed quite stable. 'Ejected' bodies either leave the asteroid domain (i.e., their semimajor axes become greater than 1 or less than 0.55) or approach Jupiter closer than 10 present Jovian radii. In the latter case, the $a(t)$ terminates in a filled circle. Figure 10 gives a histogram of the

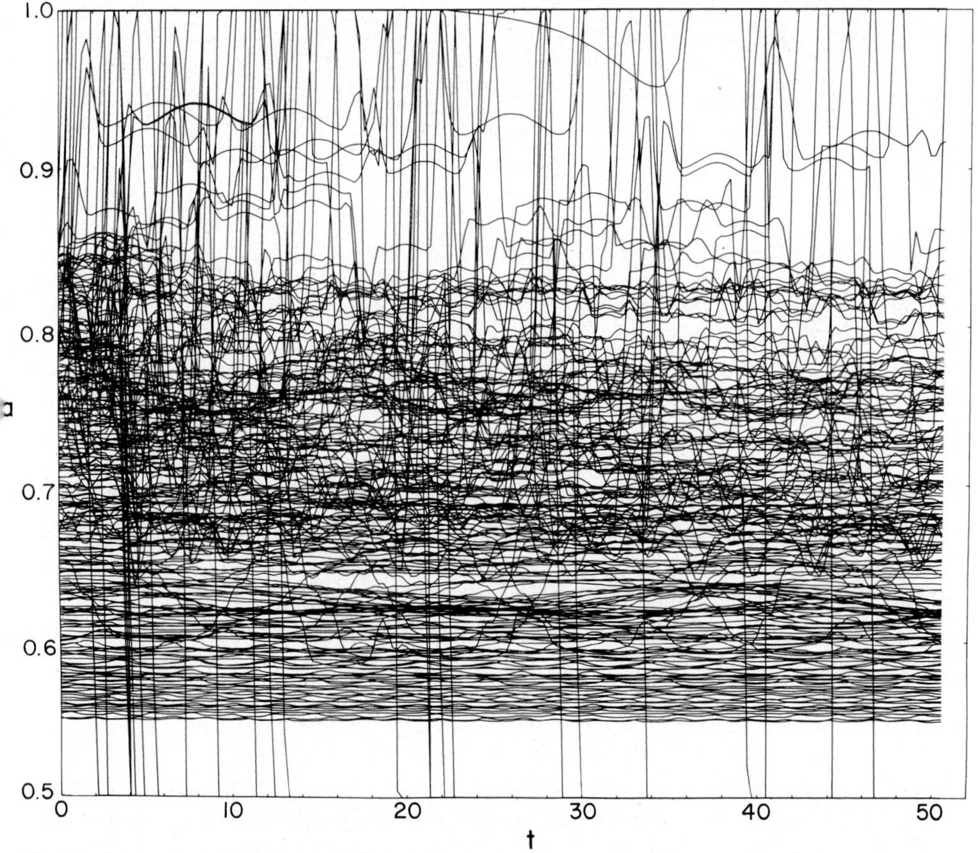

Fig. 7. Semimajor axis vs time in periods of Jupiter (eccentricity = 0.06) for a set of asteroids with initial eccentricities chosen at random between 0.0 and 0.3.

surface density at various times. The evolution of the surface density is shown in a different presentation in Figure 11. The evolution proceeds very rapidly for the first 50 Jupiter periods and then markedly slows down.

If Jupiter had been in a circular orbit, the theory of the 'restricted three-body problem' would have delineated the escapers from the nonescapers by their value of the

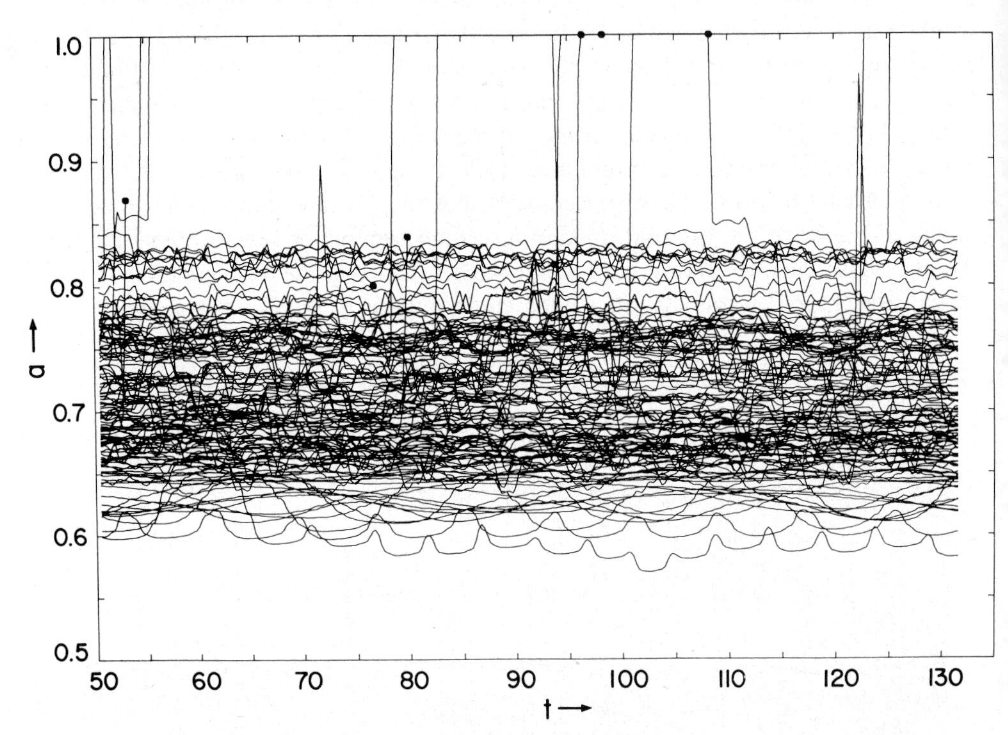

Fig. 8. A continuation, with a slightly different scale, of Figure 7 to 130 revolutions of Jupiter. Solid dots indicate objects that have approached Jupiter to closer than 10 radii and have been dropped from the calculation.

Jacobi integral. Figure 12 is a plot of the initial semimajor axes vs the initial eccentricity of the 260 asteroids except those for which $0.55<a<0.60$. The open circles denote those asteroids that subsequently escaped. In Figure 13, only the escapers are shown, and curves of constant Jacobian are given for reference.

The Jacobi constant J, in Sun-centered rotating coordinates, is given by Equation (1):

$$J=\tfrac{1}{2}r^2+\frac{1-\mu}{r}+\frac{\mu}{s}-\mu r\cos\phi+\tfrac{1}{2}\mu^2-\tfrac{1}{2}(\dot{r}^2+r^2\dot{\phi}^2), \tag{1}$$

where r denotes the Sun-asteroid distance, s denotes the Jupiter-asteroid distance, ϕ is the angle between the asteroid and Jupiter as seen from the Sun, and μ is the ratio of Jupiter's mass to the sum of the masses of Jupiter and the Sun. The sum of the masses, gravitational constant, and Jupiter's semimajor axis have been taken as unity.

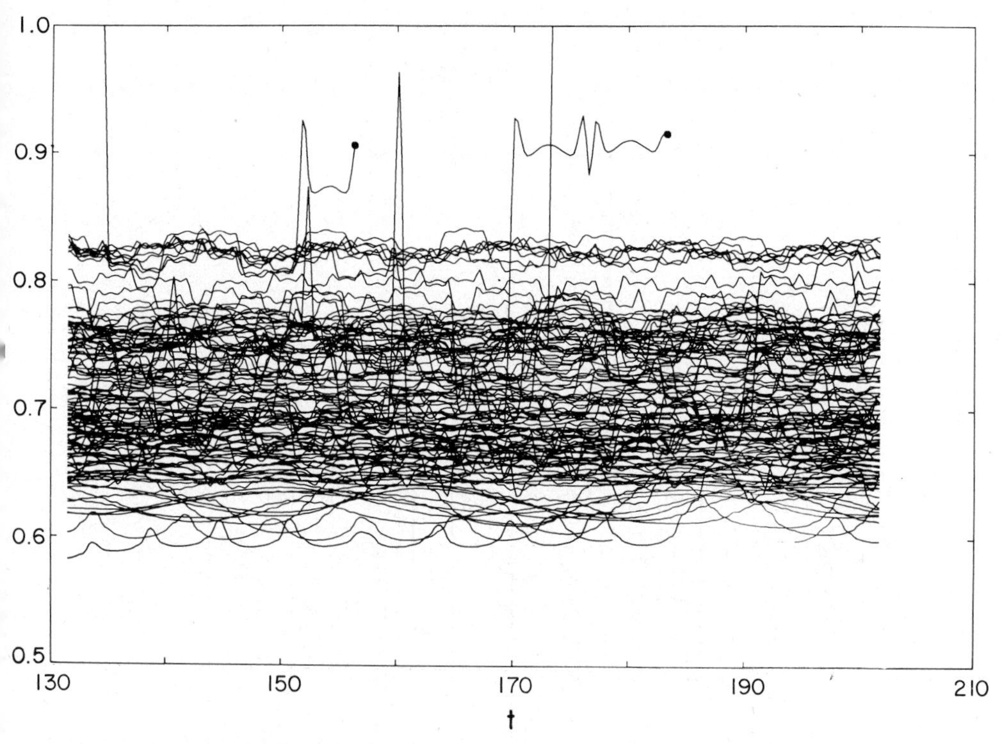

Fig. 9. Figure 8 extended to 200 revolutions of Jupiter. The region between the resonances at 2/3 and 3/4 is now largely clear, although objects can apparently librate stably at these resonances.

Let F denote J with the velocity terms set equal to zero. That is,

$$F = \tfrac{1}{2}r^2 + \frac{1-\mu}{r} + \frac{\mu}{s} - \mu r \cos\phi + \tfrac{1}{2}\mu^2. \tag{2}$$

As the velocity terms are positive-definite, motion is possible only where $F \geqslant J$. F assumes its minimum value of 1.51997 at the L_2 point, where $r = 0.932287$ (and where $\phi = 0$, and $s = (1-r)$).

We can approximately include the effect of Jupiter's eccentricity by calculating the instantaneous L_2 point when Jupiter is at perihelion. Figure 14 gives the quantities necessary to derive Equation (3), which balances the gravitational forces of the Sun and Jupiter at the distance r from the Sun, viz.:

$$GM_{\odot}/r^2 = (r - \mu R)\,\Omega^2 + GM_J/(R-r)^2, \tag{3}$$

where Ω is the Keplerian angular velocity of Jupiter at perihelion. With the substitution $x \equiv r/R$, Equation (2) can be reduced to

$$(x-\mu)\,(1+e) + \frac{\mu}{(1-x)^2} - \frac{1-\mu}{x^2} = 0, \tag{4}$$

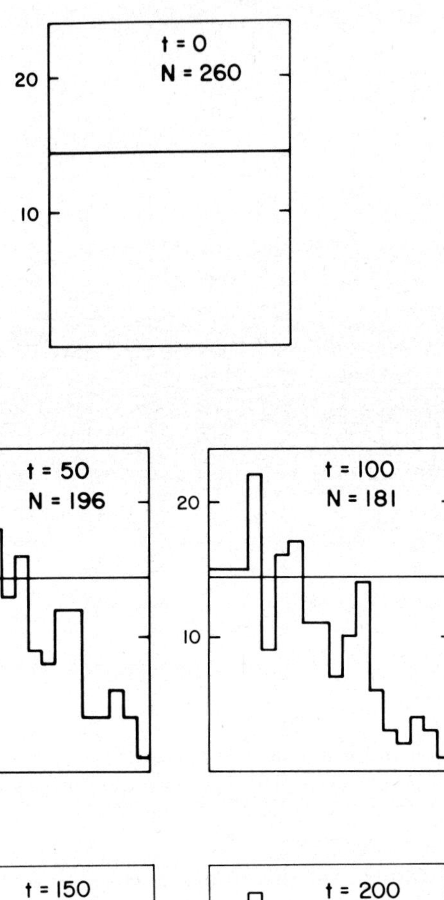

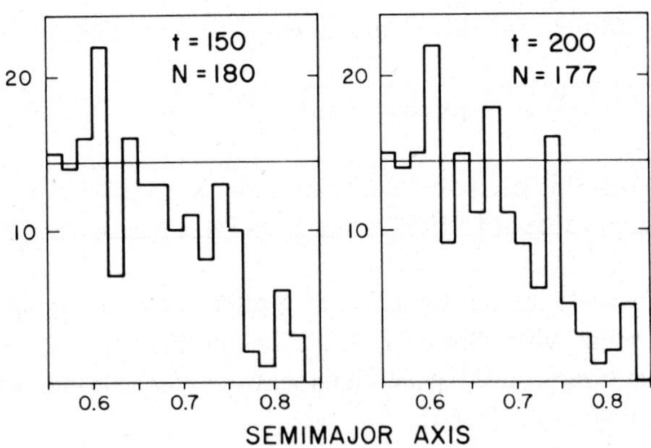

Fig. 10. Histograms of surface number density at different times.

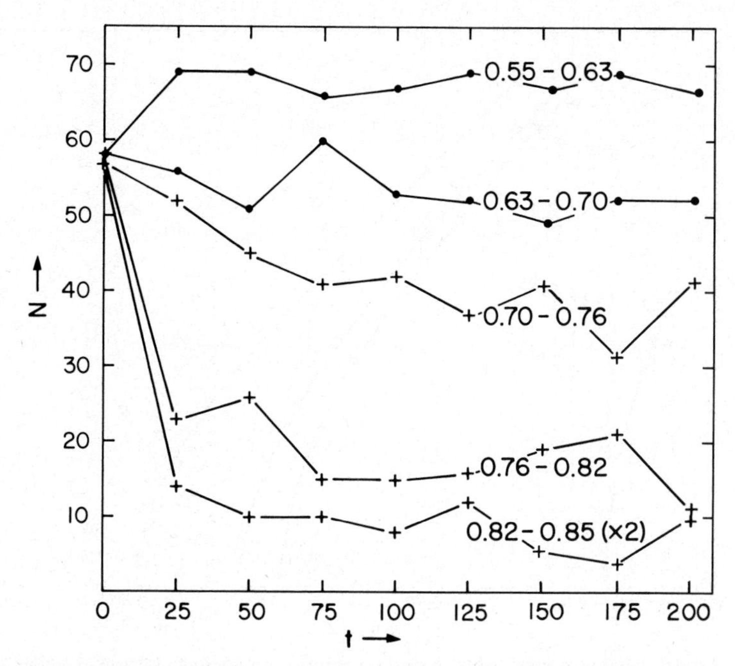

Fig. 11. Surface number density N as a function of time in revolutions of Jupiter for five ranges in semimajor axis.

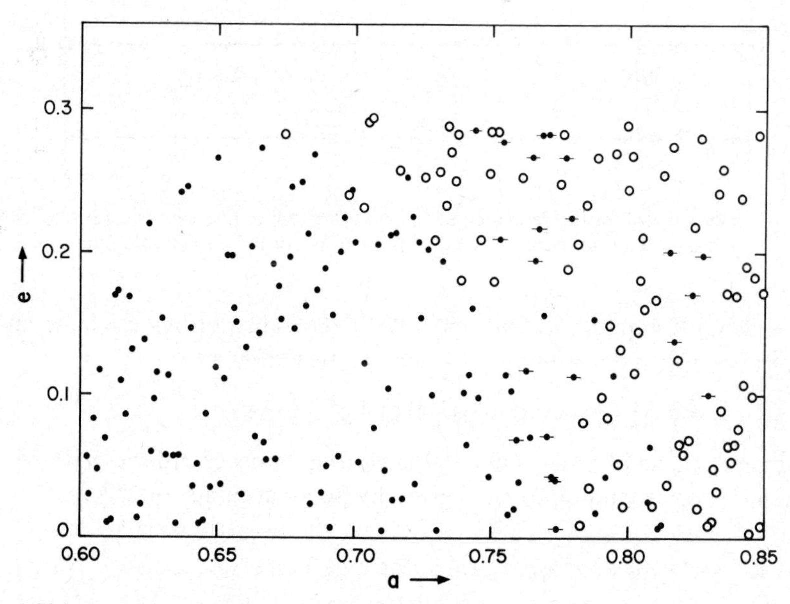

Fig. 12. Initial eccentricity and semimajor axis for bodies discussed in Section 2 in the range $0.6 < a < 0.85$. Open circles represent objects that have escaped during 200 revolutions of Jupiter; solid circles have not. Horizontal lines through a solid dot indicate a librating body.

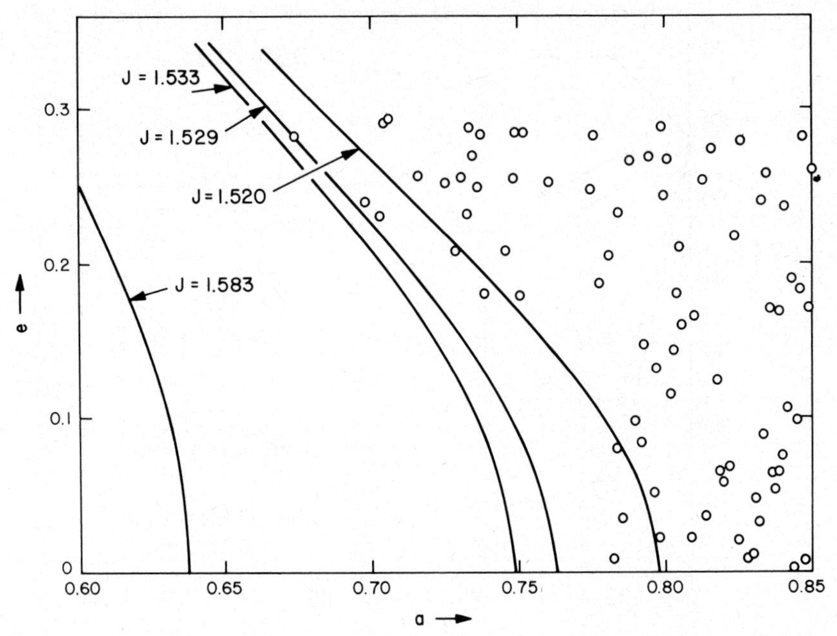

Fig. 13. Circles again give objects that have escaped. Curves for four different Jacobi constants are included.

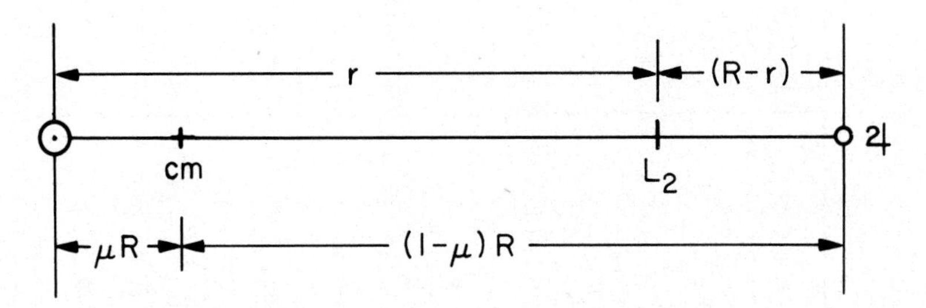

Fig. 14. Geometry useful for the derivation of Equation (2). Jupiter is assumed to move in an eccentric orbit and its 'instantaneous L_2 point' to lie at a distance r from the Sun.

where e denotes Jupiter's eccentricity. With the substitutions $\xi = 1 - x$ and $v^3 = (1/3)\,[\mu/(1-\mu)]$, and assuming $\xi \ll 1$, this can be written as

$$v^3 = \xi^3\left[1 + \tfrac{1}{3}e + v^3(1+e)\right] - \tfrac{1}{3}e\xi^2 + \xi^4 + \tfrac{4}{3}\xi^5 + \cdots. \qquad (5)$$

Letting $\alpha = \frac{1}{3}(e/v)$ be of order unity and neglecting terms of order e or v, the root of Equation (5) corresponding to L_2 is given by (to an accuracy of 2%)

$$\frac{\xi}{v} = 1 + \tfrac{1}{3}\alpha + \tfrac{1}{9}\alpha^2 + \tfrac{2}{81}\alpha^3 + \cdots \quad \text{for} \quad \alpha \lesssim 1.5$$

$$\frac{\xi}{v} = \alpha + \frac{1}{\alpha^2} - \frac{2}{\alpha^5} + \frac{4}{\alpha^8} + \cdots \quad \text{for} \quad \alpha \gtrsim 1.5.$$

The solution of Equation (4) (with $e=0.06$) for the instantaneous L_2 point gives $r=0.870652$, and the value of F, as calculated from Equation (2), is $F=1.53329$. $J=1.529$ corresponds to the largest Jacobi constant to escape after 200 Jupiter periods, and $J=1.583$ corresponds to the Jacobian that bounds the asteroids in the PLS, as is shown in Figure 15. (At large distances from Jupiter, a good approximation to J is given by $J=\frac{1}{2}a+\sqrt{[a(1-e^2)]}$.)

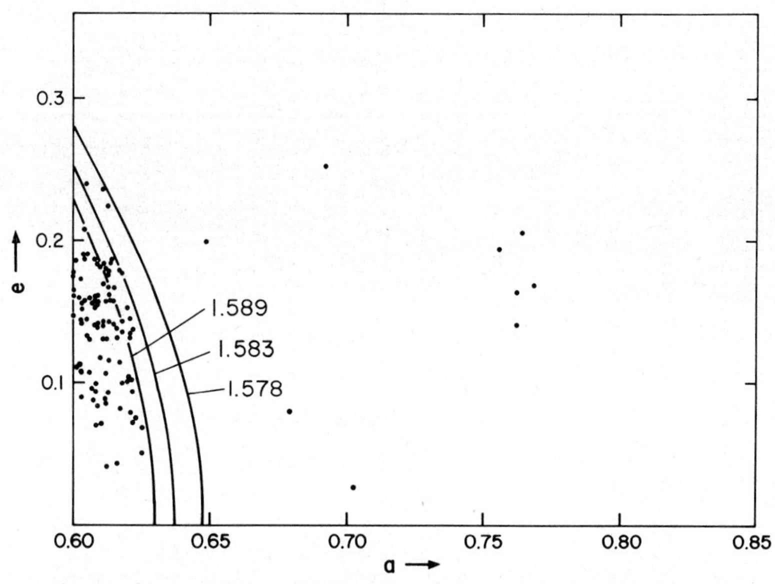

Fig. 15. PLS asteroids with three values of the Jacobi constant for reference.

Figure 16 shows the Jacobi constant as a function of time for the interval 50–130 Jupiter periods. For reference, J is 1.53 at the 2/3 resonance and 1.58 at the 1/2 resonance. If we assume the evolution of the asteroids to be a random walk in 'Jacobian space', then Jupiter has thus far eaten away those asteroids for which J is less than 1.53 and the solar system has retained asteroids with J greater than 1.58. The central problem, which we have not answered in this preliminary survey, is the following: Will Jupiter eventually eat away another 0.05 (3%) units of the Jacobi constant? Typical oscillations of J (Figure 16) are of the order of 10^{-3}. If this is taken as the order of magnitude of a random impulse, then we would expect to wait $\sim 10^5$ Jupiter periods in order that J be changed by 0.05. However, we suspect that this reasoning is faulty and, instead, that stable orbits are stable just because successive impulses are not independent but correlated. In fact, we expect that interior to the 1/2 resonance, a perturbed Jacobi integral may be valid for all time and that this integral breaks down somewhere exterior to the 1/2 resonance. We shall return to this question in a later study.

We note that the asteroids in the neighborhood of the 2/3 resonance, with the exception of the stable librators, were removed from that neighborhood in a few thousand years. This explains, we think, why there still remains a concentration of asteroids there. There was no possibility for the librators, which have large excursions in eccentricity and semimajor axis, to collide with nonlibrators, because the nonlibrators escaped almost immediately. This is in sharp contrast with the situation at the 1/2 resonance, where a large body of asteroids interior to the 1/2 resonance remained to

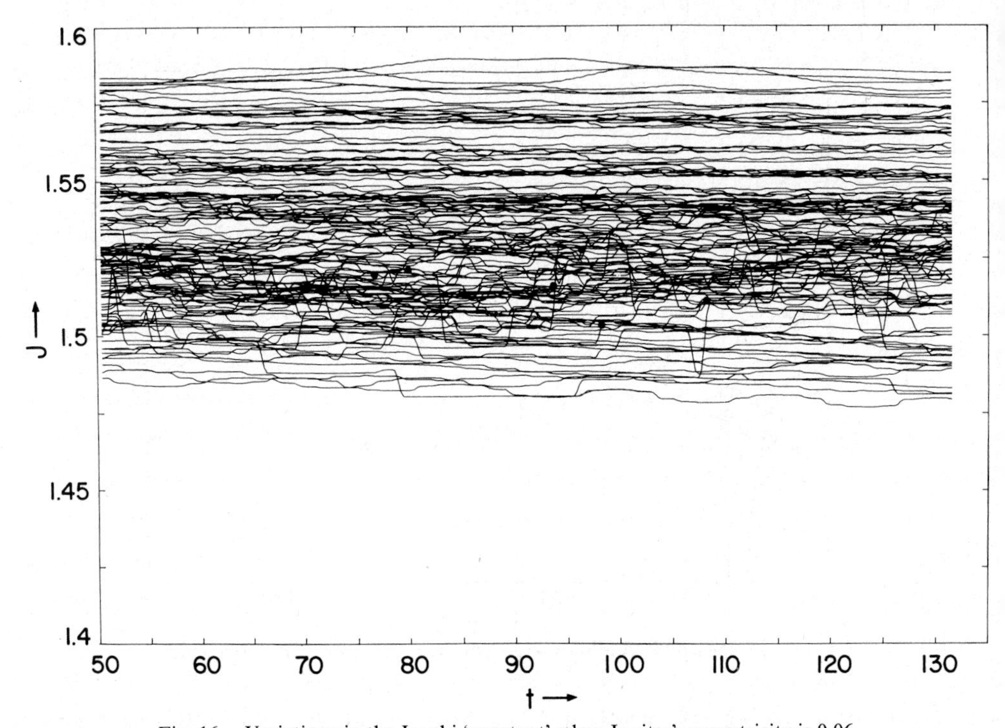

Fig. 16. Variations in the Jacobi 'constant' when Jupiter's eccentricity is 0.06.

collide with librators (see, for example, Figure 5). Thus, at the 1/2 resonance, there is a clean gap rather than a family of bodies. This provides a possible and simple explanation for the curious feature that at some resonances (2/3, 3/4, 1/1) a concentration of bodies exists, while others (1/3, 2/5, 3/7, 1/2) present a gap in the space density.

3. The Region Between Jupiter and Saturn

One-hundred asteroids were started with semimajor axes distributed from 1.1–1.75 Jupiter units with uniform surface density and with eccentricities randomly chosen between 0.0 and 0.1. Saturn, introduced only in this calculation, has a semimajor axis of 1.83 Jovian units; its mass was assumed to be 30% of Jupiter's and its eccentricity was taken as 0.08 – near its maximum value. Figures 17, 18, and 19 show the semimajor axes vs time for the intervals 0–160, 160–330, and 330–500 Jupiter periods. Figure 20 presents the surface number density – in the form of a histogram – at

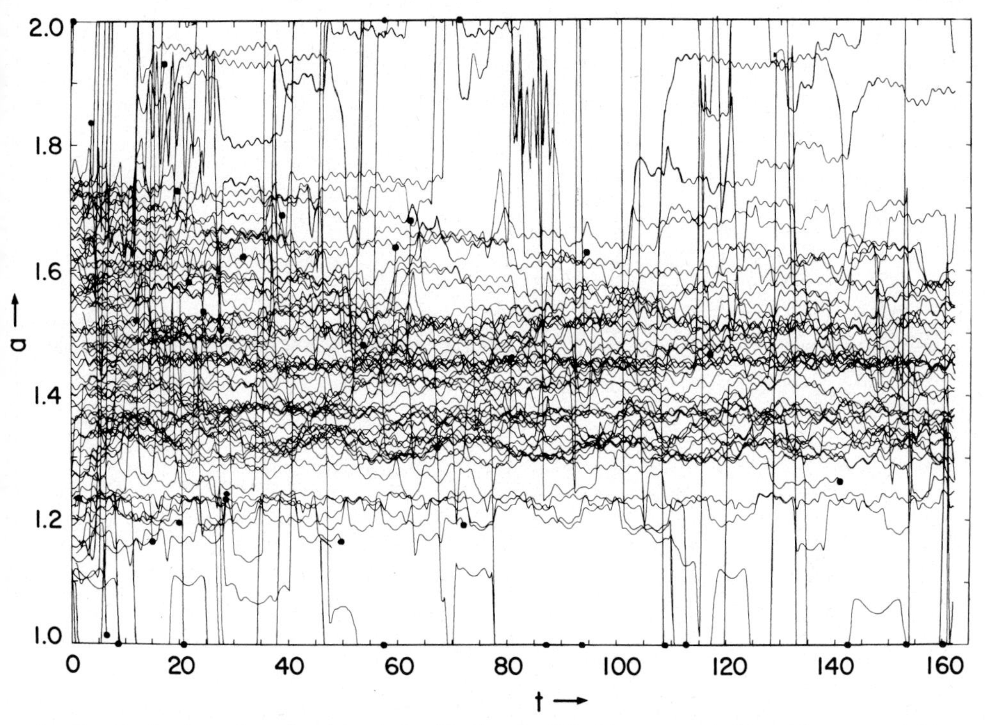

Fig. 17. Semimajor axis vs time in periods of Jupiter for 100 objects started with random eccentricities between 0.0 and 0.1. All other orbital elements, except semimajor axis, are also selected at random.

various times. Inspection of the trajectories leads us to estimate that only the bands at 1.30 and 1.45 Jupiter units (6.8 and 7.5 AU) are possibly stable. Numerous asteroids injected inside Jupiter's orbit in our first study were found to librate, thereby avoiding close approaches to Jupiter. The concentration of bodies at $a \cong 0.825$ in Figure 8, for example, owes its existence to apparently stable librators at the 3/4 resonance with Jupiter. Between Jupiter and Saturn, however, bodies that librate permanently appear to be absent, presumably because of the combined effects of two major planets.

4. Conclusions

Figure 21 compares the initial and final surface densities of the computer-simulated asteroids, where the rectangles outline the initial surface density.

We conclude that even if the Jupiter–Saturn region were initially populated with asteroids, objects there would, with the possible exception of bodies in two narrow bands, be ejected in a few thousand years.

In the Mars–Jupiter region, the central problem remains unanswered. In the short interval simulated (200 Jupiter revolutions), the region between the 1/2 and 2/3 resonances remains well populated. Assuming that asteroids were initially present there, we have to look to Jupiter perturbations over much longer times, to collisions be-

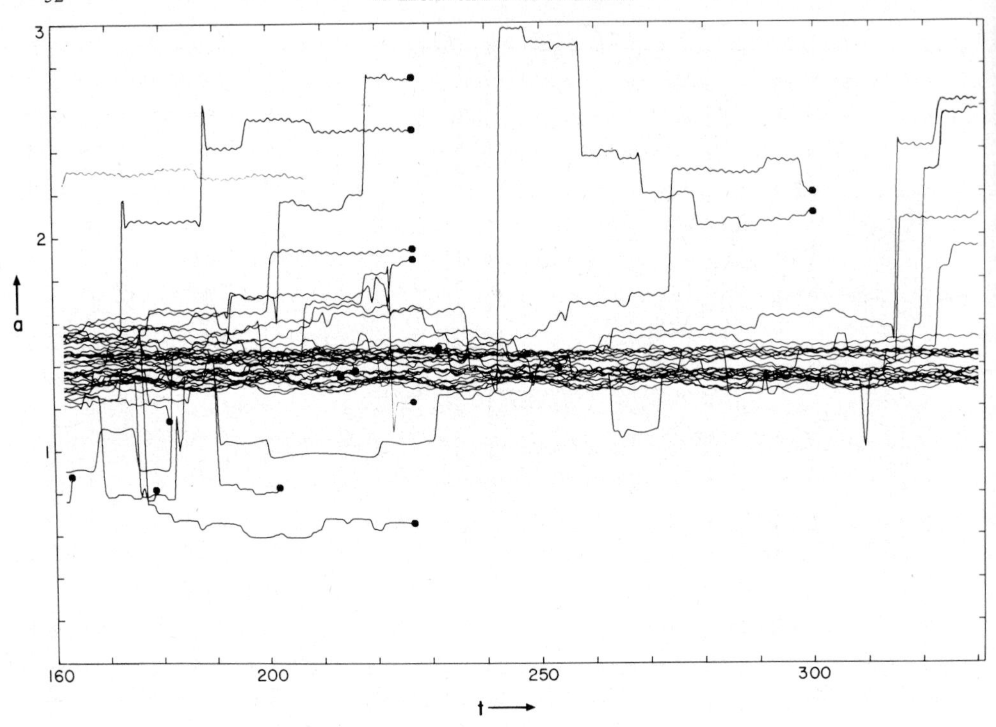

Fig. 18. Continuation of Figure 17 to 330 revolutions of Jupiter. Circles again indicate objects that have approached either planet to 10 radii.

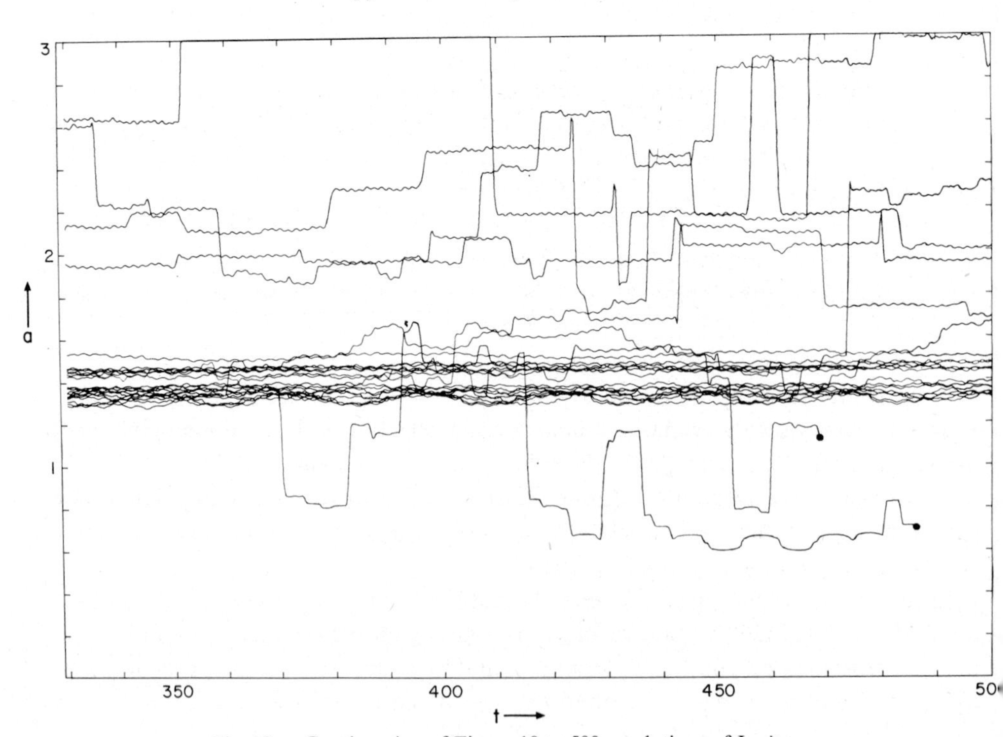

Fig. 19. Continuation of Figure 18 to 500 revolutions of Jupiter.

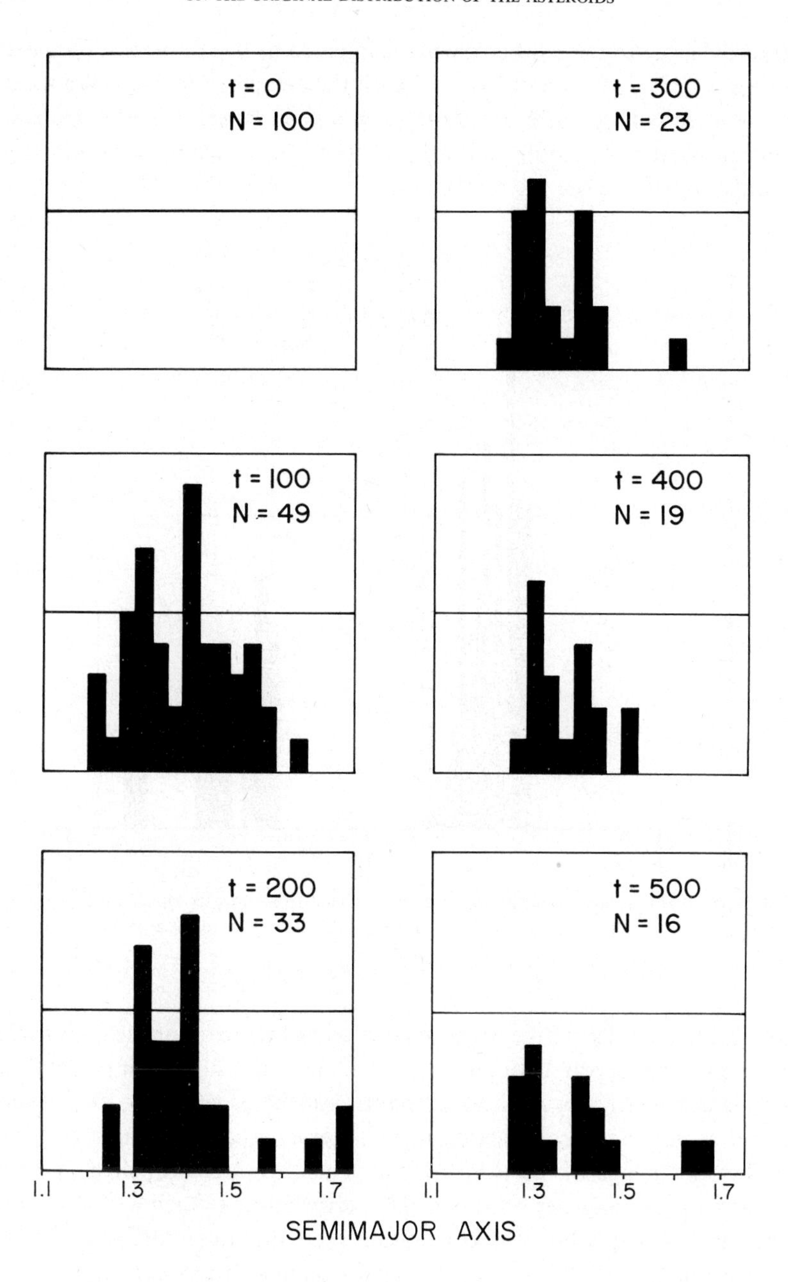

Fig. 20. Histogram of the number density of objects between Jupiter and Saturn as a function of time. Bodies near 1.3 and 1.45 may remain for some time; those near 1.65 are temporary.

tween asteroids, or to some other mechanism (such as gas drag) to depopulate that region. Figure 22 is a version of Figure 7 with the escapers removed. We see that the oscillations in semimajor axis are quite pronounced exterior to the 1/2 resonance (0.63), indicating that collisions could have been quite effective. The activity drops off markedly interior to the 1/2 resonance.

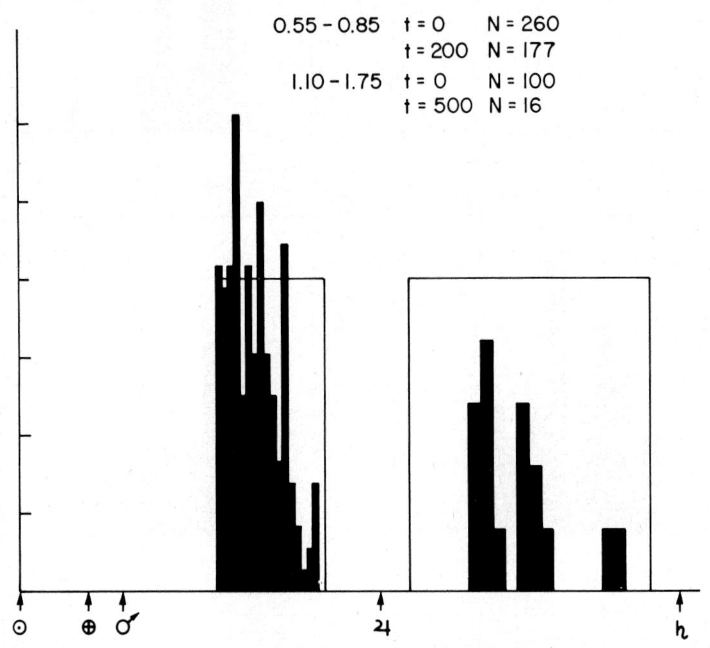

Fig. 21. Results of this survey: Number of remaining objects scaled for comparison with the initial distribution. N is the total number of bodies remaining after the indicated times in periods of Jupiter.

However, the time derivative of the approximate Jacobian for the elliptic restricted three-body problem is proportional to μe (the mass ratio of Jupiter to the Sun multiplied by the eccentricity of Jupiter), a quantity of the order of 6×10^{-5}, which indicates a time scale long in comparison to the interval studied in this survey but short with respect to the age of the solar system. Thus, we are compelled to attempt an investigation of the long-period or secular perturbations. Such a study is also suggested by the fact that, while it is known that there are no secular terms in the semimajor axis to order μ^2 (10^{-6}), the situation for higher orders is uncertain.

Times of the order of 10^6 Jupiter revolutions are not yet accessible by direct numerical integration such as was used in this study. Averaging techniques would bring such times within reach, but we strongly suspect that averaging stabilizes the equations of motion; an effect we are anxious to avoid. Thus we are led to a mixture of analytic and numerical techniques on which we will report in a subsequent communication.

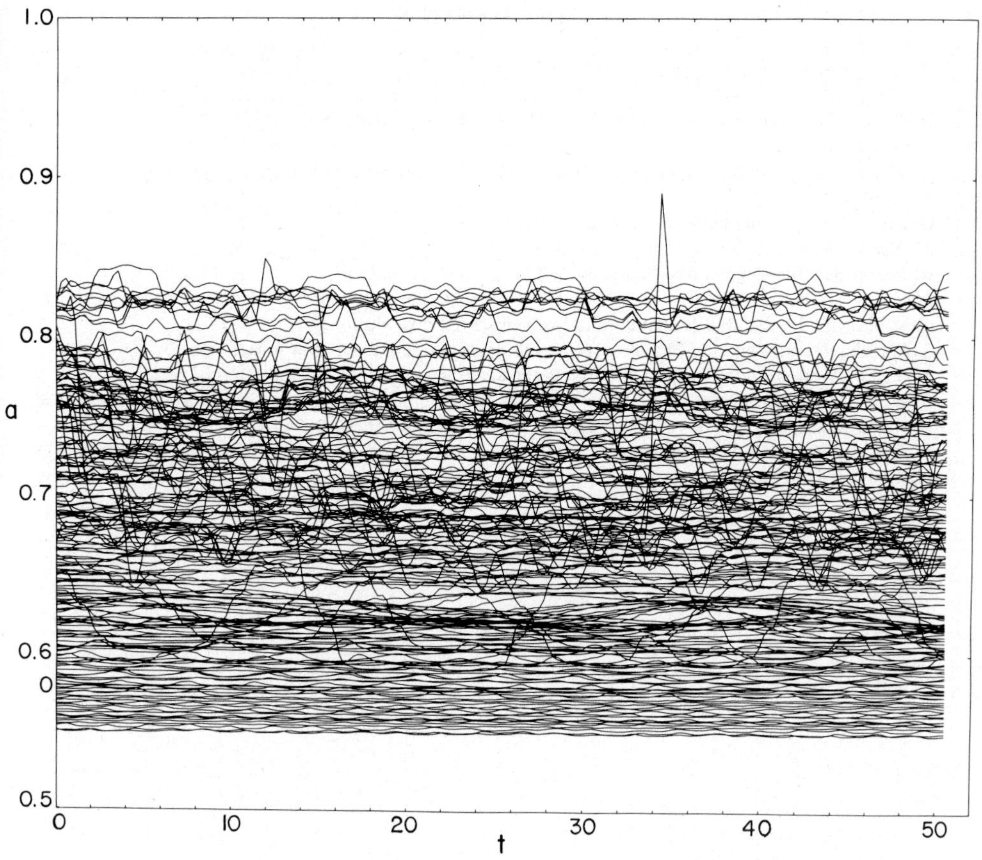

Fig. 22. Semimajor axis vs time in periods of Jupiter with escaping objects removed.

Acknowledgements

We wish to express our gratitude and regard for the competent and imaginative computer programming of our colleague Mr Rudolf Loeser. Prof. Giuseppe Colombo provided valuable insights in our many conversations. We also acknowledge the encouragement and support of Prof. Fred L. Whipple.

References

Brouwer, D. J.: 1963, *Astron. J.* **68**, 152–59.

Brown, H., Goddard, I., and Kane, J.: 1967, *Astrophys. J. Suppl. Series* **14**, 57–123.

Lecar, M., Loeser, R., and Cherniack, J. R.: 1973, in D. Bettis (ed.), *Solution of Differential Equations and Application to the Gravitational Problem*, Springer-Verlag, Berlin (in press).

van Houten, C. J., van Houten-Groeneveld, I., Herget, P., and Gehrels, T.: 1970, *Astron. Astrophys. Suppl.* **2**, 339–448.

DISCUSSION

H. Scholl: Which integration method did you use?

M. Lecar: A method which uses analytical forms for the first four time derivatives of the force.

H. Scholl: Is there any asteroid in the Hecuba gap which left the gap after 6000 yr?

M. Lecar: No.

H. Scholl: Did you use different values for the masses of Jupiter and Saturn in your computations?

M. Lecar: No.

G. Hertz: Did you integrate in rectangular coordinates?

M. Lecar: Yes.

G. Hertz: Is your original hypothesis correct?

M. Lecar: I am not sure. For the relatively short time intervals that we integrated asteroids still remained between 3.3 and 4.0. It is in conflict with observed distribution.

L. Kresák: Do you find any Apollo-type asteroids originating from your initial orbits?

M. Lecar: I don't know. We did not follow the asteroids after they left the asteroid belt.

R. H. Miller: You attributed the larger value of Jacobi constant (1.536 instead of 1.520) to the effect of Jupiter's eccentricity. Have you checked that a circular orbit for Jupiter yields a value of 1.520?

M. Lecar: In the circular problem the Jacobian is conserved to a part in 10^5 for the time intervals we integrated in this investigation.

THE ORIGIN OF COMMENSURABILITIES IN THE SATELLITE SYSTEMS

S. F. DERMOTT

Department of Geophysics and Planetary Physics, The University, Newcastle upon Tyne, NE1 7RU, England

The distribution of orbits in the satellite systems of the major planets is definitely nonrandom as there is a marked preference for commensurability among pairs of mean motions in these systems. If this preference is not the result of a formation process then it follows that since the time of satellite formation dynamical evolution has occurred. In this paper (a full version of which is to be submitted to another journal) I show that there is a bulk of evidence in favour of the hypothesis that orbital evolution has occurred as the result of tidal dissipation in the planets. This evidence can be listed as follows:

(i) In a tidally evolved satellite system there should be a linear correlation between log (orbital radius) and log (satellite mass). The expected correlation is observed among the inner satellites of Saturn and Uranus, the slope of the log-log plots being consistent with an amplitude and frequency independent Q (the tidal dissipation function). The correlation in Saturn's system must be largely due to either a formation process or possibly an early amplitude and frequency dependent tidal process, but certainly the present satellite distribution is completely consistent with the tidal hypothesis.

(ii) If Q is amplitude- and frequency-independent then the Mimas–Tethys and Enceladus–Dione resonances are stable under the action of tidal forces. This is an important result as it indicates that the necessary tidal dissipation occurred in the solid parts of the planets. Whether or not these planets have any solid parts at present is a matter of dispute.

(iii) The tidal hypothesis can account for the formation (capture into libration) of the Mimas–Tethys and Enceladus–Dione resonances.

(iv) If it is allowed that the age of the Mimas–Tethys resonance is not small ($>10^8$ yr) in comparison with that of the solar system and that orbital evolution since the time of satellite formation has been appreciable then it can be shown that tidal forces (with Q amplitude- and frequency-independent) can account for the present large amplitude of libration of this resonance.

(v) If tidal evolution in the satellite systems of Jupiter and Saturn has been appreciable then from the present orbits of Io and Mimas it can be deduced that Q (Jupiter) $\sim 1.1 \times 10^5$ and Q (Saturn) $\sim 1.2 \times 10^5$. As the masses of the two satellites differ by a factor ~ 2000 and the amplitudes of the tides they raise on their respective planets differ by a factor ~ 5000 this agreement in the Q-values is somewhat remarkable. But as the stability of the Mimas–Tethys and Enceladus–Dione resonances demands that Q is amplitude- and frequency-independent and as the chemical com-

Y. Kozai (ed.), The Stability of the Solar System and of Small Stellar Systems, 57–58. *All Rights Reserved.*

positions and structures of Jupiter and Saturn are probably similar then one should expect the Q's of these planets to be similar and thus the result supports the tidal hypothesis.

(vi) Finally, as the values of Q are so very high it would be remarkable if tidal evolution had not taken place.

It would appear that tidal forces alone can account for the observed preference for commensurability among pairs of mean motions in the satellite systems of the major planets. This does not mean that evolution due to point-gravitational forces has not occurred but it does mean that there is no need to invoke such forces to account for the above preference. These results obviously have cosmogonical importance and they may also be of interest to cosmologists. If the gravitational constant is varying inversely with the age of the Universe, T, then as Mimas must always have been above synchronous height it can be shown that $T \gtrsim 1.1 \times 10^{10}$ yr.

DISCUSSION

R. Greenberg: Your conclusions seem to depend on a very simple functional dependence of Q on amplitude and frequency. Can you show that a more general expression would yield the same results?

S. F. Dermott: My analysis, using $dn/dt = cn^x \mu^y$, has shown that appreciable orbital evolution of Mimas can be reconciled with the observed large amplitude of libration if at present $1/n(dn/dt)$ for Mimas is approximately the same as that for Tethys. This is true if the forces acting are tidal forces.

Y. Kozai: Is the mean motion of Mimas decreasing due to the tidal force by your theory? I asked this question because by analyzing data for 300 yr I found that the mean longitude of Mimas had secular acceleration.

S. F. Dermott: Yes, it is. However, the amount of change is so small and it cannot be detected from the observations.

THE ORIGIN OF THE ASTEROID RING

S. GĄSKA

Astronomical Institute, N. Copernicus University, Toruń, Poland

Abstract. From extended statistical investigations of the orbital elements of asteroids and of their dispersions it is found that the relations for some of the dispersions are stable and are regarded to be invariable. By assuming that the orbital elements derived by putting the dispersions equal to zero correspond to those of the mother planet of the asteroid ring, it is found that the mother planet is Mars.

In this paper statistical investigations are made for the orbital elements of asteroids and their dispersions. The data used are osculating elements tabulated in the *Ephemerides of Minor Planets for 1969* published by the Institute of Theoretical Astronomy at Leningrad, secular elements published by Brouwer (1950) and Jacobi's constant published by Sultanov and Ibragimov (1962).

It is well-known that the distribution function $n(\Pi)$ shown by dots in Figure 1, where n is the number and Π is the longitude of the perihelion, has the maximum at $\Pi \cong 10°$ and the minimum at $\Pi \cong 180°$. The maximum is especially prominent for $e > 0.14$ (Gąska, 1970). The distribution of points in (Π, φ)-plane $(e = \sin\varphi)$ suggests a strong correlation between Π and φ. The mean values of φ taken for arbitrary intervals of Π are shown as circles in Figure 1, which indicates that the relation $\bar{\varphi}(\Pi)$ can be represented by two straight lines. Similar results are derived even if the asteroids, for which the longitude of the perihelion is of libration with respect to those of Jupiter or Jupiter and Saturn, are rejected.

Similar statistics are performed for the two secular elements, the proper eccentricity, A, and the proper longitude of the perihelion, Π_1. In Figure 2 the distribution

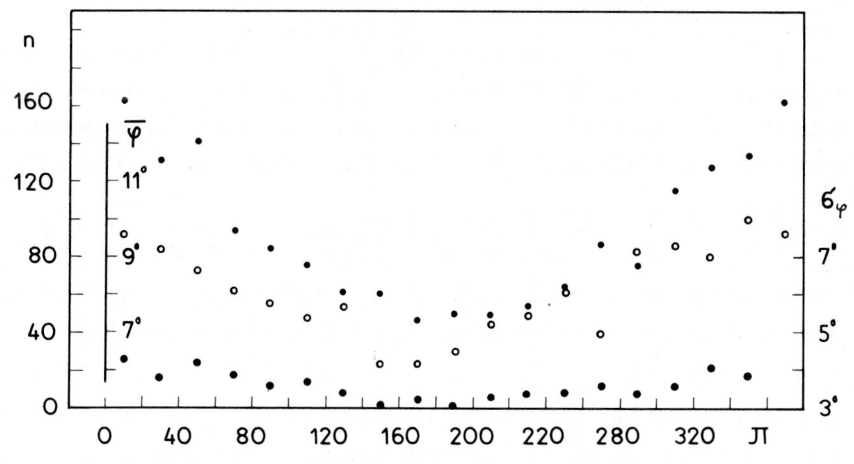

Fig. 1. The distribution function $n(\Pi)$ (dots), the mean value of φ (open circles), and the dispersions of φ (black circles) vs Π are shown.

Y. Kozai (ed.), The Stability of the Solar System and of Small Stellar Systems, 59–62.

function $n(\Pi_1)$, mean values of A, and the dispersions of A are shown as functions of Π_1 by dots, open circles, and black circles, respectively.

Comparing Figure 1 with Figure 2 we come to the following conclusions: (1) The distribution functions for Π and Π_1 are almost identical. (2) The dispersions are

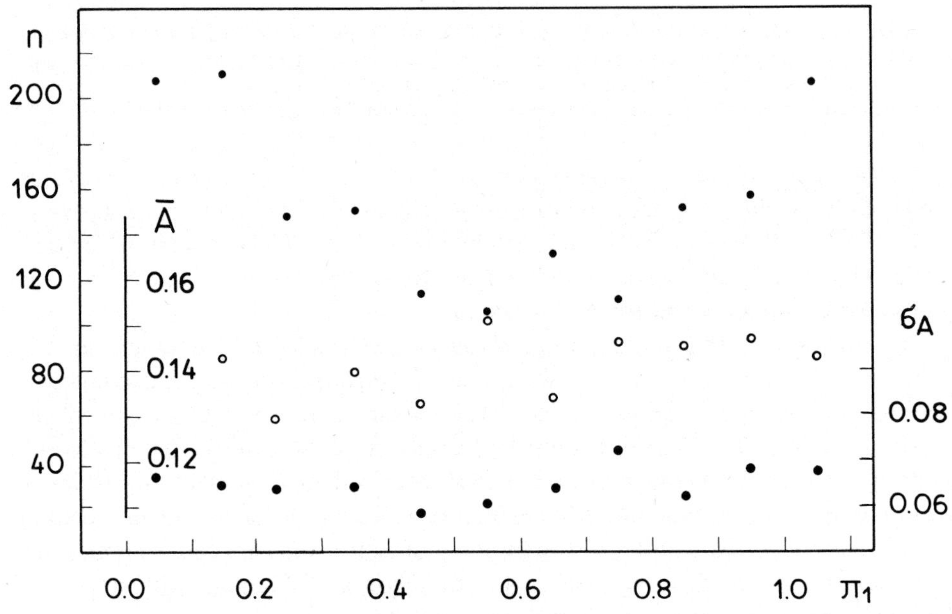

Fig. 2. The distribution function $n(\Pi_1)$ (dots), the mean value of the proper eccentricity, A (open circles), and the dispersion of A (black circles) vs the proper longitude of the perihelion, Π_1, are shown.

similar. (3) However, the relations $\bar{\varphi}(\Pi)$ and $A(\Pi_1)$ are not similar, as $\bar{\varphi}(\Pi)$ has been disturbed by the secular perturbations due to the major planets.

More interesting results are derived for the semimajor axis, a, the inclination, i, and the proper inclination, B. The relations, $a(\sigma_i)$ and $a(\sigma_B)$, dispersions σ_i and σ_B as functions of a, which are shown in Figure 3, do not lie on straight lines, and their deviations from them may be explained by Jupiter's large action due to commensurability.

Figure 4 shows similar relations, $a(\bar{i})$ and $a(\bar{B})$, the mean values of the inclination and of the proper inclination vs the semimajor axis.

By assuming that the relations $a(\sigma_i)$, $a(\sigma_B)$, $a(\bar{i})$ and $a(\bar{B})$ are linear with respect to a, the following expressions are derived by the method of least squares:

$$a = \underset{\pm 0.027}{0.221}\sigma_i + \underset{\pm 0.12}{1.82}, \qquad a = \underset{\pm 0.28}{10.81}\sigma_B + \underset{\pm 0.08}{1.87},$$

$$a = \underset{\pm 0.029}{0.222}\bar{i} + \underset{\pm 0.09}{0.90}, \qquad a = \underset{\pm 0.22}{7.22}\bar{B} + \underset{\pm 0.09}{1.63}.$$

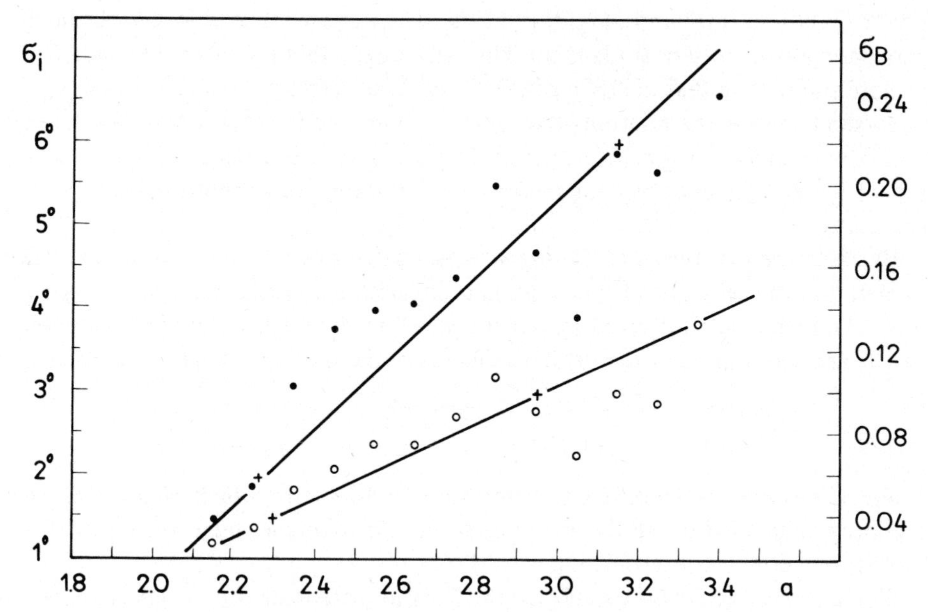

Fig. 3. The dispersion of the inclination, i (dots), and that of the proper inclination, B (open circles), vs the semimajor axis, a, are shown.

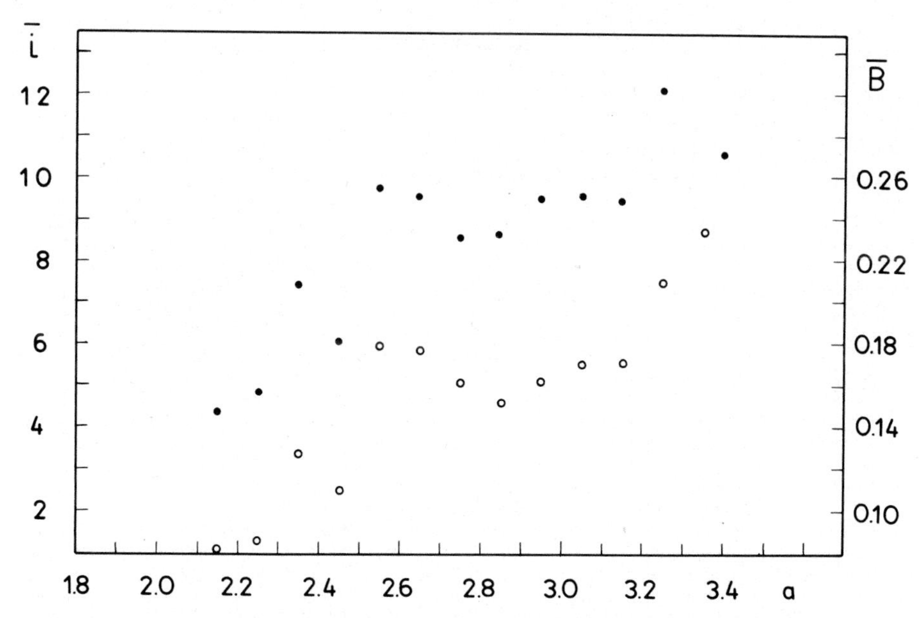

Fig. 4. The mean value of the inclination, i (dots), and that of the proper inclination, B (open circles), vs the semimajor axis, a, are shown.

For the relations $a(\sigma_i)$ and $a(\sigma_B)$ the values of a_0, the values of a, for which the dispersions vanish, are almost identical. This will suggest that the relation of type $D(\sigma_C)$ is more stable than that of $D(\bar{C})$, where C and D are arbitrary orbital elements.

By similar ways the relations $Q(\sigma_i)$, $Q(\sigma_B)$, $H(\sigma_A)$ and $H(\sigma_B)$, where $Q=a(1+e)$, and H is Jacobi's constant, are derived. Of these $H(\sigma_A)$ and $H(\sigma_B)$ cannot be represented by straight lines well, as near $H=920$ the relations are disturbed by Hill's instability.

By assuming that the asteroid ring was originated due to disintegration of a large planet, that the relations of type $D(\sigma_C)$ are invariable and that the original value of any orbital element is derived by putting $\sigma_C=0$ in the relation, the following set of values are determined as the orbital elements of the mother planet of asteroids:

$$a_0=1.85, \quad Q_0=2.24, \quad H_0=1109, \quad e_0=0.196.$$
$$\pm 0.19 \qquad \pm 0.10 \qquad \pm 9 \qquad \pm 0.043$$

These values are not so different from those for Mars, for which $H_M=1278$, while the mean value of H for all the asteroids is 943. Therefore, we may conclude that the mother planet of the asteroid ring was Mars.

Similar computations are made for the Perseid meteor stream, which is known to have originated from the Comet 1862 III. The results are:

$$a_0=24.2\ \mathrm{AU}, \quad q_0=0.979\ \mathrm{AU}, \quad e_0=0.960.$$

On the other hand they are $a=24.29$, $q=0.972$ and $e=0.960$ for the Comet 1862 III. The agreement is quite satisfactory (Gąska, 1973).

References

Brouwer, D.: 1950, *Astron. J.* **56**, 9–32.
Gąska, S.: 1970, *Bull. Toruń* **45**, 1.
Gąska, S.: 1973, *Bull. Toruń* (in press).
Sultanov, G. F. and Ibragimov, N. B.: 1962, *Katalog postojannytr Jacobi*, Baku.

ORBITS OF TROJAN ASTEROIDS

G. A. CHEBOTAREV, N. A. BELYAEV, and R. P. EREMENKO
Institute of Theoretical Astronomy, Leningrad, U.S.S.R.

Abstract. In this paper the orbital evolution of Trojan asteroids are studied by integrating numerically the equations of motion over the interval 1660–2060, perturbations from Venus to Pluto being taken into account. The comparison of the actual motion of Trojans in the solar system with the theory based on the restricted three-body problem are given.

1. Introduction

Since the orbits of Trojan asteroids are typical examples of the motion in the vicinity of the libration triangular points L_4 and L_5 in the restricted three-body problem, many investigations have been made for these asteroids, especially for libration orbits about the triangular points. In Table I the minimum angular distance of the longitudes of Jupiter and the asteroid, α_1 the maximum angular distance, α_2, and the libration period in years, T, are given for seven libration orbits based on the restricted problem of three bodies. Orbit No. 1 corresponds to the exact triangular solution, whereas No. 7 corresponds to the asymptotic solution, for which the longitude difference becomes as large as 180°.

TABLE I
Libration orbits near triangular points L_4 or L_5

No.	α_1	α_2	T
1	60°00	60°	147.42
2	51°32	70°	148.9
3	39°54	90°	153.9
4	29°79	120°	172.6
5	25°25	150°	215.7
6	24°05	170°	294.0
7	23°91	180°	∞

Applying the results to actual Trojans maximum and minimum deviations and the maximum possible deviations of the mean motion from that of Jupiter $(n-n_1)$ are obtained and are given in Table II.

The purpose of this paper is to compare the actual motion of Trojans in the Solar System with the theory based on the restricted three-body problem.

2. Evolution of Orbits of 15 Catalogued Trojans

In Table III the orbital elements of 15 Trojans given in *The Ephemerides of Minor Planets for 1972* are also tabulated. The asteroids are arranged in increasing order of

Y. Kozai (ed.), The Stability of the Solar System and of Small Stellar Systems, 63–69.

mean longitude. Ten asteroids move ahead of Jupiter, describing libration orbits around the point L_4. Five asteroids follow Jupiter (libration point L_5). Table III includes along the orbital elements the difference of the mean longitudes of these

TABLE II

Libration orbits of Trojans

Asteroid	α_1	α_2	$(n-n_1)$
588 Achilles	54°63	65°86	2″36
617 Patroclus	53°96	66°65	2″66
624 Hector	45°22	79°13	7″03
659 Nestor	49°07	73°11	5″01
884 Priam	49°71	72°20	4″69
911 Agamemnon	46°72	76°68	6″23
1143 Odysseus	51°37	69°95	3″88
1172 Aeneas	51°08	70°31	4″02
1173 Anchises	40°61	87°72	9″65
1208 Troilus	52°05	69°05	3″55
1404 Ajax	44°01	81°21	7″69
1437 Diomedes	35°64	99°61	12″83

asteroids and Jupiter $\Delta\lambda = \lambda - \lambda_1$. For this epoch $\lambda_1 = 342°8$ and the positions of libration points are $\lambda_4 = 42°8$ and $\lambda_5 = 282°8$.

For studying the orbits of Trojans a numerical integration of equations of motion has been made taking into account the perturbations by eight major planets (Venus–Pluto) covering the time interval of 400 years (1660–2060). The initial osculating elements are taken from *The Ephemerides of Minor Planets for 1972.*

Table IV summarizes the changes of orbital elements of Trojans for 400 years, and the maximum deviations $(n-n_1)$, as well as the maximum and minimum angular distances from Jupiter α_1 and α_2. Comparison of Tables II and IV demonstrates a good agreement between the theory and the real motion of Trojans. Changes of inclinations and eccentricities of the orbits are insignificant. The nodal lines are nearly stationary. The lines of apsides turn in direct direction by angles from 1°2 (1437) to 49°2 (659). During the time interval of 1660–2059 the approach of Anchises (1173) to Jupiter is the nearest; its minimum distance is $\Delta = 2.6$ AU. Small eccentricities are characteristic for the orbits of the Trojans.

Rabe (1967) suggested that in the case of the circular restricted three-body problem the limiting eccentricity for Trojans has the value $e = 0.19$; when allowance is made for Jupiter's orbital eccentricity this limit increases up to $e = 0.24$. From the catalogued Trojans the orbit of Achilles (588), with $e = 0{,}15$, has the maximum eccentricity.

In accordance with the theory for Trojans the difference of the longitudes of perihelia is also 60°, $(\pi - \pi_1) = 60°$. As $\pi_1 = 14°$, for the Trojans near the libration point L_5, the mean perihelion longitude should be equal to 74° and for those about the libration point $L_4 \pi = 314°$. Comparison with real Trojans proves that the perihelia of all known Trojans get into the interval of $\pi = 314° \pm 45°$, while the perihelia of the

TABLE III

Orbital elements of the Trojans. Epoch $T = 1950$ Nov. 15.0 ET

Asteroid	λ	$\Delta\lambda$	n	a	e	ω	Ω	π	i
1437 Diomedes	22°9	+40°1	305″4	5.130	0.048	128°9	315°3	84°2	20°6
1583 Antilochus	23°5	+40°7	292″7	5.276	0.054	186°3	221°1	47°4	28°3
588 Achilles	37°9	+55°1	299″1	5.202	0.150	128°4	316°4	84°5	10°3
624 Hector	40°4	+57°6	306″1	5.122	0.025	180°0	342°1	162°1	18°3
911 Agamemnon	47°7	+64°9	305″1	5.133	0.066	79°5	337°3	56°8	21°9
1749 Telamon	49°7	767°0	293″5	5.267	0.111	106°1	340°8	86°9	6°1
659 Nestor	51°2	+68°4	295″6	5.243	0.110	333°7	350°5	324°3	4°5
1647 Menelaus	51°7	+68°9	298″1	5.214	0.026	286°6	239°8	166°4	5°6
1143 Odysseus	53°1	+70°3	300″3	5.187	0.092	233°3	220°6	93°9	3°1
1404 Ajax	63°5	+80°7	301″9	5.170	0.113	57°6	332°3	29°9	18°1
884 Priam	269°8	−72°9	298″3	5.211	0.120	331°6	301°0	272°7	8°9
1172 Aeneas	273°2	−69°6	300″6	5.184	0.103	46°4	246°8	293°2	16°7
1208 Troilus	282°3	−59°5	302″6	5.161	0.093	292°9	48°0	340°9	33°7
617 Patroclus	287°9	−54°9	298″8	5.206	0.142	303°8	43°9	347°7	22°1
1173 Anchises	290°8	−51°9	308″4	5.097	0.137	30°9	284°2	315°1	7°0

TABLE IV

Orbital evolution of Trojans for 400 years

Asteroid	α_1	α_2	$(n-n_1)$	Δe	$\Delta\pi$	$\Delta\Omega$	Δi
1437 Diomedes	40°.1	93°.0	11″.6	−0.004	+ 1°.2	−1°.9	−0°.3
1583 Antilochus	37°.6	84°.8	8″.9	+0.013	+ 8°.7	−1°.1	−0°.1
588 Achilles	54°.7	66°.1	3″.2	0.000	+25°.9	−1°.1	0°.0
624 Hector	47°.4	81°.4	7″.4	−0.026	+19°.0	−1°.3	−0°.2
911 Agamemnon	43°.5	76°.0	6″.0	+0.007	+10°.0	−0°.9	−0°.1
1749 Telamon	48°.2	71°.9	6″.2	−0.005	+24°.4	−1°.9	0°.0
659 Nestor	52°.0	70°.0	4″.1	+0.030	+49°.2	−1°.8	0°.0
1647 Menelaus	56°.2	69°.0	4″.0	−0.033	+38°.0	−1°.4	+0°.1
1143 Odysseus	52°.1	70°.4	4″.6	−0.009	+20°.9	−1°.5	+0°.1
1404 Ajax	46°.5	81°.9	8″.7	+0.020	+24°.4	−2°.2	+0°.1
884 Priam	54°.2	73°.1	5″.2	+0.026	+31°.2	−2°.4	0°.0
1172 Aeneas	51°.5	69°.6	4″.6	+0.016	+24°.7	−1°.0	−0°.1
1208 Troilus	48°.4	69°.2	3″.9	−0.003	+12°.4	−0°.1	−0°.1
617 Patroclus	54°.8	63°.6	2″.2	−0.011	+21°.3	−0°.8	0°.0
1173 Anchises	43°.6	89°.4	10″.4	+0.010	+29°.2	−3°.5	+0°.1

three Trojans about the libration point L_4 (624, 659 and 1647) lie beyond the limit of $\pi = 74° \pm 45°$.

According to the orbital inclination, the Trojans may be divided into two groups (see Table V). In this table, besides i, the value $Z_0 = a \sin i$ is given, which indicates the height (in AU) over the plane of the ecliptic to which the Trojans can rise.

Table V gives also the size of Trojans. These dimensions have been determined

TABLE V

Distribution of Trojans by the angle of inclination

Asteroid	Z_0 (AU)	i	d (km)
1208 Troilus	2.86	33°.7	43
1583 Antilochus	2.50	28°.3	43
617 Patroclus	1.96	22°.1	57
911 Agamemnon	1.92	21°.9	65
1437 Diomedes	1.80	20°.6	57
624 Hector	1.61	18°.3	71
1404 Ajax	1.61	18°.1	34
1172 Aeneas	1.49	16°.7	52
588 Achilles	0.94	10°.3	52
884 Priam	0.80	8°.9	41
1173 Anchises	0.62	7°.0	37
1749 Telamon	0.55	6°.1	23
1647 Menelaus	0.52	5°.6	20
659 Nestor	0.41	4°.5	45
1143 Odysseus	0.28	3°.1	52

from their brightness by the formula (Combes, 1971):

$$\log d(\mathrm{km}) = 3.592 - 0.2g.$$

It is interesting to note that among the Trojans with high orbital inclinations the asteroids of larger dimensions prevail.

3. Orbital Evolution of 15 Noncatalogued Trojans

In 1970 the Palomar and Leiden observatories completed an extensive collective work (van Houten *et al.*, 1970) on taking the photographs and determining the orbits of faint minor planets. Among the discovered asteroids 15 typical Trojans were detected whose orbits are nearly circular. The orbital elements of these Trojans are given in Table VI. Photographs of Trojans were taken in the vicinity of libration point

TABLE VI

Orbital elements of 15 noncatalogued Trojans. Epoch 1950 Nov. 15.0 ET

Astroid	λ	$\Delta\lambda$	n	a	e	ω	Ω	π	i
9607	34°.7	51°.9	307″.6	5.105	0.111	256°.2	34°.1	290°.4	5°.0
6629	49°.3	66°.5	312″.4	5.053	0.008	299°.6	162°.1	101°.7	4°.2
6020	57°.1	74°.3	297″.0	5.226	0.093	90°.3	189°.0	279°.3	1°.4
4523	59°.2	76°.4	305″.8	5.125	0.050	129°.5	208°.5	338°.0	0°.9
4139	64°.0	81°.2	306″.7	5.115	0.003	309°.5	2°.7	312°.2	17°.6
2008	68°.6	85°.9	302″.8	5.159	0.112	167°.6	197°.2	4°.8	16°.8
6844	69°.0	86°.2	301″.1	5.178	0.104	24°.0	17°.6	41°.6	8°.2
4572	69°.1	86°.4	304″.7	5.138	0.058	42°.0	16°.0	58°.0	9°.3
4655	70°.6	87°.9	297″.7	5.217	0.032	93°.5	5°.8	99°.3	17°.1
6591	71°.1	88°.3	292″.8	5.276	0.043	57°.0	18°.6	75°.6	7°.4
4596	72°.1	89°.3	294″.9	5.251	0.070	331°.0	43°.9	14°.8	4°.0
6581	72°.3	89°.5	291″.6	5.291	0.031	54°.3	350°.6	45°.0	4°.9
6540	72°.6	89°.6	296″.6	5.231	0.059	67°.3	24°.0	91°.3	9°.1
6541	73°.3	90°.5	297″.3	5.223	0.088	263°.4	152°.9	56°.3	8°.1
2706	73°.3	90°.6	313″.4	5.042	0.120	355°.9	98°.8	94°.7	1°.2

L_4 and, therefore, all 15 asteroids are in the orbit ahead of Jupiter. The lack of Trojans with great inclinations is explained by the fact that the photographs were taken in the region of the ecliptic. Nevertheless, in the case of faint Trojans it is also possible to distinguish two groups of asteroids with considerable and moderate inclinations (see Table VII). In this case the largest asteroids will again get into the first group.

Only for 9 asteroids out of 15 the perihelion longitude comes into the interval of $\pi = 74° \pm 45°$.

The lines of nodes are so located that they form two groups enclosed within the intervals of 350°.6–43°.9 (9 asteroids) and 152°.9–208°.5 (5 asteroids). The asteroid 2706 has a longitude of node $\Omega = 98°.8$ which is very close to the longitude of Jupiter's node $\Omega_1 = 100°.1$. However, the lacuna 208°.5–350°.6 is completely filled with 13 catalogued Trojans whose longitudes of node are enclosed within the limits 220°.6–350°.5. Thus,

TABLE VII

Distribution of Trojans according to the angle of inclination

Asteroid	Z_0 (AU)	i	d (km)
4139	1.57	17°.6	13.5
4655	1.53	17°.1	17.0
2008	1.50	16°.8	30.0
4572	0.84	9°.3	11.0
6540	0.82	9°.1	11.5
6844	0.74	8°.2	6.8
6541	0.73	8°.1	11.5
6591	0.67	7°.4	14.0
9507	0.45	5°.0	16.0
6581	0.44	4°.9	15.5
6629	0.38	4°.2	13.5
4596	0.36	4°.0	14.0
6020	0.13	1°.4	10.0
2706	0.11	1°.2	6.2
4523	0.08	0°.9	13.5

in the distribution of the line of nodes we observe (Table VIII) a lacuna 48°.0–152°.8 wide, into which gets only one asteroid (2706). The middle of the lacuna, $\Omega = 100°.4$, exactly coincides with the longitude of Jupiter's node. To the descending node longitude of Jupiter ($\Omega_1 = 280°.1$) corresponds the second lacuna of 246°.8–301°.0, in which one asteroid, Anchises (1173), is located. This planet has the greatest angular velocity of the line of nodes ($\Delta\Omega = 3°.5$).

Table IX provides the general characteristic of orbital evolution of faint Trojans

TABLE VIII

Distribution of Trojans by the longitudes of nodes

Asteroid	Ω	Asteroid	Ω
4139	2°.7	4523	208°.5
4655	5°.8	1143 Odysseus	220°.6
4572	16°.0	1583 Antilochus	221°.1
6844	17°.6	1647 Menelaus	239°.8
6591	18°.6	1172 Aeneas	246°.8
6540	24°.0	1173 Anchises	284°.2
9507	34°.1	884 Priam	301°.0
4596	43°.9	1437 Diomedes	315°.3
617 Patroclus	43°.9	588 Achilles	316°.1
1208 Troilus	48°.0	1404 Ajax	332°.3
2706	98°.8	911 Agamemnon	337°.3
6541	152°.9	1749 Telamon	340°.8
6629	162°.1	624 Hector	342°.1
6020	189°.0	659 Nestor	350°.5
2008	197°.2	6581	350°.6

for 400 years. The eccentricity value of asteroid 4139 has passed through zero; the longitude of perihelion, therewith, has changed from $\pi = 187\overset{\circ}{.}7$ to $\pi = 312\overset{\circ}{.}2$. The perturbations of orbital inclinations and eccentricities are insignificant.

TABLE IX

Orbital evolution of 15 noncatalogued Trojans

Asteroid	α_1	α_2	$(n-n_1)$	Δe	$\Delta\pi$	$\Delta\Omega$	Δi
9507	51°.9	96°.1	11″.5	+0.004	+50°.7	−3°.5	0°.0
6629	42°.8	103°.7	13″.8	−0.022	−26°.1	−2°.9	+0°.1
6020	52°.7	74°.9	4″.5	+0.002	+61°.6	−2°.5	+0°.1
4523	40°.9	84°.0	6″.5	+0.029	+64°.9	−3°.7	+0°.1
4139	32°.5	93°.0	10″.1	−0.018	—	−2°.3	−0°.1
2008	41°.5	87°.8	9″.4	+0.036	+38°.4	−2°.2	+0°.2
6844	50°.3	86°.4	9″.5	+0.020	+26°.5	−2°.7	−0°.1
4572	37°.6	92°.0	10″.0	+0.005	+10°.7	−2°.6	0°.0
4655	45°.0	87°.9	9″.6	−0.018	− 2°.1	−2°.2	+0°.1
6591	38°.7	88°.3	10″.9	−0.008	− 1°.5	−3°.1	0°.0
4596	42°.1	89°.3	8″.8	+0.035	+28°.9	−2°.5	0°.0
6581	36°.2	93°.2	12″.5	+0.009	−26°.4	−4°.1	0°.0
6540	46°.1	89°.8	9″.7	−0.016	+15°.1	−2°.5	0°.0
6541	47°.7	90°.5	9″.9	+0.009	+22°.1	−2°.3	+0°.2
2706	37°.6	121°.0	16″.9	−0.014	+29°.7	+1°.3	0°.0

References

Combes, M. A.: 1971, *L'Astronomie*, Décembre, p. 413.

Rabe, E.: 1967, *Astron. J.* **72**, 10.

van Houten, C. J., van Houten-Groeneveld, I., Herget, P., and Gehrels, T.: 1970, *Astron. Astrophys. Suppl.* **2**, 339–448.

THE ROLE OF SATURN'S OBLATENESS IN THE MIMAS–TETHYS RESONANCE

R. GREENBERG
Lunar and Planetary Laboratory, University of Arizona, Tucson, Ariz. 85721, U.S.A.

Saturn's satellites Mimas and Tethys appear to be involved in a unique resonance. All orbit–orbit resonances, by definition, have the satellites' conjunction librating about some specific longitude. Equivalently, their orbital periods, measured relative to that longitude, are commensurable. Most orbit–orbit resonances are of the eccentricity-type; the conjunctions librate about the longitude of an apse of one orbit. The Mimas–Tethys resonance is of the inclination type.

These satellites have mean motions in a ratio of nearly 2:1. They have significant inclination and nearly circular orbits. Their masses are comparable so that their mutual perturbations are important. The masses contrast with, for example, the Titan–Hyperion resonance where the effect of Hyperion on Titan is negligible. The Mimas–Tethys conjunction librates about the midpoint of the longitudes of the satellites' nodes on Saturn's equatorial plane. This configuration holds true even as the nodes precess at rates on the order of a revolution per year (Brouwer and Clemence, 1961).

While the thrust of recent research has been toward explaining the origin of resonances, it is worthwhile to examine the physical mechanism which currently maintains this unique Mimas–Tethys case. The object is to explain this resonance in a physical way; the mathematical analysis has already been extensive and quite successful in describing the properties of this satellite interaction (e.g. Allan, 1969).

The typical mathematical analysis of this resonance is a study of the behavior of the resonance variable

$$\Psi \equiv 4\lambda_T - 2\lambda_M - \Omega_T - \Omega_M,$$

where λ is the mean longitude and Ω is the longitude of the ascending nodes. The longitude of conjunction of the two satellites is given by $\xi \equiv 2\lambda_T - \lambda_M$ and the average longitude of the ascending nodes is $\Omega_{AVG} \equiv (\Omega_T + \Omega_M)/2$. Clearly the resonance variable can be interpreted as twice the difference between the longitude of conjunction and the average longitude of the nodes. The second derivative is evaluated by Lagrange's planetary equations, ignoring short-period terms in the disturbing function and ignoring high orders of small quantities, leaving

$$\ddot{\Psi} = -C I_M I_T \sin \Psi,$$

where the quantity C is positive. Thus Ψ is stable at 0, i.e. conjunction is stable at the average of the nodes.

As outlined here, the analysis appears to hold for any reference plane from which the inclinations are small, not only for the equatorial plane. But changing the reference

Y. Kozai (ed.), The Stability of the Solar System and of Small Stellar Systems, 71–76.

plane slightly can drastically alter Ω_{AVG}. Why does the Mimas–Tethys conjunction librate about Ω_{AVG} measured on the equatorial plane instead of on some other reference plane? The answer to this question is that from the point of view of the satellites' dynamics, the equatorial plane is special because of Saturn's oblateness.

In fact the oblateness is included implicitly in the analysis that I have described here. Some of the short-period terms that were neglected from the disturbing function contained

$$\Psi_1 \equiv 4\lambda_T - 2\lambda_M - 2\Omega_M = \Psi - \Omega_M + \Omega_T$$

and

$$\Psi_2 \equiv 4\lambda_T - 2\lambda_M - 2\Omega_T = \Psi + \Omega_M - \Omega_T$$

as arguments of cosines. These arguments have short periods compared to Ψ only because of the rapid precession of Ω_T and Ω_M due to the oblateness of Saturn.

If we are to understand clearly the physical basis of this resonance we must be able to separate the role of the oblateness from the role of the mutual perturbations of the satellites. We can begin by first investigating the properties of inclination-type resonances with a model that does not include oblateness.

Consider the circular orbits of two satellites projected on a celestial sphere centered on their spherical primary (Figure 1). Let the north pole be defined by the total orbital angular momentum of the two satellites. Suppose the orbital mean motions are in a ratio of 2/1 so that conjunction occurs repeatedly at the same meridian a few degrees after the satellite's mutual node as shown here. Since the mutual inclination is small, virtually no orbital energy is exchanged at conjunction by gravitational interaction.

However, energy *is* exchanged before and after conjunction. It is evident from Figure 1 that the gravitational effects are greater before conjunction because the

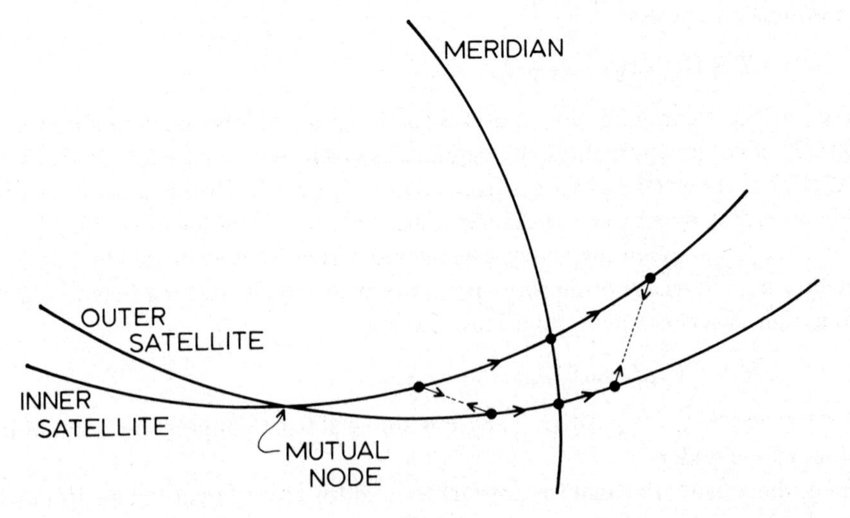

Fig. 1.

satellites are closer to one-another than they are after conjunction; moreover, the lines of force are directed more nearly along the direction of motion before than after conjunction.

Thus the net effect of the gravitational interaction is to transfer energy from the outer satellite to the inner one as occurs before conjunction. This energy transfer slows the angular velocity of the inner satellite, lowering the mean motion ratio below the commensurable value 2/1. Thus the next conjunction occurs further away from the mutual node of the satellites. Similarly if conjunction occurs before the mutual node, the net effect of the gravitational interaction is to cause conjunction to regress away from the mutual node.

Since there are two mutual nodes 180° apart, conjunction is stable at either of the two longitudes 90° from the satellites' mutual nodes. A more rigorous mathematical analysis has confirmed this result (Greenberg, 1973).

This type of resonance has two characteristics in common with eccentricity-type resonances. First, the stable condition is a 'mirror-configuration'. In other words, the velocities of the primary and its two satellites, measured relative to their center of mass, are normal to a plane containing all three masses. And second, conjunction is stable at that longitude where the orbits are furthest apart.

The inclination aspect of the Neptune–Pluto resonance is consistent with this description in that conjunction does librate about the longitude 90° from the planets' mutual node (Williams and Benson, 1971). However, the Mimas–Tethys resonance does not appear to be consistent with this description. Let us consider what happens to our model if we make the primary oblate. The ascending node of each satellite on the equatorial plane of the primary now precesses due to the oblateness. But it can be shown that the mutual perturbations of the two satellites still tend to draw conjunction toward the longitudes 90° from the mutual nodes. For want of a better name, we shall call these longitudes 'quasi-stable' since they are stable if the nodes do not precess.

We need to answer the following question: Why should conjunction librate about the midpoint of the nodes on the equatorial plane (as observed in the Mimas–Tethys case) instead of librating about the 'quasi-stable' longitude toward which it is drawn by mutual perturbations?

Let us compare the behavior of the various longitudes with time (Figure 2). Actually, the independent variable in Figure 2 is $\Delta\Omega$, the difference between the two ascending nodes on the equator. This quantity varies monotonically, and for our purposes, nearly linearly with time. For the Mimas–Tethys case, variation of $\Delta\Omega$ by 2π takes place in about a year.

All longitudes plotted in Figure 2 are measured from the longitude of the node of the inner satellite. The dotted lines represent the mid-point of the longitudes of ascending nodes. This longitude is a multi-valued function as shown. In this figure, the observed libration in the Mimas–Tethys case would be indicated by an oscillation of the conjunction longitude about a dotted line with a period of about 70 yr.

We also have plotted on this figure the quasi-stable longitudes. These have been calculated by straightforward spherical geometry and are shown as the solid wavy curves. These are the longitudes toward which the mutual gravitational effects tend to draw conjunction. The curves shown here were calculated for a case with the orbital inclinations equal to those in the Mimas–Tethys case. If these inclinations were equal,

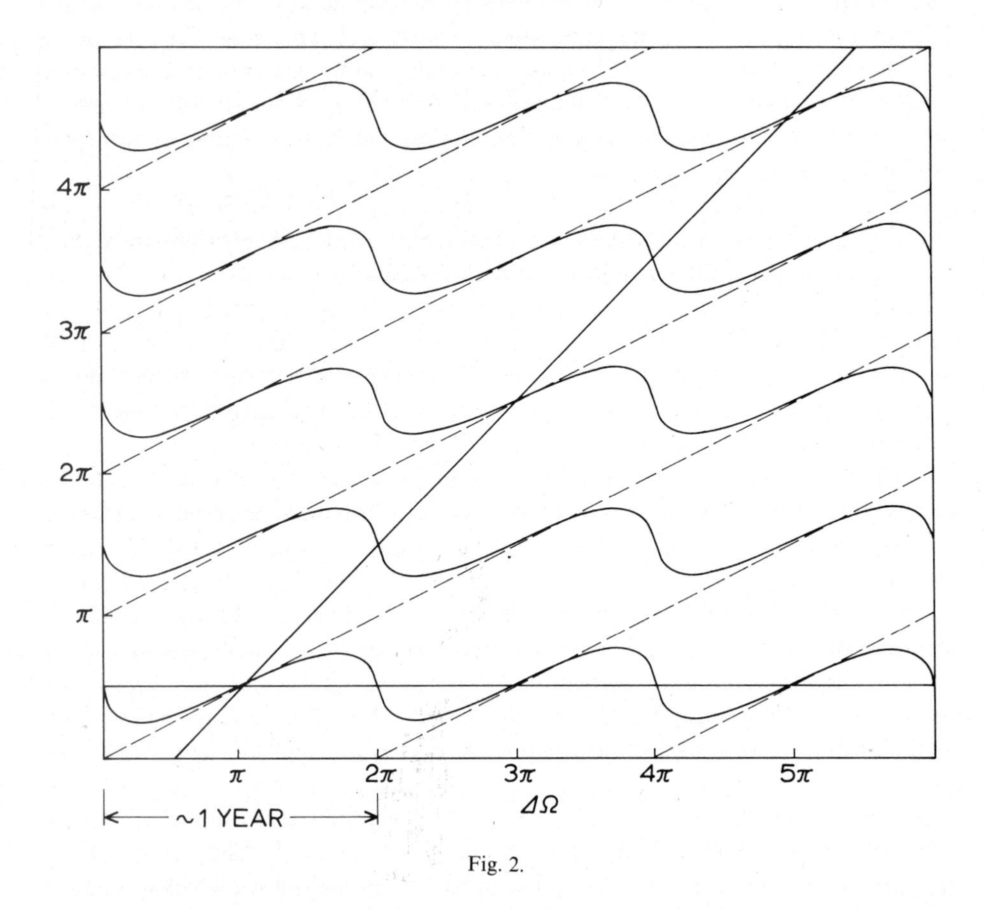

Fig. 2.

the curves would each be a sharp saw-tooth pattern with discontinuities at integer multiples of 2π. Our problem would be solved because in this case the quasi-stable longitudes would always be at the midpoint between equatorial nodes (Roy and Ovenden, 1955).

However, in the more realistic case where the inclinations are not equal, the curves of Figure 2 represent the quasi-stable longitudes. Suppose conjunction occurs at the midpoint between the equatorial nodes at an instant when the longitude is quasi-stable (e.g. at $\Delta\Omega = \pi$). At the quasi-stable longitude the mutual gravitational interaction does not accelerate conjunction one way or the other. Now suppose the ratio of the mean motions of the two satellites is such that the conjunction longitude follows one of the dotted lines.

Immediately the mutual interaction tends to pull conjunction down toward the quasi-stable longitude. However, the time scale for response to this pull is quite long. (You will recall that the libration period for the Mimas–Tethys case is about 70 yr.) Before conjunction can respond significantly it has moved along the dotted line to a region where the mutual interaction tends to pull conjunction up toward the nearest quasi-stable longitude. As conjunction follows a dotted line, it experiences exactly as much pull up as down, so that the net effect is zero. In this sense, we see that conjunction at the midpoint of the equatorial nodes is an equilibrium configuration.

Is this a stable configuration? Suppose conjunction moves just below a dotted line. Most of the time it is pulled upwards toward the quasi-stable longitude so that the net effect over several years is to draw conjunction toward the dotted line. In this way we can show heuristically that conjunction is in stable equilibrium on the dotted line.

In summary, we have found that the instantaneous influence of mutual perturbation is to draw conjunction toward the quasi-stable longitudes, but that the average effect, due to the planet's oblateness, is to draw conjunction toward the *observed* stable longitudes.

I would like to point out that the mathematical theory of this resonance indicates that conjunction might also be stable at either satellite's *descending* node on Saturn's equator. Such longitudes can be represented in Figure 2 by the solid straight lines. Note that the net acceleration along either of these lines toward quasi-stable longitudes is zero, in perfect agreement with the mathematical theory.

In a sense this physical interpretation has not gone beyond the mathematical treatment in terms of revealing new properties of the Mimas–Tethys resonance. However, it is important to understand the mechanism maintaining the present resonance if we are to investigate successfully the origin of the resonance. In this regard, I would like to point out one more possible interpretation of this figure.

We can regard the solid curves as representing, at any instant in time, the positions of minima on a sinusoidal potential field governing the behavior of the conjunction longitude. As time goes on, the potential topography alternately follows the average nodes and then jumps back 180° quite suddenly. In the case of the observed Mimas–Tethys resonance, the sudden jumps occur too quickly for the system to respond. However, this effect may have played a role in the events that first locked these two satellites in resonance, when the amplitude of libration was nearly 180°. In this context, an understanding of the role of Saturn's oblateness is essential.

References

Allan, R. R.: 1969, *Astron. J.* **74**, 497.

Brouwer, D. and Clemence, G. M.: 1961, in G. P. Kuiper and B. M. Middlehurst (eds.), *Planets and Satellites*, Univ. of Chicago Press, p. 31.

Greenberg, R. J.: 1973, *Monthly Notices Roy. Astron. Soc.* **165**, 305.

Roy, A. E. and Ovenden, M. W.: 1955, *Monthly Notices Roy. Astron. Soc.* **115**, 296.

Williams, J. G. and Benson, G. S.: 1971, *Astron. J.* **76**, 167.

DISCUSSION

G. Hertz: How does the difference of the longitudes of the ascending nodes of Tethys and Mimas behave? And how is the effect of Titan?

R. Greenberg: To this approximation each longitude of the ascending node varies at a constant rate. Thus, their difference also varies with a constant rate. The discrepancy between this approximation and the true behavior of the difference is too small to affect the qualitative description of the resonance mechanism. Similarly, the effect of Titan would not change the basic mechanism which I have described.

STABILITY OF ASTEROIDAL MOTION IN THE HECUBA GAP

H. SCHOLL and R. GIFFEN

Astronomisches Rechen-Institut, Heidelberg, Germany

There exist gaps in the frequency distribution of the osculating mean motions of the asteroids. In these gaps, the mean motions of the asteroids are commensurable to Jupiter's mean motion.

Recently, Giffen (1973) studied the stability of commensurable motion in the Hecuba gap where the ratio of the asteroid's mean motion to that of Jupiter is 2:1. Giffen's model was based on the plane, elliptic restricted three-body problem which means that e_A, $e_J \neq 0$, $i_A = i_J = 0$ and $m_A = 0$. The dimension of the corresponding phase space is 5. Poincaré developed a system of canonical variables that are suitable for the study of the secular behaviour of commensurable motion in the elliptic restricted three-body problem. This system was then modified slightly by Schubart. The Hamiltonian $H(\dots t)$ of this system depends on the mean anomaly which is a function of the time. If one removes that part of the Hamiltonian H which depends on the time then the new Hamiltonian H^* is constant and therefore is a nonclassical integral of the motion. Schubart obtained such a nonclassical integral by averaging the Hamiltonian H over the commensurable period. The resulting averaged Hamiltonian H^* depends then on four variables, that is, $H^* = H^*(a, e, \sigma, g)$, where σ is the critical argument. For a given value of H^*, the motion of an asteroid can be described in a three-dimensional phase space, $H^*(a, e, \sigma)$.

To study the stability of the motion in the Hecuba gap, Giffen used the method of invariant curves. He chose a certain plane in the three-dimensional phase space. Each time when the orbit crosses this plane, the point of intersection is plotted. Figure 1 shows these plotted points for an orbit with $e = 0.30$. The points are plotted when the eccentricity is close to its minimum. All the points lie on a curve. That means that each time when the orbit in the phase space crosses the designated plane, the intersection point will lie on this invariant curve. Therefore, apparently, for this orbit there exists an isolating integral which indicates that the motion of an asteroid in the Hecuba gap with eccentricity $e = 0.30$ is stable. Giffen found that all the other high eccentric orbits he computed have invariant curves.

In Figure 2, points for a low eccentric orbit, $e = 0.14$, are plotted. The points do not lie on an invariant curve. They have the tendency to fill the whole area. This orbit is 'ergodic'. Here and in the following 'ergodic orbits' mean 'wild' orbits. It is not stated that the orbits are ergodic in the exact mathematical sense.

More examples show that the small eccentric orbits in the Hecuba gap are all ergodic. 140 points are plotted here which corresponds to a run over 60000 yr. The points of the first 30000 yr fall in the area on the left side and the points of the second 30000 yr fall in the area on the right side of Figure 2. This indicates that on the aver-

Y. Kozai (ed), The Stability of the Solar System and of Small Stellar Systems, 77–79. All Rights Reserved.

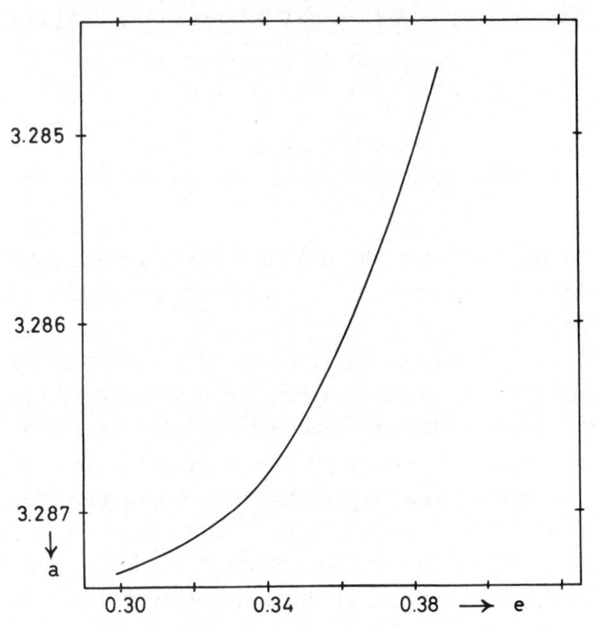

Fig. 1. Plots of intersections of orbits with a plane in the phase space, where the eccentricity is close to its minimum ($e=0.30$).

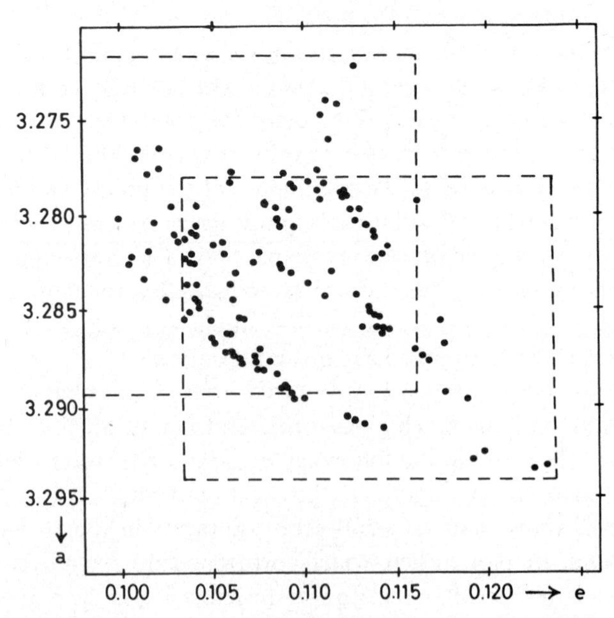

Fig. 2. Plots of intersections of orbits with a plane for $e=0.14$.

age the eccentricity increases. This fact leads Giffen to the following hypothesis: If there was an asteroid with small eccentricity in the Hecuba gap, then its eccentricity increased on the average. If the eccentricity was large, the collision probability with asteroids close to the Hecuba gap became larger, so that the asteroid left the Hecuba gap after collisions. This procedure could explain why there are no asteroids with small eccentricity in the Hecuba gap.

To test this hypothesis, we made computations over a longer period of time, over 200000 yr instead of 60000 yr. The results show that for the high eccentric orbits the invariant curves do not dissolve. For small eccentric orbits, the eccentricity increases but then decreases. On the average, it oscillates. No slow increase on the average in the eccentricity was found.

In addition, according to a suggestion from Froeschlé, we studied the divergence of orbits which start with slightly different initial values. The high eccentric orbits diverge very slowly while the small eccentric orbits diverge quickly. This result also indicates that the small eccentric orbits are ergodic and the high eccentric orbits are nonergodic. However, after 200000 yr, the eccentricity of the ergodic orbits did not increase as was expected by the hypothesis. But this is not yet a disproof of the hypotheses because it is possible that the small eccentric orbits develop by a diffusion process, which has been described by Froeschlé and Scheidecker (1974), into higher eccentric orbits and then escape from the gap because of collisions. In collaboration with Froeschlé we now are testing this new hypothesis and we also are testing if the invariant curves for the high eccentric orbits dissolve very slowly and become therefore also ergodic.

If there were asteroids with small eccentric orbits in the Hecuba gap and if these asteroids left the gap because of perturbations of Jupiter which increased their eccentricities so that collisions occured, then this process must have taken a much longer time than 10^5 yr.

Acknowledgements

The authors would like to express their gratitude to Dr J. Schubart, Astronomisches Rechen-Institut Heidelberg, for his helpful advice and for many valuable discussions. The calculations were carried out on the IBM 360/44 of the University of Heidelberg's Rechenzentrum.

References

Froeschlé, C. and Scheidecker, J.-P.: 1974, this volume, p. 297.
Giffen, R. B.: 1973, *Astron. Astrophys.* **23**, 387–403.

SECULAR PERTURBATIONS FOR ASTEROIDS BELONGING TO FAMILIES

Y. KOZAI and M. YUASA
Tokyo Astronomical Observatory, Mitaka, Tokyo, Japan 181

The first-order theory of secular perturbations for an asteroid indicates that the secular motions of the longitudes of the perihelion and of the ascending node have an equal absolute value with opposite signs, that is, $\dot{\Pi} = -\dot{\Theta} = b$, where b is a function of the semimajor axis and has the disturbing mass as a factor, if the eccentricity and the inclination are sufficiently small so that their squares are of the order of the disturbing mass and if there is no commensurable relation between the mean motions of the asteroid and the disturbing planet. Thus $\Pi + \Theta$ is a stable quantity according to the first-order theory, therefore, the origin of families of asteroids has been discussed by using the data for distribution of $\Pi + \Theta$ as well as those for dispersions of other orbital elements of asteroids belonging to a family.

However, $\Pi + \Theta$ is not so stable even if short-periodic terms are subtracted, since the eccentricity and the inclination are not generally so small and the second-order terms are not so small as one estimates by a formal expression (Kozai, 1953).

The secular terms of the second-order of the disturbing mass and of the fourth degree of the eccentricity and the inclination can be included in the disturbing function if there is no commensurable relation by either von Zeipel's or Lie-Hori's method easily. Although the differential equations including these terms are no more linear, they can be solved approximately (Kozai, 1953; Yuasa, 1973).

The results show that the mean secular motion of $\Pi + \Theta$ take the following form:

$$\dot{\Pi} + \dot{\Theta} = a_1 v^2 + 2a_2 v\mu + a_3 \mu^2 + m' a_4,$$

where μ and v are, respectively, the proper eccentricity and the proper inclination of the asteroid, m' is the disturbing mass, and a_j $(j = 1, \ldots, 4)$ are functions of the semimajor axis and have a factor of m'. Therefore, for a fixed value of the semimajor axis the equation $\dot{\Pi} + \dot{\Theta} = 0$ can be expressed by a quadratic curve in (μ, v)-plane. If the values of μ and v of asteroids belonging to a family are plotted as well as the quadratic curve, it is found that there is a tendency that the points are not so far from the curve, that is, $\dot{\Pi} + \dot{\Theta}$ takes a small value for an asteroid which belongs to any family.

There is also a very interesting relation between $\Pi_1 + \Theta_1$ and their time derivatives. In Figure 1 the values of $\Pi_1 + \Theta_1$ which are free from short-periodic perturbations are plotted against their time derivatives for asteroids of Flora family ($\alpha = a/a' = 0.43$). If points in left-side between 0 and 2π are moved vertically by 2π upwards and those in right-side are moved downwards by 2π, it is found that all the points line on a straight line. The slope of this line indicates that $T = 9.98 \times 10^5$ yr ago $\Pi_1 + \Theta_1$ took the same value for the asteroids. We cannot easily claim that this is the age of Flora family, since there is still ambiguity to determine this number.

Y. Kozai (ed.), Stability of the Solar System and of Small Stellar Systems, 81–82.

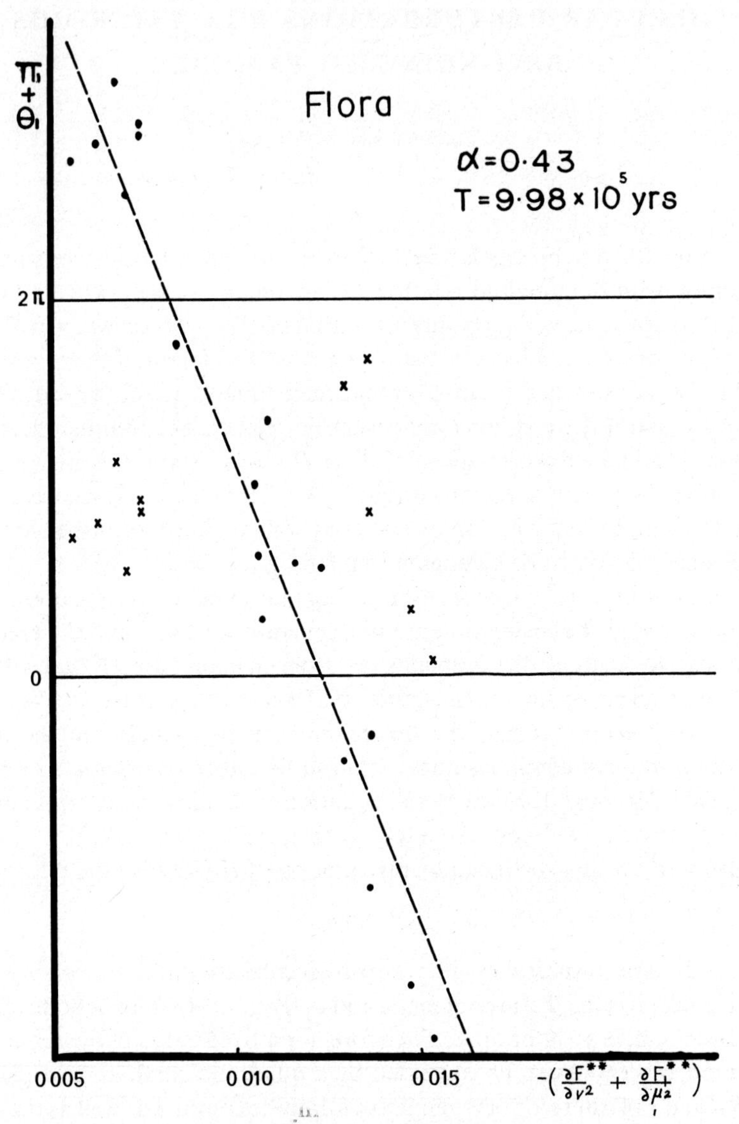

Fig. 1. $\Pi_1 + \Theta_1$ against their time derivatives for asteroids of Flora family. Points × are moved vertically by 2π to

For Coronis, Eos and Themis families $T = 3.7 \times 10^6$, 6.3×10^5 and 4.6×10^5 yr, respectively, are derived.

References

Kozai, Y.: 1953, *Publ. Astron. Soc. Japan* **6**, 41–66.
Yuasa, M.: 1973, *Publ. Astron. Soc. Japan* **25**, 399–445.

THE FORMATION OF DISKS BY INELASTIC COLLISIONS OF GRAVITATING PARTICLES. APPLICATIONS TO THE DYNAMICS OF THE SATURN'S RING AND TO THE FORMATION OF THE SOLAR SYSTEM

A. BRAHIC

Observatoire de Paris-Meudon, Université de Paris VII, France

Abstract. We integrate numerically the evolution of a three-dimensional system of particles with finite dimensions, which bounce inelastically upon each other. The particles are subjected to the attraction of a central mass; their mutual attraction is neglected. This model is used to study the evolution of Saturn's ring. The first results are presented: such a collision mechanism can flatten very quickly the Saturn's ring and the system tends towards a final equilibrium state.

Collisions between macroscopic objects are surely a common feature of the Universe: these objects may range from interstellar clouds to solid bodies in the solar system. We would like to know in what way such collisions have affected the evolution of the system in which they occur.

Poincaré (1911) showed qualitatively that, in a nebula consisting of a number of bodies undergoing inelastic collisions, a central condensation will be formed, the system as a whole will flatten into a plane perpendicular to the initial angular momentum vector and the orbits of the bodies will become more circular.

Since the time of Descartes, Laplace and Poincaré, the idea of a primordial chaotic nebula has played a central role in cosmogonic speculation.

Many researchers invoke collisions between discrete bodies or 'flocules' to explain the formation of the solar system 'disk' and of the planets (MacCrea, 1960; Alfvén and Arrhenius, 1970a, b; Urey, 1972).

Collisions have certainly played an essential role in the formation and evolution of Saturn's ring (Jeffreys, 1947; Cook *et al.*, 1972). Many critical questions still remain unanswered:

Is the ring a few thousand millions years old and was it, in that case, created from circulating material which was flattened through inelastic collisions and which could not condense into a satellite because it is within the Roche limit? We may in this case ask why the ring does not extend uniformly to the planet itself.

Or is the ring younger than this and the debris of a satellite disrupted by tidal forces? Now, if such a satellite were to pass near to Saturn on an hyperbolic orbit, it would certainly be broken up, but in this case the pieces would not remain around the planet. It is difficult to envisage a satellite spiralling inward since, in this particular case, tidal forces would tend rather to push the satellite out. Note that if a satellite did get broken up inside the Roche limit and the debris remained in orbit, then since the rocks have some degree of cohesion, their dimensions would be about one hundred times the observed thickness of the ring (Bobrov, 1970).

Y. Kozai (ed.), The Stability of the Solar System and of Small Stellar Systems, 83–93.

Does the ring consist of one layer of bodies, or more? Since the last century, many researchers have suggested that the luminosity of the rings as a function of their inclination indicates the presence of several layers which throw shadows on each other; on the other hand, collisions would rapidly reduce the system to one layer only. It has been shown in the case of the lunar surface that one layer of material could be sufficient to generate such a 'cat's eye' effect.

What is the characteristic size of the material? What is its nature? Is it frost, boulders or anything else...?

There has been some analytic analysis made for systems of orbiting colliding particles, based on the Boltzmann equation, but in view of the approximations which have to be introduced, its usefulness in the present context is debatable.

With the help of Michel Hénon, I am at present developing a general programme for this kind of problem. It is a numerical simulation of a gravitating system of particles with inelastic collisions. Such a programme has many potential applications, for instance the dynamics of Saturn's ring, the formation of the solar system, the evolution of the nuclei of galaxies (Spitzer and Saslaw, 1966; Spitzer and Stone, 1967; Sanders, 1970), the formation of galaxies (Brosche, 1970).

Calculations of this kind have so far only been carried out by Ulam (1968), who was interested in the nuclei of galaxies, and by Trulsen (1972a, b, c), who studied the dynamics of jet streams and Saturn's ring. Corresponding numerical experiments (Alder and Wainwright, 1959, 1960; Rahman, 1964; Verlet, 1967, 1968) have simulated significant progress in molecular dynamics.

The technical details of the rather intricate calculations will be given in a paper in preparation. The principal difficulty is to know whether two particles will in fact collide.

A very similar approach has also been used by us in the case of the formation of galaxies. The results are presented elsewhere (Brahic, 1973).

The time scale t_c of the evolution of the systems is of the order of the mean time between collisions for one particle. Therefore, it is inversely proportional to the number N of particles and to their geometrical cross-section, which is itself proportional to the square of the radius R of a particle.

In fact, if we change either N or R, we simply make the evolution go slower and faster. Consequently, a proper choice for N and R permits us to follow realistically the evolution of our system using a minimum of computer time.

Results have been obtained for the simplest case in which particle orbits are Keplerian around a central mass point. Positions and velocities at any given time are obtained from Kepler's equation. In a collision, the normal component is multiplied by a coefficient k which lies between 0 and 1.

The kinetic rotational energy of the bodies has been neglected. Each collision is assumed to be instantaneous.

The initial conditions were set up by selecting at random the six elements of the Keplerian orbit in such a way that trajectories were all ellipses lying between two spheres centred on the central mass point and with inclinations lying between 0 and some maximum value. We have assumed that clouds on hyperbolic trajectories escape at once and, for technical reasons, that clouds near to the center are captured by the center of mass.

In our experiments, a particle undergoes on the average one collision per revolution: a higher frequency and therefore a smaller mean free path would change the

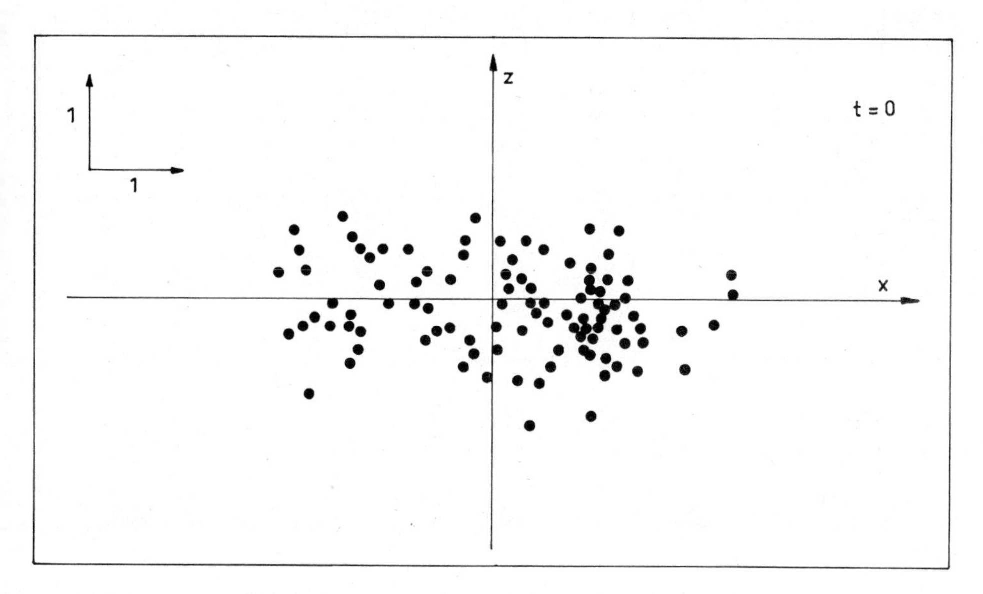

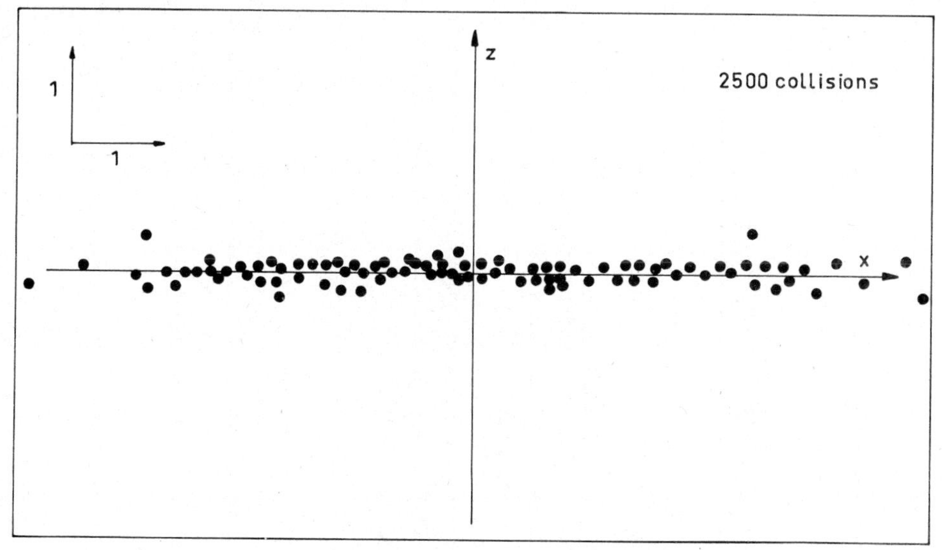

Fig. 1. Projections onto a plane parallel to the initial angular momentum vector at the initial time and at the end of the computation. The number N of clouds is equal to 100. Each cloud has the same radius $R=0.07$ and the same mass. The coefficient of elasticity k is equal to 0.1. The maximum value of the initial inclinations of the orbits is equal to 0.5 rad. After 2500 collisions, no cloud has escaped and twelve out of 100 clouds have fallen on the central body of radius 0.1.

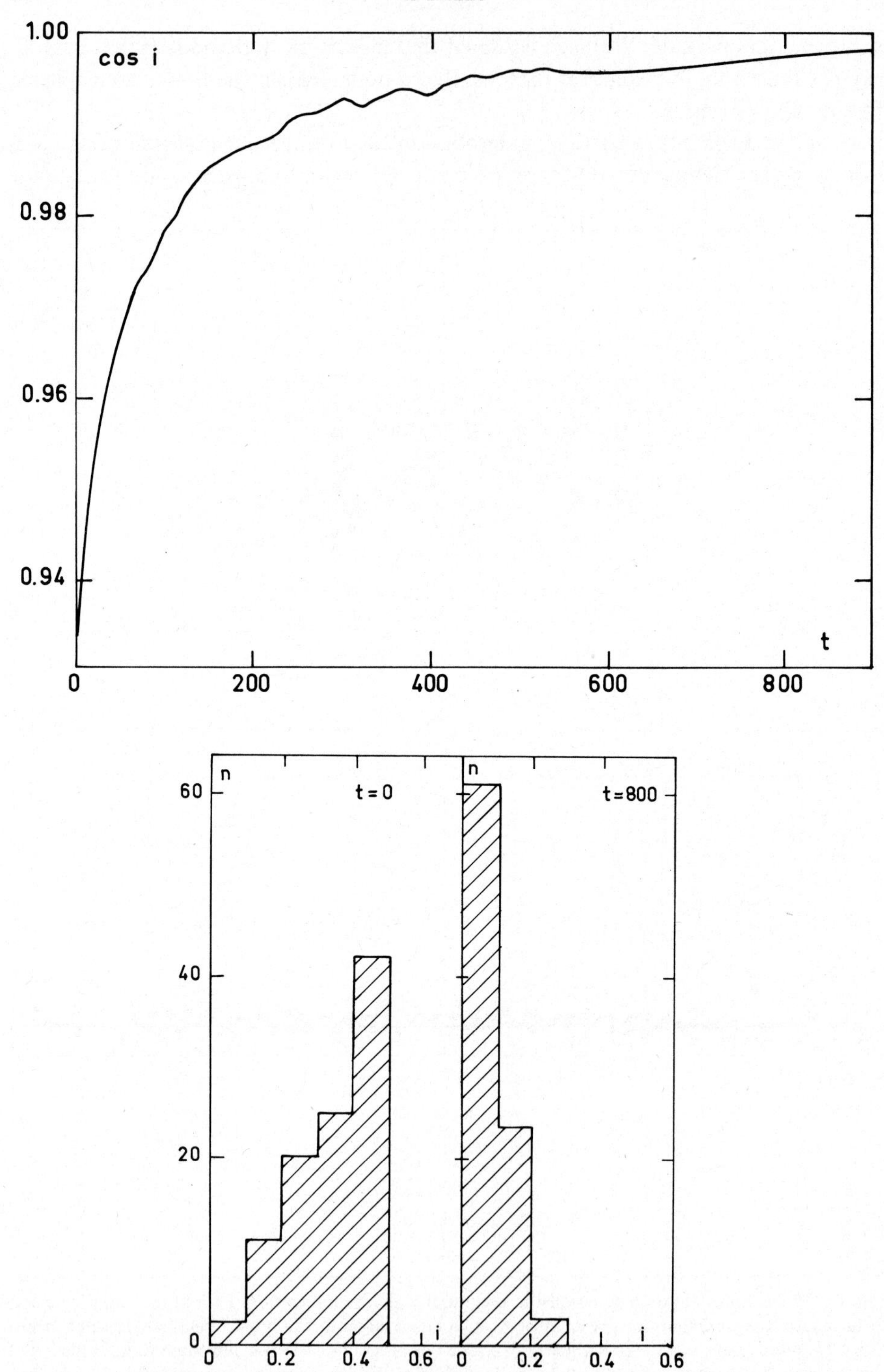

Fig. 2. Top: the variation as a function of time of the mean cosines of the inclination i. Bottom: the histogram of i at the initial time and at the end of the computation. The unit of time corresponds approximately to the mean time necessary to one cloud to turn of one radian on its orbit. $k = 0.1$.

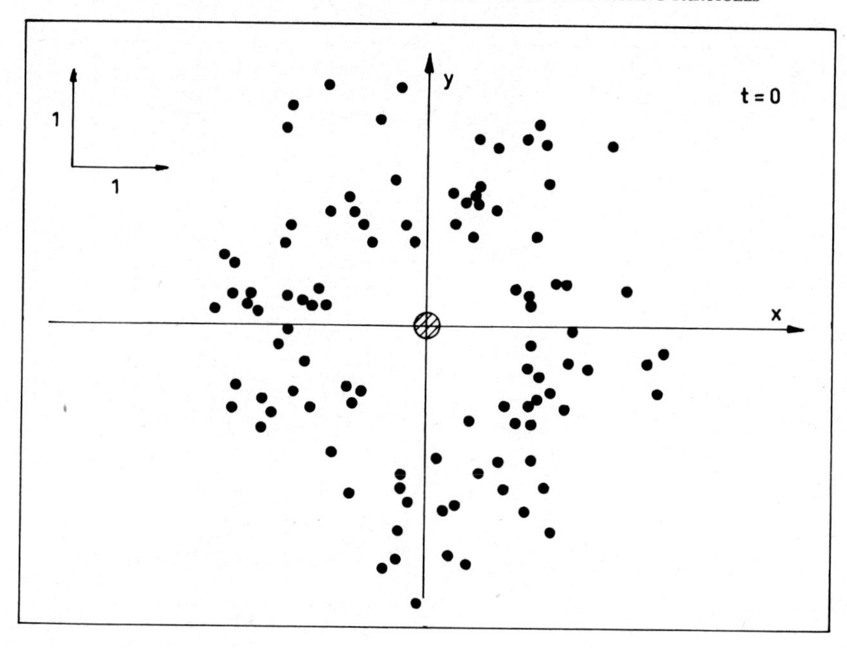

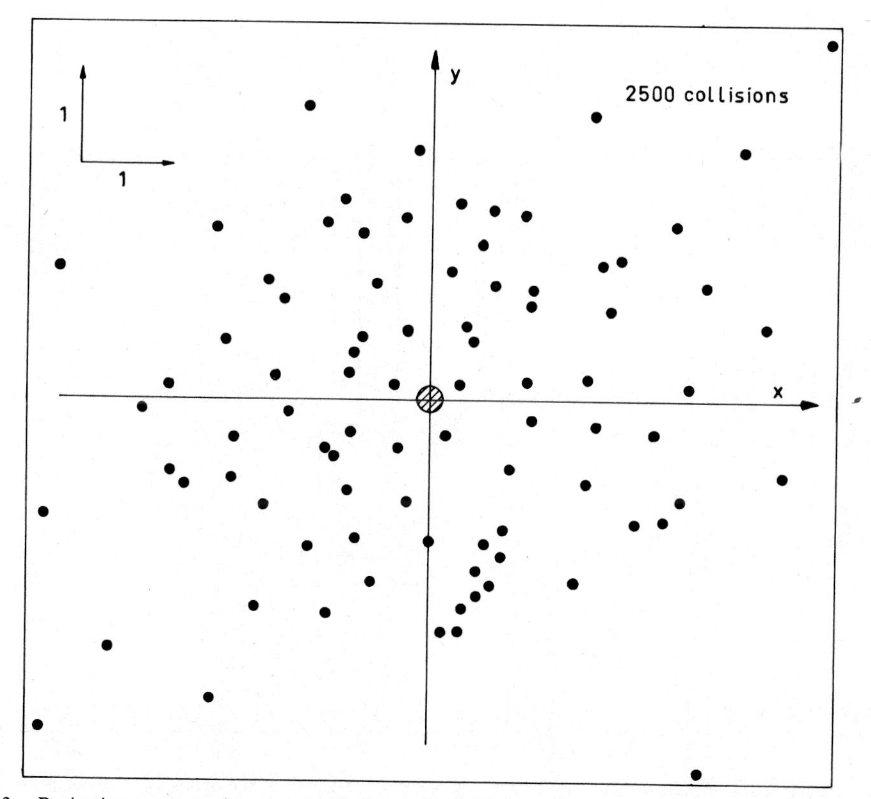

Fig. 3. Projections onto a plane perpendicular to the initial angular momentum vector. Initial trajectories are all ellipses lying between two spheres of radius 1 and 3 respectively centred on the central mass point. $k=0.1$.

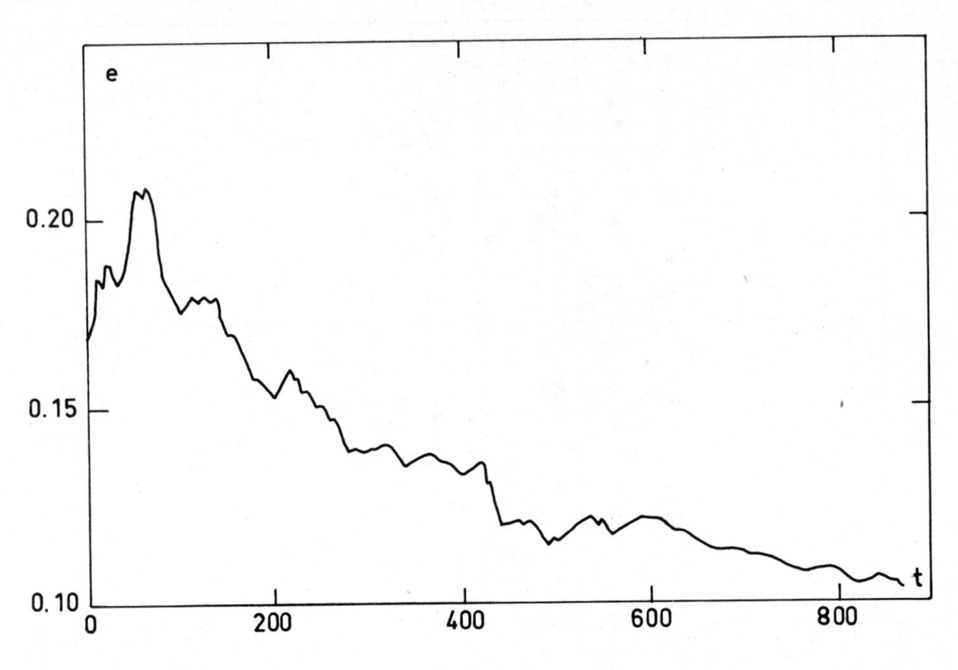

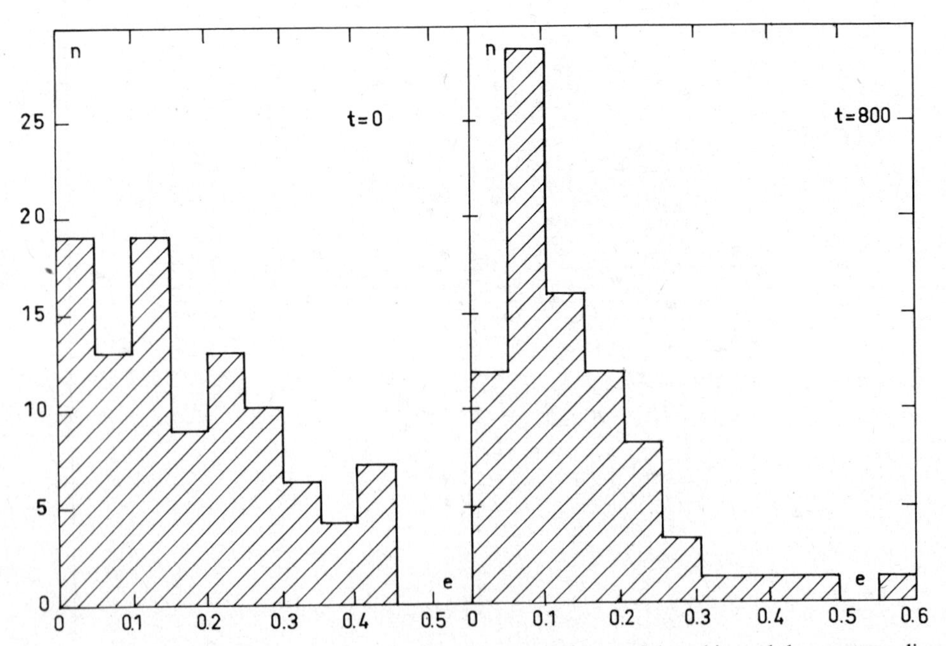

Fig. 4. The variation as a function of time of the mean excentricity of the orbits and the corresponding histograms. $k=0.1$.

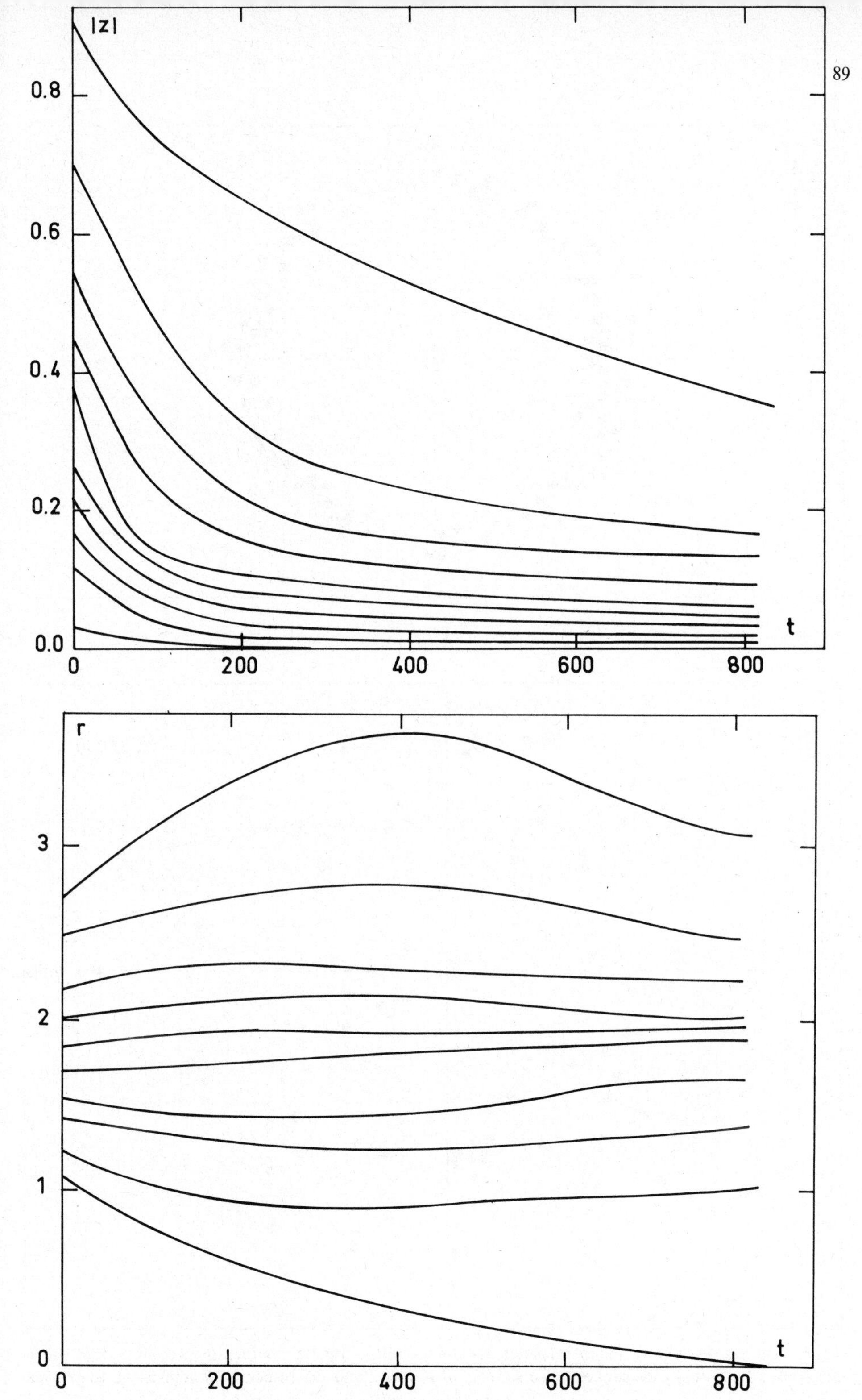

Fig. 5. Top: the ordinate represents the distance $|z|$ to the reference plane. At each time, the N bodies are sorted in order of increasing $|z|$ and divided into ten groups. The curves represent the mean value of $|z|$ for each group. Bottom: same figure for the distance r to the center.

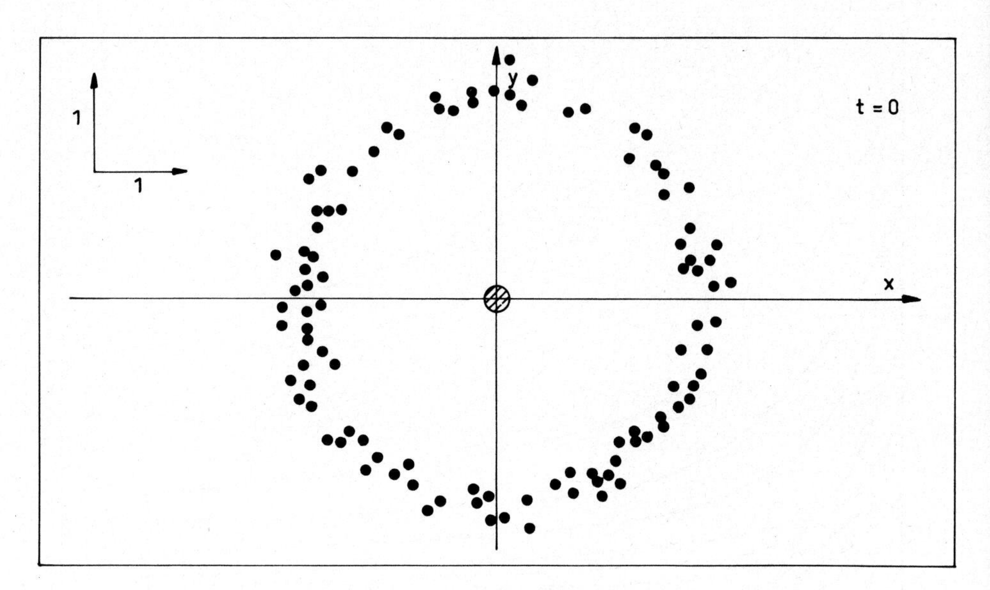

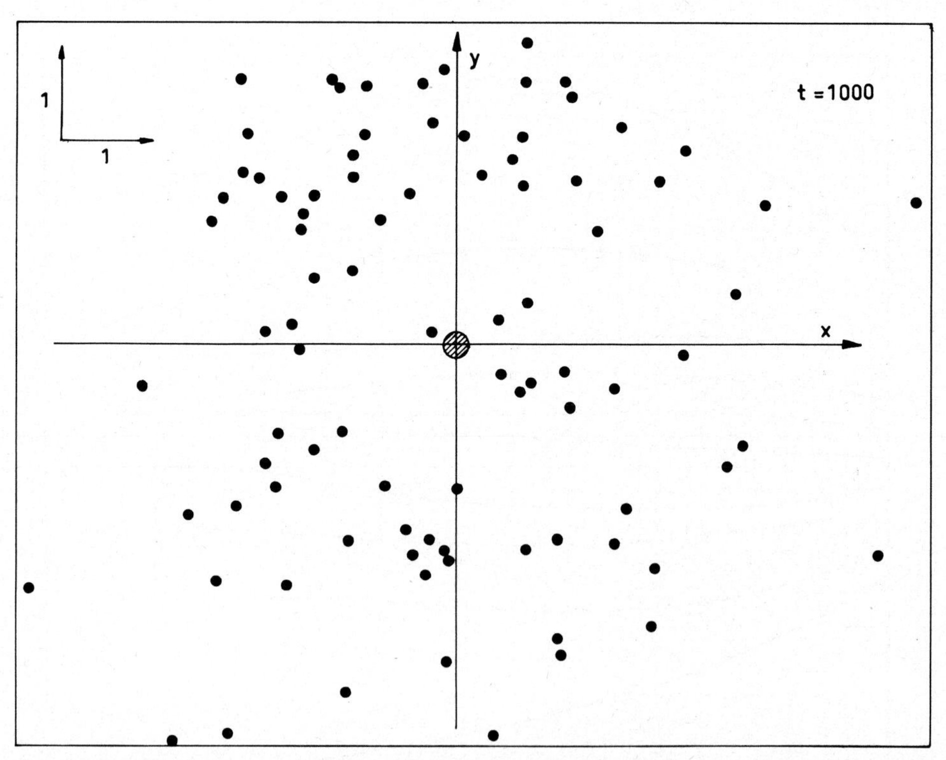

Fig. 6. Projections onto a plane perpendicular to the initial angular momentum vector. Initial trajectories are all ellipses lying between two spheres of radius 2 and 2.5 respectively centred on the central mass point. $k = 0.3$.

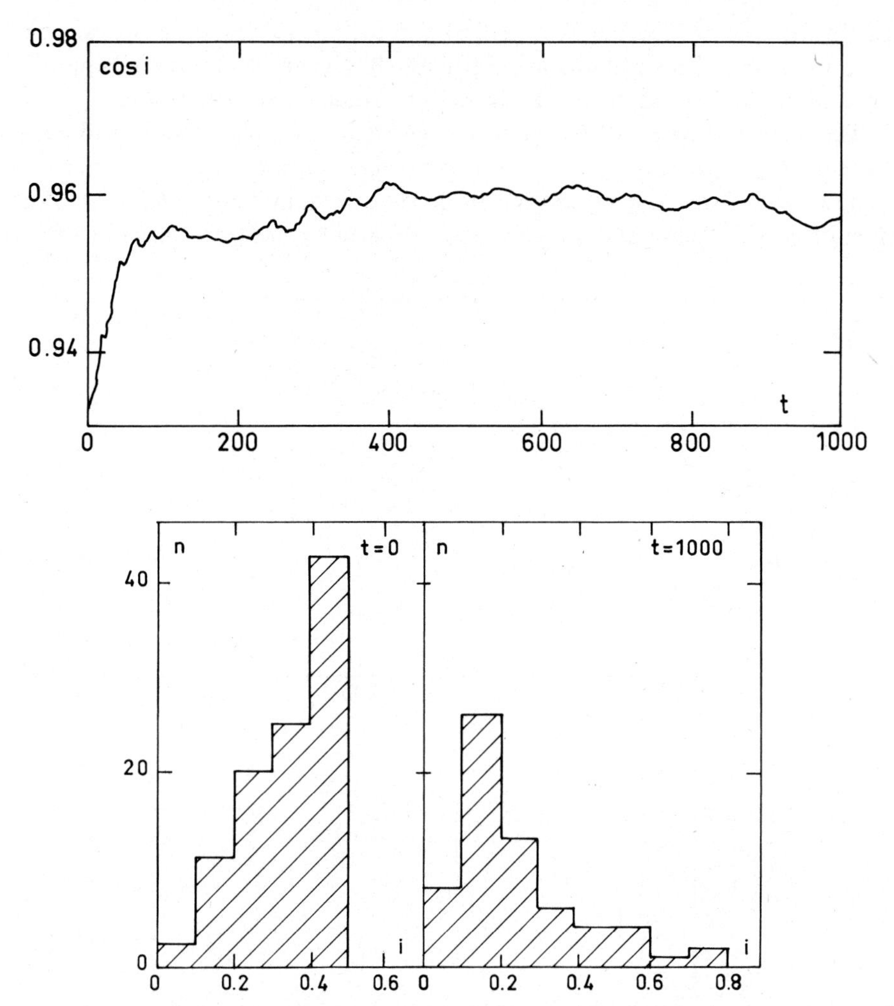

Fig. 7. Top: the variation as a function of time of the mean cosines of the inclination i. Bottom: the histogram of i at the initial time and at the end of the computation. $k=0.7$.

physics because then the effect of each collision would be only local and the system would behave like a gas. On the other hand, a smaller frequency would increase computing time without changing fundamentally the physics.

Figures 1, 2, 3, 4 and 5 show an example of the evolution of the system. We can see on Figure 1 that the system has flattened considerably after 2500 collisions, but it is not yet completely flat. Figure 2 shows the same flattening in a more quantitative way. Figure 3 shows a spread on radius: the system extends. We can see on Figure 4 that, at first, the excentricities increase because a thermal equilibrium is established between radial and vertical velocities. After that the orbits tend to become more and more circular. Figure 5 shows the flattening and the concentration towards the centre in a more detailed way. Figures 6, 7, 8 show the evolution of the system for a different case. In the case of Figure 6, the smallest and the largest initial distances from the centre are 2 and 2.5 instead of 1 and 3 (Figure 3). The bodies are initially more concentrated in a thinner shell and the spread on radius is very important.

Alfvén and Arrhenius (1970a, b) have suggested that a planet is formed out of a jet stream. A jet stream is a group of bodies moving in similar elliptical orbits around a central body. Now, a given body will necessarily return to the point at which it last suffered a collision and since, moreover, partially inelastic collisions will tend

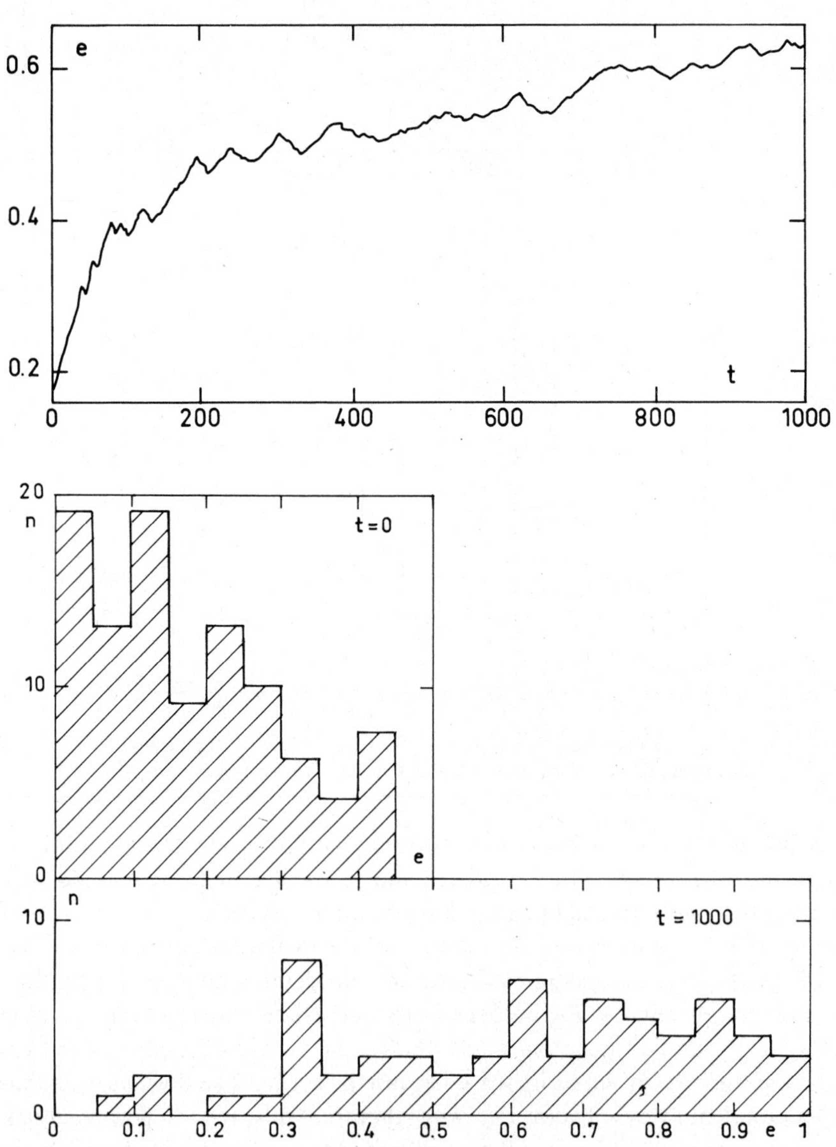

Fig. 8. The variation as a function of time of the mean eccentricity of the orbits and the corresponding histograms. $k=0.7$.

to make orbits increasingly similar, it seems not unlikely that trajectories will become 'focused', and the circulating bodies will form themselves into a jet stream stable against external perturbations.

This theory of the formation of planets is very attractive; however, it is by no means clear that such a stream can actually be formed out of a more uniform distribution. After a collision between two bodies A and B, if a third body intervenes and collides either A or B, A and B do not collide again. Experiment does not indicate a focusing process.

In the case of Figures 7 and 8, the collisions are more elastic than in the case of Figures 2 and 4. The flattening process is not very efficient if the coefficient of elasticity is too large and the orbits are less circular.

Of course, this first model must be perfected. But the results already suggest (a) that the collision rate decreases with time and the system tends towards a final equilibrium state, and (b) that such a mechanism can flatten the system. The flattening is already important when each particle has suffered about ten collisions. It is a very fast evolution since probably each particle in the case of Saturn has suffered about one collision per day.

Subsequent models will include fragmentation and coalescence during each collision, and also the use of a more realistic central field taking into account the important flattening of Saturn. The effect of a satellite can also be easily included.

References

Alder, B. J. and Wainwright, T. E.: 1959, *J. Chem. Phys.* **31**, 459.

Alder, B. J. and Wainwright, T. E.: 1960, *J. Chem. Phys.* **33**, 1439.

Alfvén, H. and Arrhenius, G.: 1970a, *Astrophys. Space Sci.* **8**, 338.

Alfvén, H. and Arrhenius, G.: 1970b, *Astrophys. Space Sci.* **9**, 3.

Bobrov, M. S.: 1970, in A. Dollfus (ed.), *Surfaces and Interiors of Planets and Satellites*, Academic Press, London, New York, p. 377.

Brahic, A.: 1974, in J. R. Shakeshaft (ed.), 'The Formation and Dynamics of Galaxies', *IAU Symp.* **58**, in press.

Brosche, P.: 1970, *Astron. Astrophys.* **6**, 240.

Cook, A. F., Franklin, F. A., and Palluconi, F. D.: 1973, *Icarus* **18**, 317.

Jeffreys, H.: 1947, *Monthly Notices Roy. Astron. Soc.* **107**, 260.

MacCrea, W. H.: 1960, *Proc. Roy. Soc. London* **A256**, 245.

Poincaré, H.: 1911, *Leçons sur les hypothèses cosmoniques*, Hermann, Paris, Chap. 5, p. 86.

Rahman, A.: 1964, *Phys. Rev.* **136A**, 405.

Sanders, R. H.: 1970, *Astrophys. J.* **159**, 1115.

Spitzer, L. and Saslaw, W. C.: 1966, *Astrophys. J.* **143**, 400.

Spitzer, L. and Stone, M. E.: 1967, *Astrophys. J.* **147**, 519.

Trulsen, J.: 1972a, *Astrophys. Space Sci.* **17**, 241.

Trulsen, J.: 1972b, *Astrophys. Space Sci.* **17**, 330.

Trulsen, J.: 1972c, *Astrophys. Space Sci.* **18**, 3.

Ulam, S. M.: 1968, *Bull. Astron.* **3**, 265.

Urey, H. C.: 1972, *Astrophys. Space Sci.* **16**, 311.

Verlet, L.: 1967, *Phys. Rev.* **159**, 98.

Verlet, L.: 1968, *Phys. Rev.* **165**, 201.

STATIONARY SOLUTIONS OF THE AVERAGED THREE-BODY PROBLEM AND SOME PROBLEMS OF PLANET MOTION STABILITY

G. A. KRASINSKY

Institute of Theoretical Astronomy, Leningrad, U.S.S.R.

Abstract. A general theory of stationary solutions of the averaged N-body problem is briefly described. Numerical results in some particular cases (general three-body problem in the nonresonant case and restricted circular problem both in the nonresonant and resonant ones) are presented. Some applications into the problem of planetary stability are developed.

1. Introduction

At present the significance of the averaged problems of celestial mechanics for investigating planet motions is clearly understood. To a certain degree the stationary solutions of the averaged problems determine the general features of the planet motions and that is why their study is of great importance. If we put aside the problem of convergence, the stationary solutions can be considered as a natural extension of Poincaré's periodic solutions to the case of arbitrary resonances (and to the nonresonant case as well) for the N-planet problem. In recent years a number of works were devoted to the averaged restricted problem of three bodies and its stationary solutions; some important results were obtained by Jefferys and Standish (1966, 1972), Kozai (1962, 1969), Lidov (1962). Other works dealt with applications of the plane stationary solutions to the stability problem for resonant asteroids (Schubart, 1964, 1968; Sinclair, 1969; Marsden, 1970). At the same time the corresponding analysis of the general three-body problem has not yet been developed to such a degree. In this case some results for the problem of the critical inclinations were obtained by von Zeipel (1898, 1901) and Jefferys and Moser (1966). Recently Lieberman (1971) constructed plane stationary solutions by a converging method. For the resonant case some periodic solutions of a type which is called 'trivial' in this paper were constructed by Poincaré (1892–1895), but they by no means exhaust the rich set of the existing periodic solutions. In Krasinsky (1972) an attempt was made to work out a general mathematical theory of the stationary solutions of the averaged N-planet system and to give a unified treatment of all these results which might seem rather heterogeneous. Here we shall use the terminology by Krasinsky (1972). By the averaged system we mean the system obtained after eliminating all short-periodic terms from the original Hamiltonian of the N-planet system by von Zeipel's method. Taking into account arbitrary high (but finite) powers of a small parameter μ (which is of the order of the disturbing masses) we may consider the averaged Hamiltonian as represented by a convergent series. The variant of von Zeipel's method proposed in Krasinsky (1973) enables us to preserve all the invariant properties characterizing

Y. Kozai (ed.), The Stability of the Solar System and of Small Stellar Systems, 95–116. *All Rights Reserved.*

the Hamiltonian of the original system, at the highest approximations. Namely, at any step the averaged Hamiltonian proves to be invariant under three linear transformations; one of them being a rotation of the reference frame relative to vector $\mathbf{l}$ of the angular momentum and the others being reflections relative to two orthogonal planes. One of the planes is orthogonal to the vector $\mathbf{l}$ (invariant Laplace plane) and the other contains $\mathbf{l}$. As a result the Hamiltonian H remains independent of time in a reference frame rotating uniformly with the arbitrary angular velocity σ with respect to $\mathbf{l}$ and differs from the Hamiltonian based on the immovable reference frame by the Coriolis' term $\sigma|\mathbf{l}|$ only. The aim of our investigation is to find stationary solutions of the corresponding system (more exactly, of the system arising after substituting $l_s \to l_s + n_s t$, $s = 1, \ldots, N$, where l_s are the mean longitudes of the averaged system, and n_s are the mean motions). The problem may be reduced to finding extrema of H on the hypersurface determined by the area integral,

$$|\mathbf{l}| = c, \tag{1}$$

under the condition that the invariant plane is chosen as a reference one. The Lagrange factor of this conditional extremum problem coincides with the unknown angular velocity σ. If we restrict ourselves to symmetrical solutions only, we have to seek extrema H relative to nonangular variables, angular variables (the longitudes of the perihelia and nodes and the critical arguments) having to be put equal to certain values according to symmetry conditions. In particular, the critical arguments are equal either to 0 or π. As $\mu \to 0$ the problem reduces to finding the extrema of the averaged perturbation function $[R]$ on the hypersurface (1) relative to the eccentricities and inclinations only. We shall use the scheme of classification of the symmetrical stationary solutions, given in Table I (Krasinsky, 1973).

TABLE I

Classification of the stationary solutions of the N-planet problem

	Solutions of the 1st kind	Solutions of the 2nd and 3rd kind			
		Trivial solutions		Nontrivial solutions (of the 3rd kind)	
		Plane solutions (of the 2nd kind)	Space solutions (of the 3rd kind)	Positive type solutions	Negative type solutions
Averaged values of the eccentricities and inclinations	$e = i = 0$	$e \neq 0$ $i = 0$	$e = 0$ $i \neq 0$	$e \neq 0$ $i \neq 0$	$e \neq 0$ $i \neq 0$
Averaged values of the perihelion arguments	–	–	–	$0, \pi$	$-\frac{1}{2}\pi, \frac{1}{2}\pi$

The extremal treatment of the stationary solutions gives the opportunity to prove easily the existence of 'trivial solutions' (for which the arguments of the perihelia are undetermined) at any values of the area constant c in (1), i.e. for any values of the eccentricities or inclinations. In fact, the integral (1) which is considered as a function of the eccentricities $e_1, \ldots, e_N$ and inclinations $i_1, \ldots, i_N$, determines a hypersurface which is homeomorphic to a $2 \times N$-dimensional sphere. Hence, existence of at least two trivial plane solutions providing $[R]$ with the minimum and the maximum is evident (if $[R]$ has no singularities on (1)). In the same way existence of two space trivial ('circular') solutions may be established if the resonances are of the odd orders or absent at all. But in this case one of these solutions corresponds to the zero value of σ; it is a well-known solution of the 'first kind' (i.e. plane and 'circular') related to a reference plane which does not coincide with the common plane of the planet orbits.

It is important to investigate the dependence of the trivial solutions on the area constant c. In Krasinsky (1972) two equations were deduced:

$$g^{+}(c)=0, \tag{2a}$$

$$g^{-}(c)=0, \tag{2b}$$

which determine 'bifurcational' values of c. If any of Equations (2) is fulfilled a non-trivial solution branches from the trivial solution under consideration. If Equation (2a) is fulfilled, the resulting nontrivial solution is of positive type, otherwise this solution is of negative type. Existence of the bifurcational values is intimately connected with stability of the corresponding trivial solution. If the parameter c passes its bifurcational value, then a pair of the characteristic exponents vanishes and the imaginary exponents become real (or on the contrary). Let c_0 be the value of c corresponding to zero eccentricities and inclinations. Then, if c^* is the smallest of the bifurcational values (at $c_0 > c$) the stationary trivial solution under consideration being stable at $c_0 > c > c^*$ becomes unstable as $c < c^*$. The value c^* will be called the critical one. If we consider the space trivial solutions the corresponding inclinations will be called critical. The notion of critical eccentricity may be introduced in the same way.

The main objective of this paper is to present results of the numerical calculations illustrating the general theory for restricted and general three-body problem. For the restricted problem, the stability of stationary values of the critical arguments is investigated as well. In particular, a new type of stationary solutions with librational motion of the critical argument is constructed. For these solutions (which are of the space trivial type) the close approaches of 'asteroid' with 'Jupiter' are impossible, even for commensurability 1:1.

2. The General Three-Body Problem in the Non-Resonant Case. Bifurcational and Critical Values

As it was mentioned above, constructing the stationary solutions may be reduced to finding extrema of the averaged perturbation function $[R]$ on the hypersurface (1). In the nonresonant case $[R]$ depends on the mutual inclination I, the angles, φ_1 and φ_2, of the eccentricities, e_1 and e_2, the arguments of the perihelia, g_1 and g_2 (which are referred to the common line of the orbital plane intersection), and the difference $\Omega_1 - \Omega_2$ of the nodes, Ω_1 and Ω_2 (indices 1 and 2 relating to inner and outer planets, respectively). If Laplace's invariant plane is chosen as the reference one, the area integral (1) may be written in the following form:

$$\beta_1 \sqrt{a_1} \cos\varphi_1 \cos i_1 + \beta_2 \sqrt{a_2} \cos\varphi_2 \cos i_2 = c, \tag{3}$$

where i_1, i_2 are the inclinations, a_1, a_2 are the semimajor axes, m_1, m_2 are the planet masses, $\beta_j = km_0 m_j / \sqrt{(m_0 + m_j)}$, $j = 1, 2$, m_0 is the mass of the central body.

Other two area integrals give

$$\beta_1 \sqrt{a_1} \sin i_1 = \beta_2 \sqrt{a_2} \sin i_2, \tag{4}$$

$$\Omega_1 - \Omega_2 = \pi. \tag{5}$$

As the nodes on the invariant plane coincide, we have

$$I = i_1 + i_2. \tag{6}$$

The function R must be calculated at certain values of the angular variables. Namely, according to formula (5) the difference $\Omega_1 - \Omega_2$ has to be equal to π and for the perihelion arguments it is necessery to set either $g_1 = 0, \pi$; $g_2 = 0, \pi$ (positive type solutions), or $g_1 = -\frac{1}{2}\pi, \frac{1}{2}\pi$; $g_2 = -\frac{1}{2}\pi, \frac{1}{2}\pi$ (negative type solutions). The choice of the values g_1 and g_2 is not arbitrary but determined by a condition of positivity of the extremal values e_1 and e_2. If e_1 (or e_2) proves to be negative we always can add π to the corresponding value of the perihelion argument and thus change the sign e_1 (or e_2). It seems convenient to put $g_1 = g_2 = 0$ (for the positive type solutions) or $g_1 = g_2 = \frac{1}{2}\pi$ for the negative type solutions), and search extrema $[R]$ at $-1 < e_1, e_2 < 1$. Writing down the equations for finding a conditional extremum we have

$$\begin{aligned} -\sin I \frac{\partial [R]}{\partial \cos I} &= -\sigma \beta_1 \sqrt{a_1} \sin i_1 \cos\varphi_1, \\ -\sin I \frac{\partial [R]}{\partial \cos I} &= -\sigma \beta_2 \sqrt{a_2} \sin i_2 \cos\varphi_2; \end{aligned} \tag{7}$$

$$\begin{aligned} \frac{\partial [R]}{\partial \varphi_1} &= -\sigma \beta_1 \sqrt{a_1} \cos i_1 \sin\varphi_1, \\ \frac{\partial [R]}{\partial \varphi_2} &= -\sigma \beta_2 \sqrt{a_2} \cos i_2 \sin\varphi_2, \end{aligned} \tag{8}$$

where σ is a Lagrange factor.

Further, we use nondimensional parameters β and δ:

$$\beta = \frac{\beta_1 \sqrt{a_1}}{\beta_2 \sqrt{a_2}} = \frac{n_2 m_1}{n_1 m_2 \alpha}, \qquad \delta = c/(\beta_1 \sqrt{a_1} + \beta_2 \sqrt{a_2}) \quad (\alpha = a_1/a_2).$$

First we consider conditions of existence of the plane trivial solutions. As it was noticed above there exist at least two such solutions if $[R]$ has no singularity on the surface,

$$\beta_1 \sqrt{a_1} \cos\varphi_1 + \beta_2 \sqrt{a_2} \cos\varphi_2 = c. \tag{9}$$

The singularities may correspond either to the case of the unit value of the eccentricities or the case of intersection of the orbits. Both of these situations cannot take place under conditions

$$a_1(1+e_1) < a_2(1-e_2), \qquad e_1 < 1, \quad e_2 < 1,$$

which are fulfilled if the inequalities,

$$\alpha < \tfrac{1}{2}\left(1 - \sqrt{1 - ((1+\beta)\,\delta - \beta)^2}\right), \qquad \delta > \max(1, \beta)/(1+\beta), \tag{10}$$

hold. It may be proved that, if eccentricities are small (more exactly if $\beta_1 \sqrt{a_1} + \beta_2 \sqrt{a_2} \sim c$), the longitudes of the perihelia for some of these solutions coincide with each other and that for others they differ by π.

In order to construct the space trivial solutions we must put $\varphi_1 = \varphi_2 = 0$ in Equations (7) and (8); then Equations (7) will be satisfied identically, and only one of Equations (8) will remain independent. Finding σ in terms of I from (8) we have

$$\sigma = \frac{\partial [R]}{\partial \cos I} \frac{\sqrt{a_1}\, \beta_1 \cos i_1 + \sqrt{a_2}\, \beta_2 \cos i_2}{\beta_1 \beta_2 \sqrt{a_1 a_2}}. \tag{11}$$

Thus, the space trivial solutions exist for arbitrary mutual inclinations I. Any solutions with small initial eccentricities will be sometimes in a vicinity of the corresponding space trivial solution. That is why finding the critical inclinations (which are the upper bound of the inclinations of the stable stationary 'circular' orbits) is a problem of great interest. As the critical inclinations belong to the set of bifurcational inclinations they can be found by equating the Jacobian of system (8) (relative to φ_1 and φ_2) to zero. Using the notations,

$$[R] = \frac{k^2 m_1 m_2}{a_2} \left[\frac{a_2}{\Delta}\right], \qquad d_{ij} = \left.\frac{\partial^2 [a_2/\Delta]}{\partial \varphi_i \varphi_j}\right|_{\substack{\varphi_1 = 0 \\ \varphi_2 = 0}}, \qquad \nu = \frac{\partial [a_2/\Delta]}{\partial \cos I},$$

we have

$$d_{ij} = a_{ij} \cos(g_i - g_j) + b_{ij} \cos(g_i + g_j), \quad i, j = 1, 2$$

(coefficients a_{ij}, b_{ij}, ν are found in Krasinsky (1973). Hence, the equations for finding the bifurcational inclinations are expressed as follows:

$$\begin{vmatrix} a_{11} \pm b_{11} - \nu \cos i_1 (\cos i_2 + \beta \cos i_1), & a_{12} \pm b_{12} \\ (a_{21} \pm b_{21})\, \beta, & \beta(a_{22} \pm b_{22}) - \nu \cos i_2 (\cos i_2 + \beta \cos i_1) \end{vmatrix} = 0.$$

Expressing the left sides of these equations in terms of $\cos I$ (by means of (4) and (6)) we find

$$\begin{vmatrix} a_{11} \pm b_{11} - \nu(\beta + \cos I), & a_{12} \pm b_{12} \\ \beta(a_{21} \pm b_{21}), & \beta(a_{22} \pm b_{22}) - \nu(1 + \beta \cos I) \end{vmatrix} = 0. \tag{12}$$

Now we consider the bifurcational and critical inclinations for small α. If the ratio of the planet masses is finite as $\alpha \to 0$, according to its definition β tends to zero too. It is

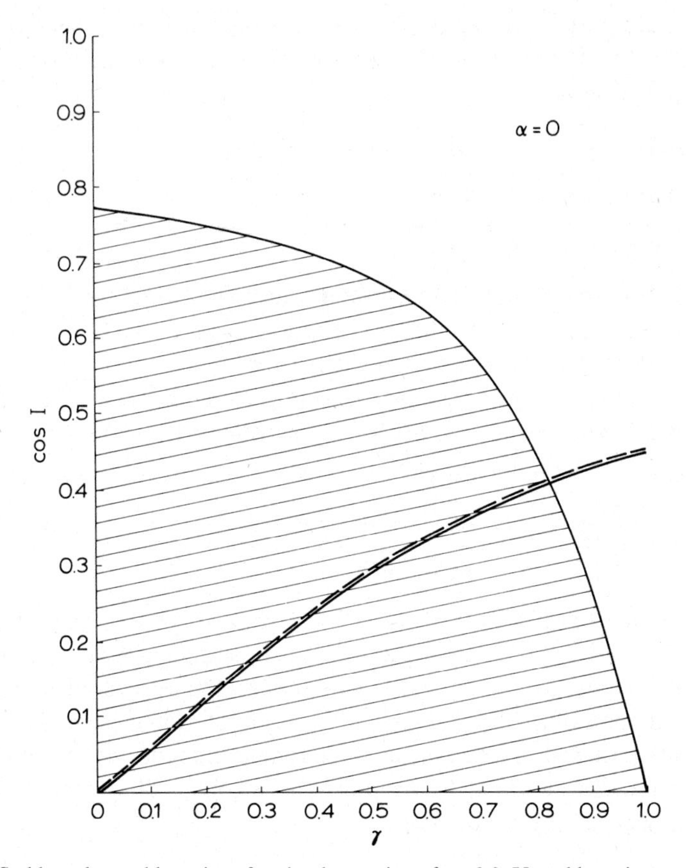

Fig. 1. Stable and unstable regions for circular motion of $\alpha = 0.0$. Unstable regions are hatched.

more convenient to consider α and β as independent parameters for any α. Writing down only the lowest terms in α for the coefficients a_{ij}, b_{ij} and v we have

$$a_{11} = a_{22} = \tfrac{1}{2}\alpha^2(-1 + 3\cos^2 I),$$
$$b_{11} = b_{22} = \tfrac{15}{8}\alpha^2 \sin^2 I, \qquad v = -\tfrac{3}{4}\alpha^2 \cos^2 I,$$
$$a_{12} = a_{21} = b_{12} = b_{21} = 0.$$

Thus, as $\alpha \to 0$ the equation determining the bifurcational inclinations for the negative type solution becomes

$$(5\cos^2 I - 3 + \beta\cos I)(5\beta\cos^2 I - \beta + 2\cos I) = 0.$$

For the positive type solutions we have

$$(1 + 2\beta\cos I)(5\beta\cos^2 I - \beta + 2\cos I) = 0.$$

Roots of these equations in the interval $(-1, 1)$ are the following (negative $\cos I$ corresponding to the retrograde orbites):

$$\cos I_1^- = \cos I_1^+ = \sqrt{\frac{1}{5}+\frac{1}{25\beta^2}} - \frac{1}{5\beta},$$

$$\cos I_2^+ = \cos I_2^- = -\sqrt{\frac{1}{5}+\frac{1}{25\beta^2}} - \frac{1}{5\beta}$$

$$\cos I_3^+ = -\frac{1}{2\beta}, \quad 2 \leqslant \beta \leqslant \infty, \tag{13}$$

$$\cos I_3^- = \sqrt{\frac{3}{5}+\frac{\beta^2}{100}} - \frac{\beta}{10}, \quad 0 \leqslant \beta \leqslant \infty,$$

$$\cos I_4^- = -\sqrt{\frac{3}{5}+\frac{\beta^2}{100}} - \frac{\beta}{10}, \quad 0 \leqslant \beta \leqslant 2.$$

Here plus and minus sign mark the roots for which bifurcation into the positive

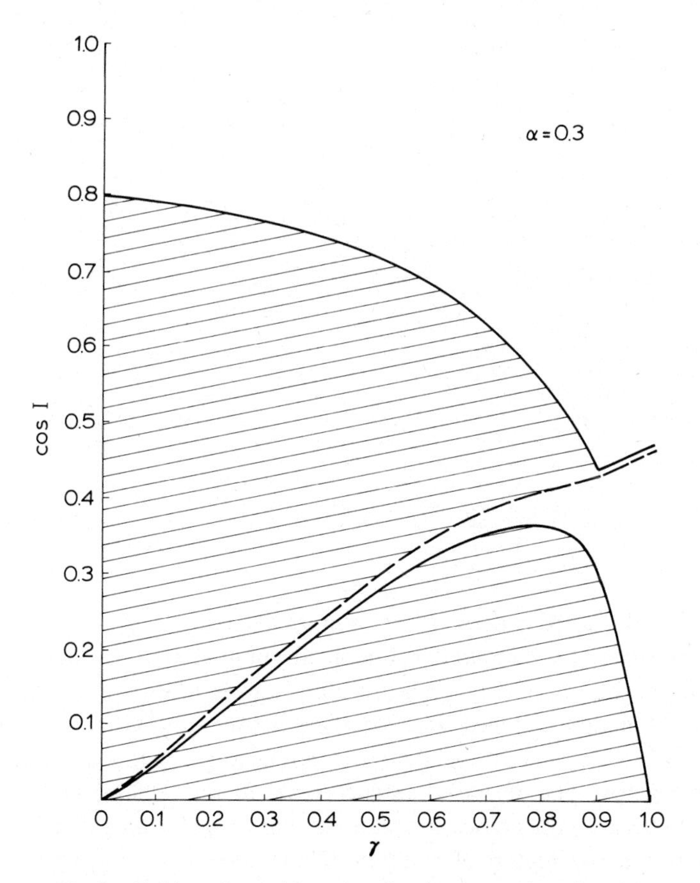

Fig. 2. Stable and unstable regions for circular motion of $\alpha = 0.3$.

or the negative type solutions takes place. The roots $I_1^- = I_1^+$, $I_2^- = I_2^+$ are double ones of the equation for the characteristic exponents and though a pair of these exponents becomes equal to zero at $I = I_1^-$ or $I = I_2^-$ the stability of the space trivial solution is

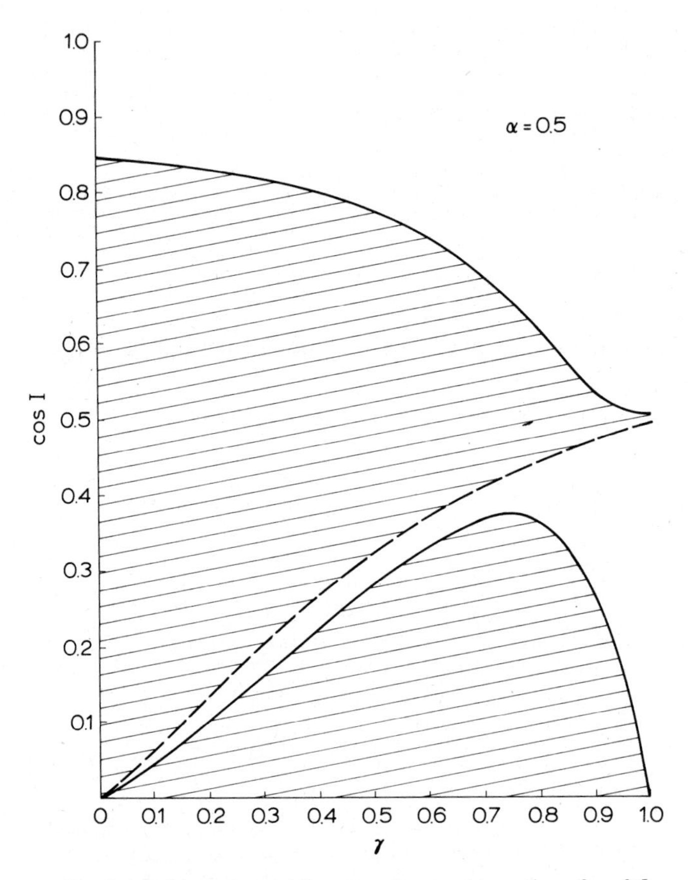

Fig. 3. Stable and unstable regions for circular motion of $\alpha = 0.5$.

preserved. For the graphic demonstration it seems convenient to use the parameter $\gamma = \beta/(1+\beta)$ instead of β. The subdivision of the phase plane $(\cos I, \gamma)$ into domains of stability and unstability of the circular motion is given in Figures 1–5 for several α's. The unstable domain is hatched. The boundary between stable and unstable domains is drawn by a broken or solid line if it corresponds to the points of bifurcation of the positive or negative type. If $\alpha \neq 0$ the structure of the phase plane undergoes a qualitative change because the coinciding (at $\alpha = 0$) lines now become divergent and there arises a new narrow region of stability. These results were obtained by solving Equation (12) numerically.

3. The Averaged Restricted Circular Three-Body Problem in the Nonresonant Case

In the nonresonant case the averaged restricted circular problem is integrable and the stationary solutions determine the topological structure of the phase plane (e, g) at the different values of the 'area integral',

$$\cos\varphi \cos I = c, \tag{14}$$

(which is an integral only for the averaged system). In as much as the averaged

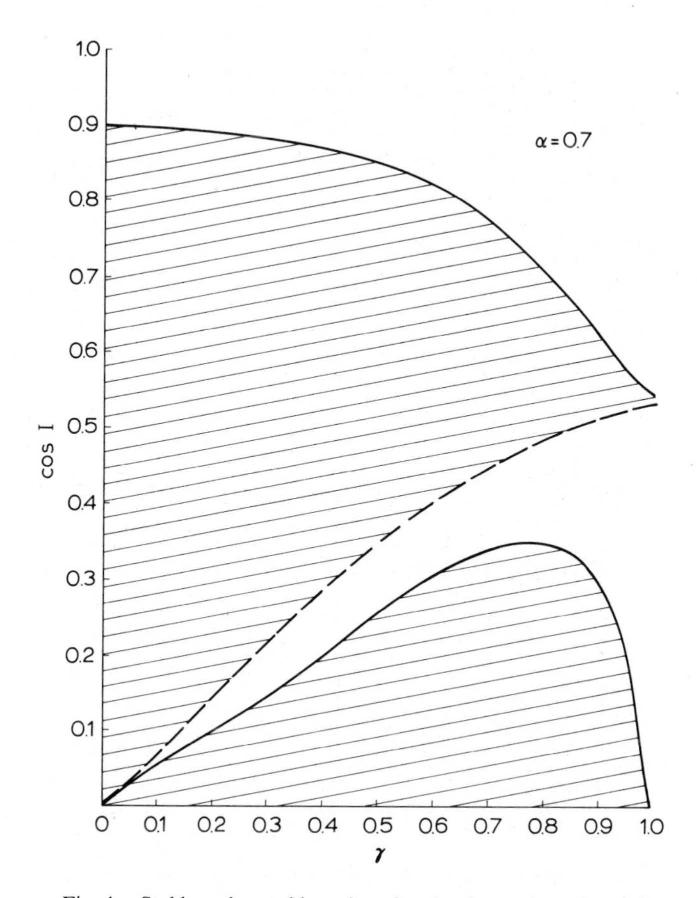

Fig. 4. Stable and unstable regions for circular motion of $\alpha = 0.7$.

perturbation function depends on $\cos^2 I$ the region of the direct and retrograde orbits are similar to each other and we can consider orbits with the direct motion only (for which $\cos I \geqslant 0$). The equations to find the bifurcational points for the inner variant of the problem may be deduced from (12) as $\beta \to 0$:

$$a_{11} \pm b_{11} - \nu \cos I = 0.$$

For the outer variant $\beta \to \infty$ we have the equation

$$a_{22} \pm b_{22} - \nu \cos I = 0.$$

In accordance with (13) we find for the inner case the following values of the bifurcational values (they are critical ones too) as $\alpha \to 0$:

$$\cos I = \sqrt{\tfrac{3}{5}}.$$

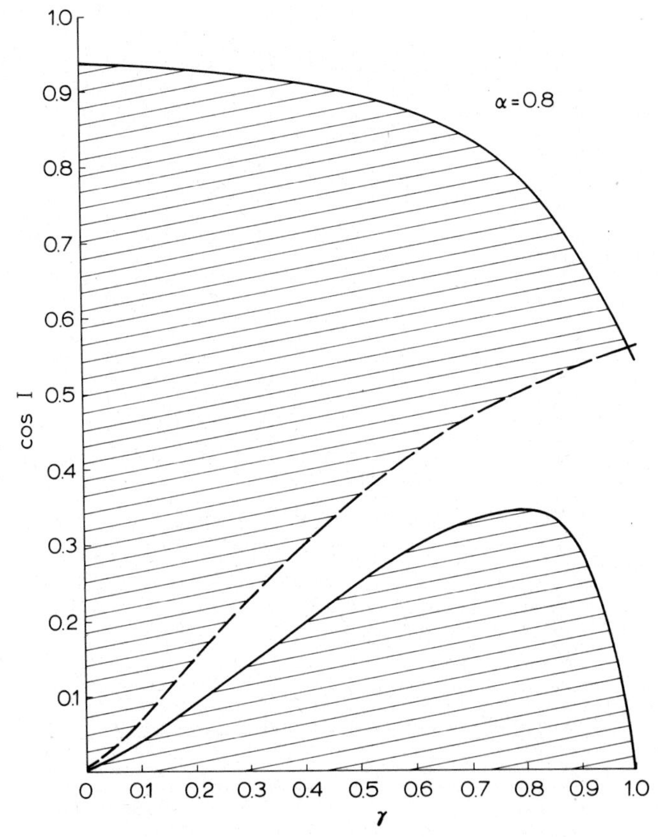

Fig. 5. Stable and unstable regions for circular motion of $\alpha = 0.8$.

For the outer case we have

$$\cos I = \sqrt{\tfrac{1}{5}}.$$

Existence of the critical inclinations in the inner variant may be proved at the arbitrary α (Krasinsky, 1972).

In this case the critical inclinations $I^*(\alpha)$ decrease with increasing α (see Table II).

Moreover, as it is proved in Krasinsky (1972) $\lim_{\alpha\to 1} I(\alpha)=0$, and the following asymptotic formula takes place:

$$\cos I^* = 1 - \tfrac{1}{4}(1-\alpha) + 0(|1-\alpha|^{3/2}\ln(1-\alpha)|).$$

For the outer problem calculations show (at $\alpha<0.95$) the existence of points of bifurcation in the positive ($I=I^+(\alpha)$) as well in the negative ($I=I^-(\alpha)$) type solutions.

TABLE II

Bifurcational inclinations for the nonresonant case

α	$\cos I$		
	Inner case		Outer case
	Bifurcations into the negative type solutions		Bifurcations into the positive type solutions
0.0	0.774595	0.447213	0.447213
0.1	0.777588	0.449883	0.449221
0.2	0.786443	0.457716	0.455186
0.3	0.800822	0.470189	0.464942
0.4	0.820211	0.486396	0.478234
0.5	0.843981	0.505978	0.494743
0.6	0.871435	0.523958	0.514116
0.7	0.901835	0.540466	0.535987
0.8	0.934368	0.550547	0.5560
0.9	0.9680	0.45013	0.5858

The critical inclination I^* for the direct orbits are determined by the relation $I^*=\min(I^+, I^-)$. Also we set $I^{**}=\max(I^+, I^-)$. The dependence of the phase plane (e, g) on the integral constant c is given in Figures 6–10. If $c>c^*=\cos I^*$, then for the inner as well as the outer problem the phase plane has a simple structure (Figure 6) from which follows the stability of the 'circular' orbits. For the inner problem at $c<c^*$, and for the outer problem at $c^*>c>c^{**}=\cos I^{**}$, the topological structure of the phase plane is the same, and it is characterized by a single stationary point. The corresponding nontrivial stationary solution is of the negative type for the inner case as well as for the outer case at $\alpha<\alpha_0=0.75\ldots$, otherwise (at $\alpha>\alpha_0$) in the last case the stationary solution is of the positive type (Figures 7 and 8). Finally, if $c<c^{**}$ then in the outer case there exist two nontrivial solutions. One of them is stable and another unstable; if $\alpha<\alpha_0$, the stable solution is of the negative type, and if $\alpha>\alpha_0$, it is of the positive type (Figures 9 and 10). It is worth mentioning that, as $c<c^{**}$, the space trivial ('circular') solution becomes stable again. A similar situation was investigated by Izsak (1962) for the satellite motion in the gravitational field of a nonspherical planet. In Figure 11 the subdivision of the phase plane $(\alpha, \cos I)$ into stable and unstable regions is given for the outer case. Broken and solid lines refer to the points of

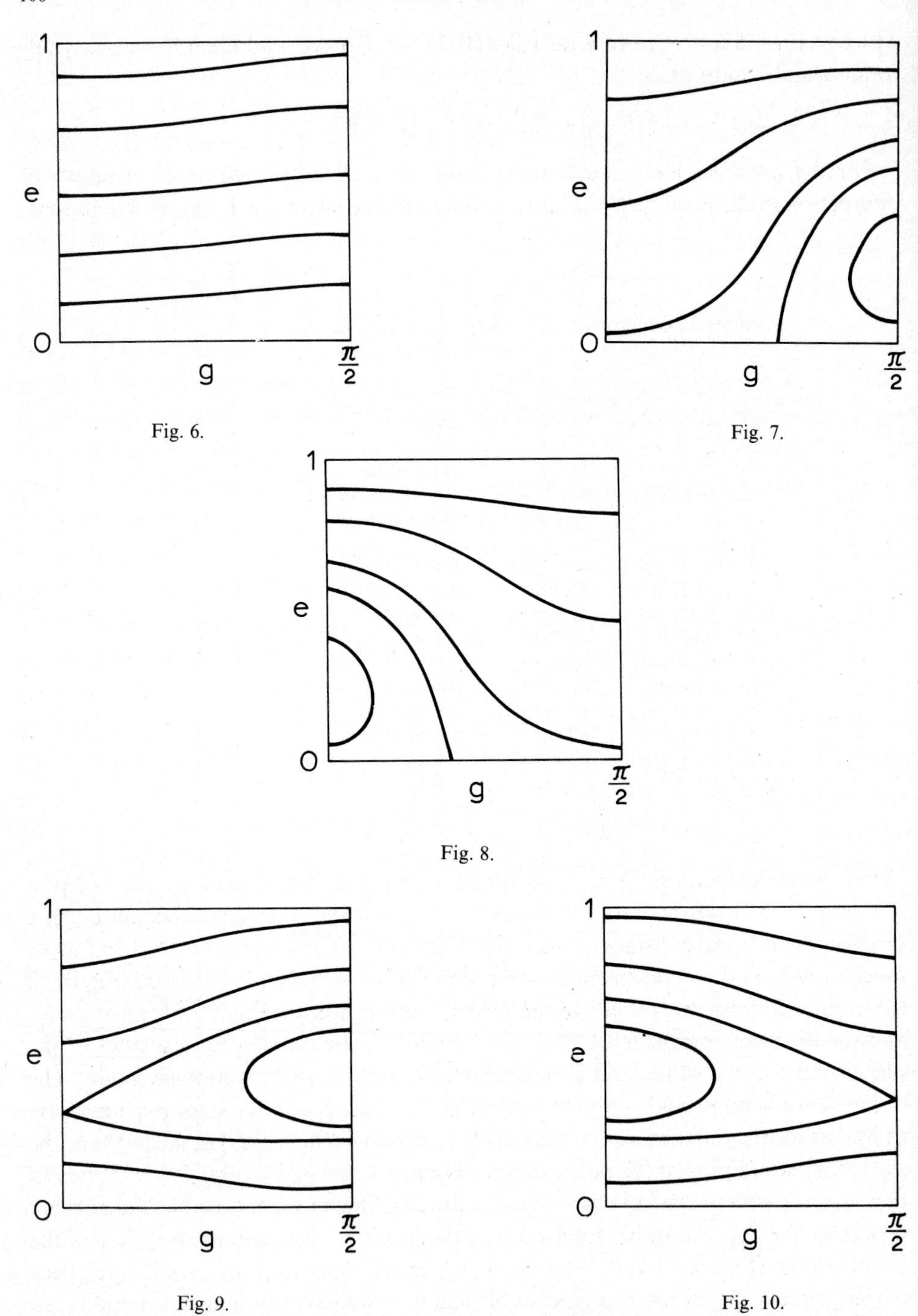

Fig. 6.

Fig. 7.

Fig. 8.

Fig. 9.

Fig. 10.

Figs. 6–10. Trajectories on (e, g)-plane for different values of c.

bifurcation into positive and negative type solutions. The unstable region is hatched.

Now we consider the inner problem again. Let $e^* = \sin\varphi^*$ and I^* be the eccentricity and inclination of the nontrivial stationary solution generated by the bifurcational

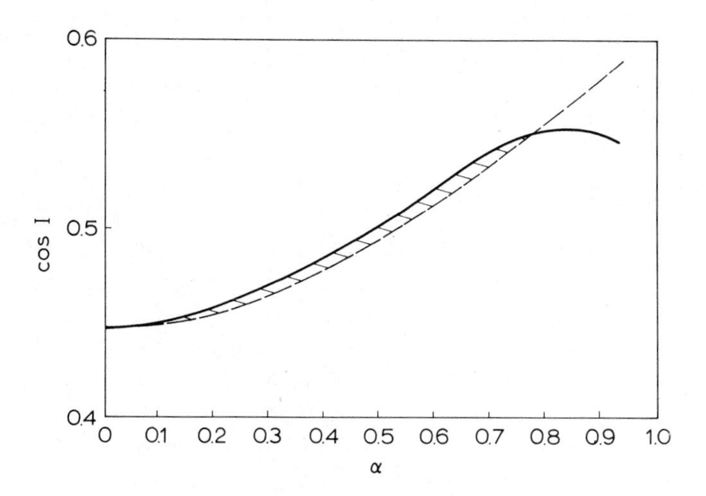

Fig. 11. Stable and unstable (hatched) regions for outer circular motion.

point and let c be the corresponding integral constant in (14). Considering the qualitative structure of the phase plane (e, g) (Figure 7) we can see that for any initial values of e_0 and I_0 (at condition $\cos\varphi_0 \cos I_0 = c$ and $\sin\varphi_0 = e_0$) there exists a moment of time t for which $e(t, e_0) \geqslant e^*$, $(e(0, e_0) = e_0)$. In particular, if $I_0 \sim \tilde{I}$, where $\cos\tilde{I} = c$, the initial value e_0 may be chosen arbitrarily small. Thus,

$$e^* = \min_{e_0} \max_{t} e(t, e_0)$$

at the condition $\cos\varphi_0 \cos I_0 = c$. That is why the problem of finding e, I on the nontrivial stationary solution is of great importance. As $\alpha \to 0$ the following relation between e, I on the stationary solution was obtained by Kozai (1962):

$$5\cos^2 I - 3\cos^2\varphi = 0. \tag{15}$$

Hence, $\varphi = \frac{1}{2}\pi$, i.e. the eccentricity of the orbit, which was circular at the initial moment and whose orbital plane was perpendicular to the orbital plane of the disturbing body, tends to unit (Lidov, 1962). It is easy to understand that this result holds true for arbitrary α (at least if $\alpha < 0.5$). And indeed, the equation determining extrema of $[R]$ on the surface (14) has the form

$$\cos\varphi \frac{\partial [R]}{\partial \cos\varphi} - \cos I \frac{\partial [R]}{\partial \cos I} = 0, \tag{16}$$

and connects e with I. If $\alpha<0.5$ the aphelion distance is less than the orbital radius of the disturbing body and $[R]$ has no singularities on the surface (14) (even for $\cos\varphi = 0$; see Krasinsky (1973)). Hence, if $\varphi = I = \frac{1}{2}\pi$, Equation (16) will be satisfied. We investigated Equation (16) numerically for several values of α. In Figure 12 the

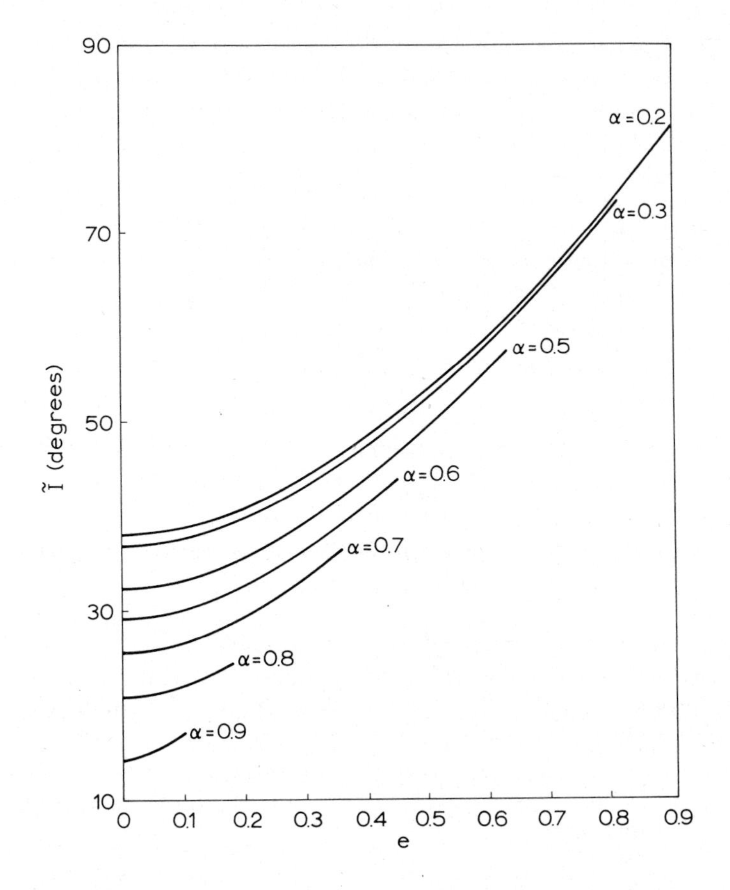

Fig. 12. Relations between the eccentricity and $\tilde{I} = \arccos c$.

dependence of e on $\tilde{I}$ ($\tilde{I} = \arccos c$) is presented. The figure shows that for $\alpha \sim 0.5$ and 0.6 (these are characteristic values for the minor planets) e increases rapidly with $\tilde{I}$. Perhaps this fact may explain the well-known peculiarity in the distributions of eccentricities and inclinations of the minor planets: large inclinations are commonly accompanied by large eccentricities. In order to illustrate this fact we plotted (Figure 13) the inclinations I and semiaxes a of all minor planets whose eccentricities are less than 0.1; the curve of the critical inclinations is drawn too. As it may be expected, all points representing the pairs (a, I) are beneath of this curve (excluding the commensurability 1:1 for which the analysis is not applicable). All inclinations are referred to the ecliptic, the mutual inclinations of the Earth and Jupiter being neglected.

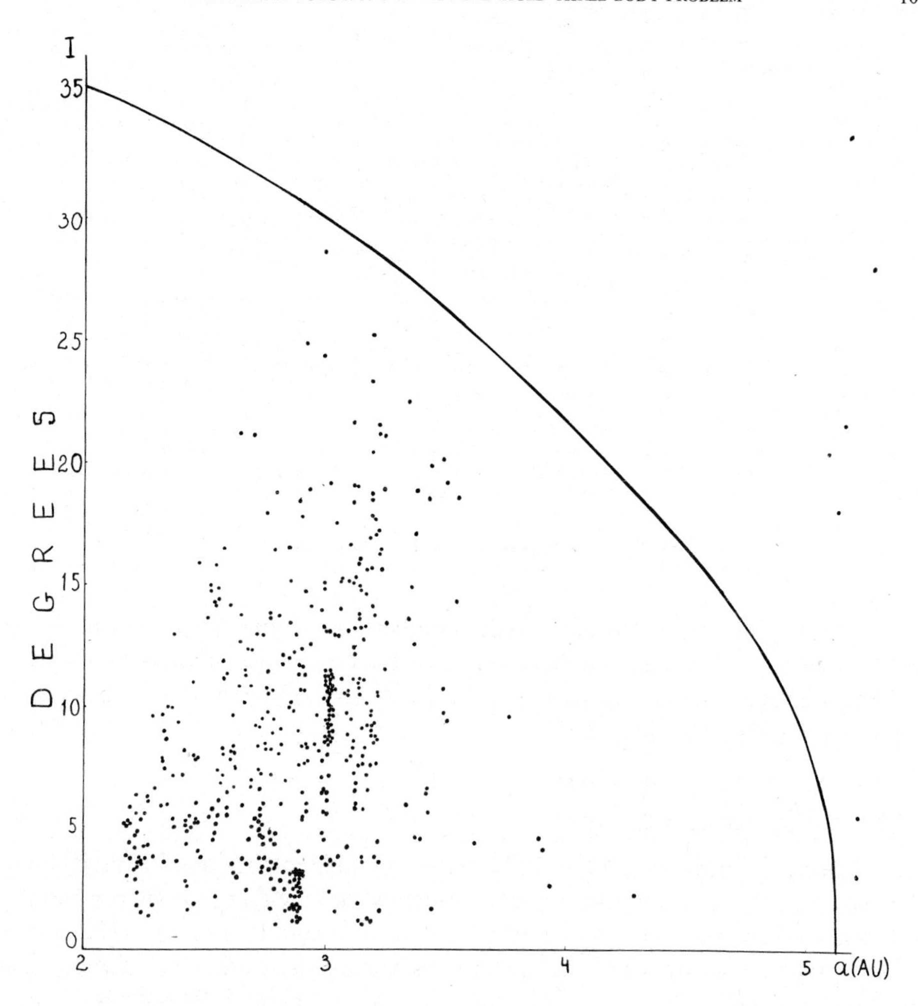

Fig. 13. Distribution of the inclinations of asteroids with respect to the semimajor axes.

In Figure 14 the dependence of e on I for the outer problem is presented for the nontrivial solutions of both positive and negative type. As $\alpha \to 0$, instead of the Kozai's relation (15), we have $\cos^2 I = \frac{1}{5}$ for any e. For $\alpha \neq 0$ we considered only moderate eccentricities; the case of large eccentricities might be useful for comet astronomy and needs further investigations.

4. The Averaged Restricted Circular Three-Body Problem in the Resonant Case

In this section the mean motions n and n' of disturbing and disturbed planets are supposed to be connected by a resonant relation

$$pn + qn' = 0,$$

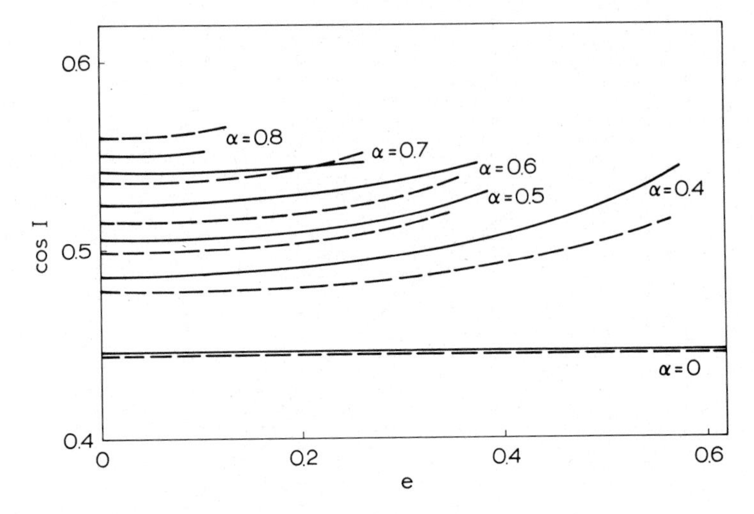

Fig. 14. Relations between e and I for outer cases.

where p and q are mutual prime integer numbers. The value $|p+q|$ will be called the order of the resonance. The averaged perturbation function in this case depends on the eccentricity e, the inclination I, the perihelion argument g and the critical argument β, whose definition is:

$$\beta=\begin{cases} pl+ql'-(p+q)\,\Omega, & \text{if} \quad I\neq 0, \\ pl+ql'-(p+q)\,\tilde{\omega}, & \text{if} \quad I=0 \end{cases} \tag{17}$$

(l and l' being the mean longitudes, Ω and ϖ the longitudes of the node and perihelion). For the stationary solution the critical argument is equal either to 0 or to π and the perihelion argument either to 0, π (the positive type solution) or to $-\frac{1}{2}\pi$, $\frac{1}{2}\pi$ (the negative type solution). The problem of constructing the stationary solutions has some peculiarities in this case because the averaged system does not possess the area integral and the averaged Hamiltonian even in a rotating coordinate system still depends on time. This dependence vanishes if we seek a stationary solution for which the corresponding mean motion is equal to $n(1+\varrho)=-n'(1+\varrho)\,q/p$, where $\varrho=-\sigma/n'$, and σ is the angular velocity of the rotating coordinate frame. In fact, we obtain the Poincaré–Schwarzschield's periodic orbits with a variable period. The equations to find the stationary solutions coincide with the conditions of periodicity and may be treated as the equations determining extrema of the averaged perturbation function $[R]$ on the surface (14) (which is not an integral one in this case). Excluding the Lagrange factor σ we again obtain Equation (15) connecting the eccentricity and the inclination on the stationary solution. This equation was investigated numerically by Jefferys and Standish (1966, 1072) and Kozai (1969). Equation (15) determines as a rule several different curves, the axis $I=0$ corresponding to the plane trivial solutions and the axis $e=0$ (for the even order $|p+q|$ of the resonance) corresponding to the space trivial solutions. According to the general theory, if the stable trivial solution passes a bifurcational point (i.e. the point of the intersection of a curve determined by Equation (15) with a coordinate axis) it becomes unstable. In the non-resonant case there exist only intersections with the axis I (the critical and bifurcational inclinations). In the resonant case the existence of intersections with axis e is proved in Jefferys and Standish (1966, 1972) and Kozai (1969). Thus, in these papers the first examples of the

TABLE III

Characteristics of the symmetric trivial periodic solutions in the restricted circular three-body problem (commensurability $pn+qn'=0$)

Oddness or evenness p, q	Type of conjunctions or oppositions	Mean values of the critical argument	Stability or instability
	Plane solutions		
$p+q$ odd	conjunction at perihelion Opposition at aphelion	0	Stability
q odd	Conjunction at aphelion Opposition at perihelion	π	Instability
$p+q$ odd	Conjunction at perihelion Opposition at perihelion	0	Stability
q even	Conjunction at aphelion Opposition at aphelion	π	Instability
$p+q$ even	Conjunction at perihelion Conjunction at aphelion	0	Instability
$p+q$ even	Opposition at perihelion Opposition at aphelion	π	Stability
	Space solutions		
$p+q$ even	Conjunction at the ascending node Conjunction at the descending node	0	Instability
	Opposition at the ascending node Opposition at the descending node	π	Stability

critical and bifurcational eccentricities are constructed. Unlike the problem of the critical inclinations the calculation of the critical eccentricities gives great difficulties because large eccentricities have to be considered. This is why there are discrepancies between corresponding numerical results by Jefferys and Standish, and by Kozai (1969).

All the nontrivial solutions may be subdivided into four groups which differ by the type (positive or negative) and values of the critical argument. This subdivision does not coincide with that by Kozai because the different definitions of the critical argument are used. For the trivial solutions the type is undetermined and they may be distinguished only by the values of the critical argument. The solutions belonging to the different groups have different types of symmetry which we are going to describe briefly. It is easy to prove the existence of two moments of time, T_1 and T_2, ($T_2 - T_1 = \frac{1}{2}T$, where $T = 2\pi|p|/(n' - \sigma)$ is the period of this periodic solution) for which the mean longitudes of the disturbed and disturbing bodies are either equal to each other or differ by π. Following Poincaré we refer to the first case as 'conjunction', to the second case as 'opposition'. At these moments the critical argument coincides with its mean value (0 or π) and the bodies are either on the line of the apsides (for the nontrivial and plane trivial solutions) or on the nodal line (for the space trivial solutions). If the nontrivial solution is the positive type, the lines of the nodes and the apsides coincide; if it is the netative type these lines are orthogonal (at the moments T_1 and T_2). In the last case the plane P containing the three bodies is orthogonal to the plane S of the disturbing body, the velocity of the disturbed body being perpendicular to P and parallel to S. This kind of symmetry was studied first by Jefferys (1965). In Table III for all virtual cases mutual positions of the bodies at T_1 and T_2 are given for the trivial solutions (depending on the evenness or oddness of p and q and the values of the critical argument). In the last column we marked whether the stationary values of the critical arguments are stable or not (for small eccentricities and inclinations), in other words whether the critical argument has a bifurcational or circular motion. The corresponding analytical proof is given in Krasinsky (1973).

In recent years the theory of the stable stationary solutions was applied to the problem of resonant asteroids. For instance, learning the commensurability 3:2 (Hilda group) we see from Table IV that for the stable plane stationary solution (which corresponds to the value π of the critical argument) the conjunctions with Jupiter take place only when the asteroid is in the perihelion. If the eccentricity is not small, the close approach of the asteroid to Jupiter does not occur. This situation was

TABLE IV

Bifurcational inclinations for resonant cases

$n:n'$	$\cos I$	
	Bifurcations into the positive type solutions	Bifurcations into the positive type solutions
3:1	0.322742	0.590852
5:1	–	0.787049
7:1	–	0.796043
9:1	–	0.790342
11:1	–	0.786699
5:3	0.585475	0.438795
7:3	–	0.802517
11:3	–	0.824638
7:5	0.668089	0.112050
9:5	–	0.787081
11:5	–	0.861418
9:7	0.111854	–
11:7	–	0.737298

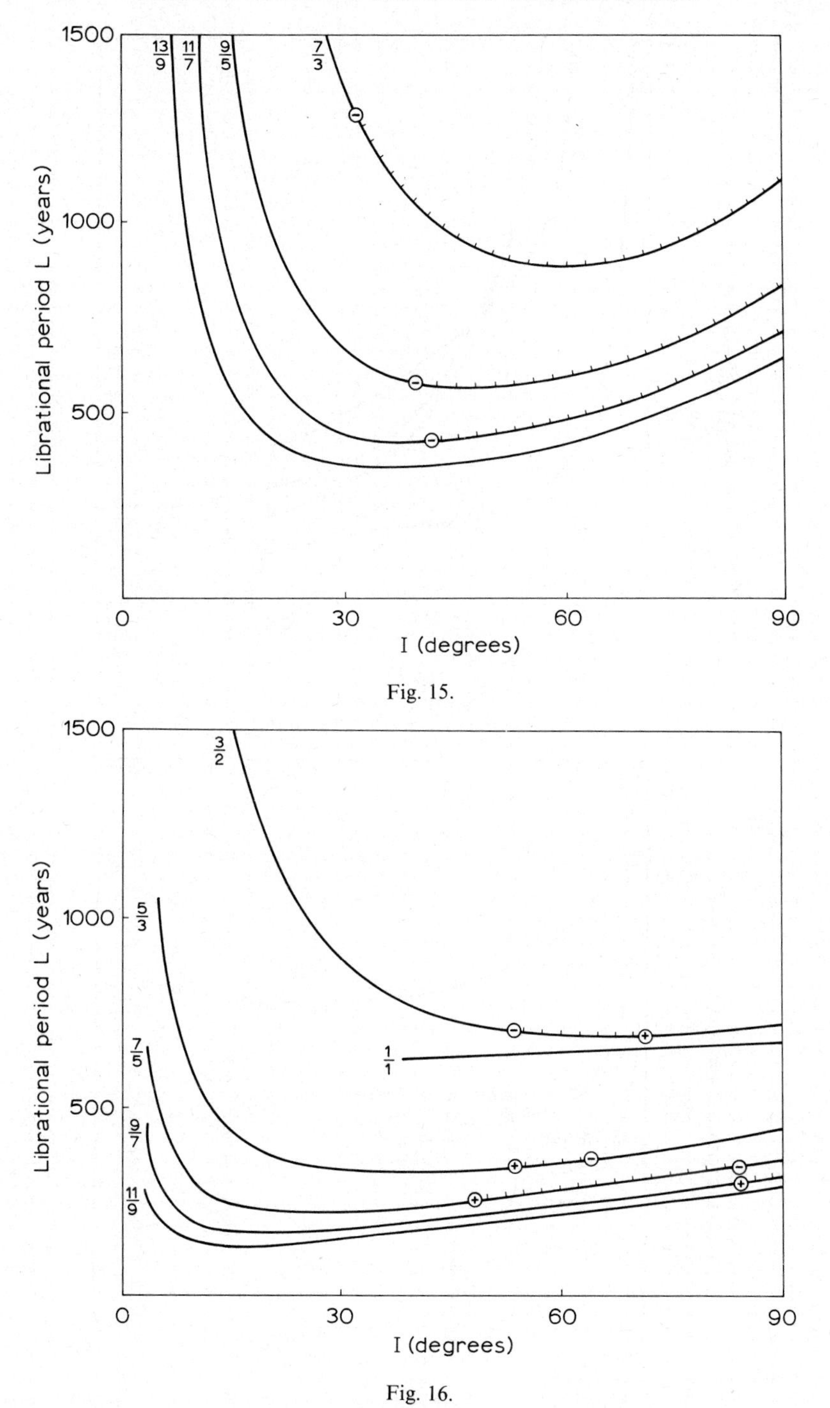

Figs. 15–16. Relations between libration periods and the inclinations for commensurable asteroids. Broken lines correspond to unstable points.

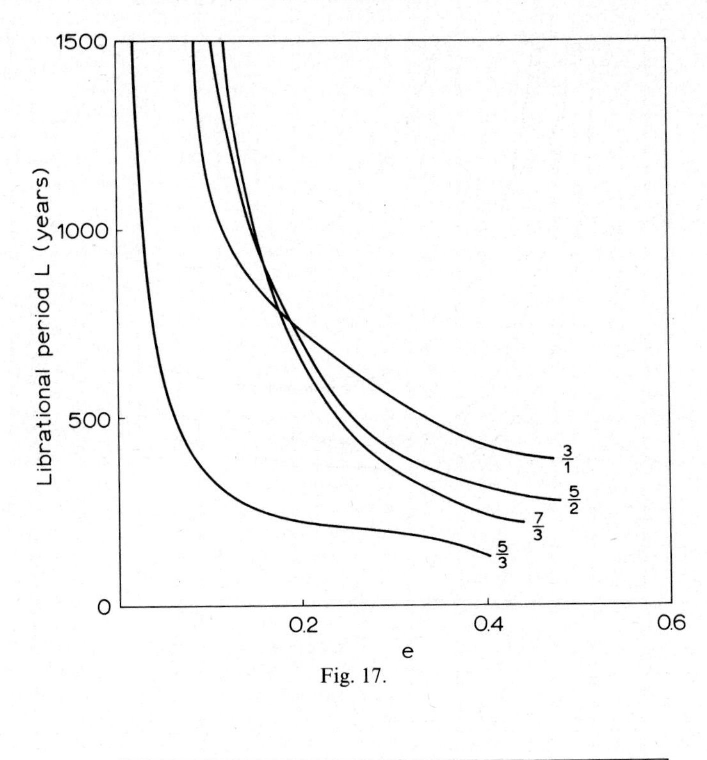

Fig. 17.

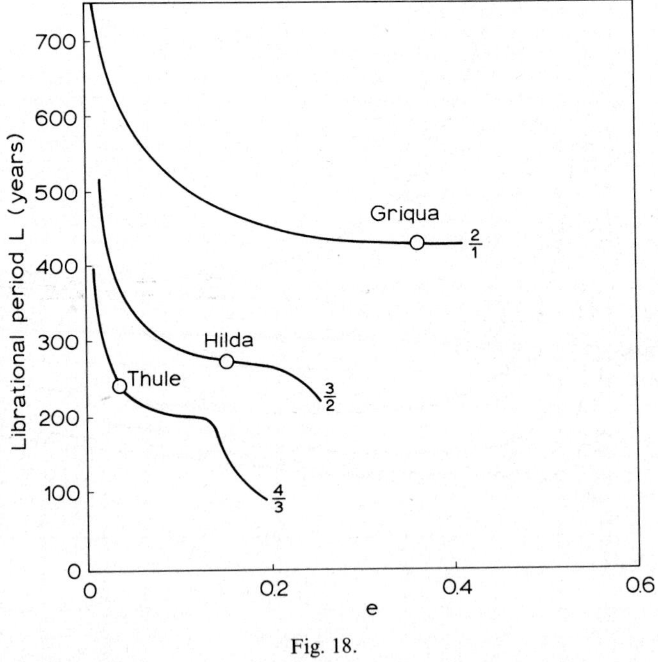

Fig. 18.

Figs. 17–18. Relations between libration periods and the eccentricities for commensurable asteroids.

studied in detail by Schubart (1964, 1968), who established that all of the asteroids of the Hilda group are in the vicinity of the stable periodic orbit (excluding two asteroids with small eccentricities). This seems to explain the stability of the Hilda group (as well as of some other groups). The space trivial solution gives another example of the libration of the critical argument and the absence of close approaches with Jupiter. The corresponding stable periodic solution (existing only if the order $|p+q|$ of the resonance is an even number) for which the mean value of the critical argument is equal to π, has no conjunctions with Jupiter at all. In the most dangerous points of the orbits (in the nodes) the longitudes l and l' of the asteroid and Jupiter differ by π. The case of commensurability 1:1 is of particularly great interest. In this case the value $l-l'$ differs from π only by small short-periodic terms. This solution is symmetrical unlike the well-known periodic solutions which correspond to the librational points L_4 and L_5. For the commensurability 1:1 the plane symmetrical solution with non-zero eccentricity exists too, but it cannot be constructed by means of the analytical method used. In Figures 15–18 we present the results of numerical calculations of the 'librational period' for the stable trivial solution in the system Jupiter–asteroid. The parts of the curves in Figures 15 and 16 marked by the broken line correspond to the region of instability of the eccentricity. The small circles with the plus or minus sign inside mark the bifurcational points corresponding to the solution of the positive or the negative type which branch from these points. The values of the bifurcational and critical inclinations are also given in Table IV. (The critical eccentricities were not considered in this work.) The numerical results may be compared in some cases with those by Kozai (1969) and Jefferys and Standish (1966, 1972). The coincidence is good enough (it is necessary to keep in mind the differences between the definitions of the critical argument).

References

Brumberg, V. A.: 1966, *Bull. Inst. Theor. Astron.* **2** (in Russian).
Izsak, I.: 1962, *Smithsonian Astrophys. Obs.*, Special Report No. 90.
Jefferys, W. H.: 1965, *Astron. J.* **70**, 393.
Jefferys, W. H. and Moser, J.: 1966, *Astron. J.* **71**, 568.
Jefferys, W. H. and Standish, E.: 1966, *Astron. J.* **71**, 982.
Jefferys, W. H. and Standish, E.: 1972, *Astron. J.* **77**, 394.
Kozai, Y.: 1962, *Astron. J.* **67**, 591.
Kozai, Y.: 1969, *Publ. Astron. Soc. Japan* **21**, 33.
Krasinsky, G. A.: 1972, *Celes. Mech.* **6**, 60.
Krasinsky, G. A.: 1973, in N. S. Samoylova-Yakhontova (eds.), *Minor Planets*, Moscow, p. 151.
Lidov, L. M.: 1962, in M. Roy (ed.), *Dynamics of Satellites*, Springer Verlag, Berlin, p. 168.
Lieberman, B.: 1971, *Celes. Mech.* **3**, 408.
Marsden, B. G.: 1970, in G. E. O. Giacaglia (ed.), *Periodic Orbits, Stability and Resonances*, D. Reidel, Dordrecht, The Netherlands, p. 151.
Poincaré, H.: 1892–95, *Les méthodes nouvelles de la mécanique céleste* **1–3**, Paris.
Schubart, J.: 1964, *Smithsonian Astrophys. Obs.*, Special Report No. 149.
Schubart, J.: 1968, *Astron. J.* **73**, 2.
Sinclair, A.: 1969, *Monthly Notices Roy. Astron. Soc.* **142**, 3.
von Zeipel, H.: 1898, *Bihang till K. Sven. Vet.-Handl.* **24**.
von Zeipel, H.: 1901, *Bihang till K. Sven. Vet.-Akad. Handl.* **26**.

DISCUSSION

F. Nahon: What is the meaning of the word 'stationary' in your title?

G. A. Krasinsky: The stationary solutions provide the averaged Hamiltonian with the extremal value. For the immovable reference frame we have no extrema of the averaged Hamiltonian excluding the case $e=i=0$, however, they exist for the rotating reference frame. By a similar way we found the conditional extrema for the immovable reference frame. This approach was initiated by Poincaré for the periodical case. We extended this approach to the case of arbitrary resonances as well as to the nonresonant case.

J. Moser: Do you have an explanation for the distribution of the orbital elements for asteroids according to your study?

G. A. Krasinsky: I have no explanation for the distribution of the perihelia, however, I can offer some explanations for the distributions of the eccentricities and the inclinations.

INTEGRABLE CASES OF SATELLITE PROBLEM WITH THE THIRD BODY AND THE OBLATE PLANET

M. L. LIDOV

Institute of Mathematics, Academy of Sciences, Moscow, U.S.S.R.

Abstract. The evolutions of satellite's orbit under the influence of the third disturbing body moving on an elliptic orbit and the oblate planet are studied by averaging the disturbing function, and integrable cases are classified in seven cases.

In this report we shall study the evolution of satellite's orbit under the influence of simultaneous perturbations of the third body, moving on an elliptical orbit, and the oblateness of the planet, around which the satellite rotates.

The problem is considered in the following assumptions, formulated by Lidov (1962):

(1) The ratio of the radius-vector of the satellite r and the radius-vector of the third body r_1 is small ($r/r_1 \ll 1$), and in the expansion of the disturbing function in the series of r/r_1 only the main term W_1 remains (Hill's approximation).

(2) The oblateness of the field, W_2, is described by the second zonal harmonic.

(3) We assume that there are no resonant relations between the frequencies of the satellite's and the third-body's rotations. In this case the disturbing function of the problem, being averaged with respect to the longitudes of the satellite and the perturbing body, λ and φ, could be used for the description of the secular evolution:

$$W = \frac{1}{(2\pi)^2} \int_0^{2\pi} \int_0^{2\pi} (W_1 + W_2) \, \mathrm{d}\lambda \, \mathrm{d}\varphi . \tag{1}$$

In the above assumptions the following formula for W is valid:

$$W = \frac{\mu_1 a^2}{8a_1^3 (1-e_1^2)^{3/2}} [6e^2 - 1 - 15e^2 \sin^2 \omega + 3 \cos^2 i \times \times (5e^2 \sin^2 \omega + 1 - e^2)] + \frac{\mu a_0^2 c_{20}}{4a^3 (1-e^2)^{3/2}} (1 - 3\cos^2 i_{\mathrm{eq}}). \tag{2}$$

Here a, e, i, ω, and Ω are the conventional notations for the Kepler's elements of the satellite's orbit. The orbital plane of the perturbing body is taken as reference; Ω is counted from the ascending node of the perturbing body's orbit on the equator of the planet; i_{eq} is an inclination of the satellite's orbit to the equatorial plane of the planet:

$$\cos i_{\mathrm{eq}} = \cos I \cos i - \sin I \sin i \cos \Omega , \tag{3}$$

where I is the angle between the orbital plane of the perturbing body and the equatorial plane, a_1 and e_1 are orbital elements of the perturbing body, a_0 is the equa-

Y. Kozai (ed.), The Stability of the Solar System and of Small Stellar Systems, 117–124.

torial radius of the planet, c_{20} is the coefficient of the second zonal harmonic, $\mu = GM$, $\mu_1 = GM_1$, G is the gravitational constant, M is the mass of the central body, and M_1 is the mass of the perturbing body.

To the disturbing function (2) corresponds the system of equations (Lidov, 1962):

$$\frac{da}{dn} = 0, \qquad \frac{d\varepsilon}{dn} = -(1-\varepsilon)\,\varepsilon^{1/2} \sin 2\omega \sin^2 i,$$

$$\frac{di}{dn} = -\frac{1-\varepsilon}{2\varepsilon^{1/2}} \sin 2\omega \sin i \cos i + \beta \frac{\sin I \sin \Omega}{\varepsilon^2} \cos i_{eq},$$

$$\frac{d\omega}{dn} = \varepsilon^{-1/2}\left[\frac{2\varepsilon}{5} + \sin^2 \omega (\cos^2 i - \varepsilon)\right] + \\ + \frac{\beta}{2\varepsilon^2}\left(1 - 5\cos^2 i - \frac{2 \sin I \cos \Omega}{\sin i} \cos i_{eq}\right),$$

$$\frac{d\Omega}{dn} = -\varepsilon^{-1/2} \cos i \left[\frac{\varepsilon}{5} + (1-\varepsilon)\sin^2 \omega\right] + \\ + \beta (\varepsilon^2 \sin i)^{-1} (\sin i \cos I + \cos i \sin I \cos \Omega) \cos i_{eq}.$$

Here instead of time t we introduced the dimensionless time n:

$$dn = \frac{15 \mu a^2}{4 a_1^3 \sqrt{\mu a} (1 - e_1^2)^{3/2}}\, dt, \tag{5}$$

$$\varepsilon = 1 - e^2.$$

β is a constant parameter, characterizing the ratio of perturbing accelerations due to oblateness and the perturbing acceleration due to the third body,

$$\beta = \frac{2 \mu a_0^2 a_1^3 (1 - e_1^2)^{3/2}\, c_{20}}{5 \mu_1 a^5}. \tag{6}$$

Besides the obvious integral $a = \text{const.}$ the system of Equations (4) always has one more integral:

$$W = \text{const.}, \tag{7}$$

where W is given in (2).

The general theory says that for integrating Equations (4) the existence of one more integral is necessary. In the general case, apparently, such integral does not exist. However, in certain special cases, Equations (4) could be integrated.

Let us enumerate the following cases:

(1) For $\beta = 0$ (the influence of oblateness is negligibly small) the full investigation of the problem was carried out by Lidov (1961) and Kozai (1962). In this case there exists one more integral $\varepsilon \cos^2 i = c_1$, and integral (7) could be simplified and rewritten in the following form:

$$(1-\varepsilon)\left(\tfrac{2}{5} - \sin^2 i \sin^2 \omega\right) = c_2.$$

To make the consideration complete we shall give, without comment, the qualitative pictures, describing the behaviour of the trajectories in the plane (ε, ω) for the three characteristic cases:

$$c_1 > \tfrac{3}{5} \text{ (Figure 1a);} \qquad 0 < c_1 < \tfrac{3}{5} \text{ (Figure 1b);}$$
$$c_1 = 0 \quad (\cos i = 0) \quad \text{(Figure 1c).}$$

(2) For $\beta = \pm\infty$ $(\mu_1 = 0)$. The solution in this popular case was obtained by many

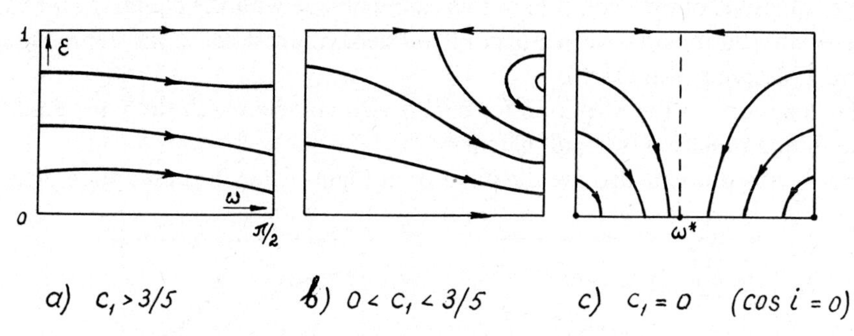

Fig. 1. ε–ω diagrams for $\beta = 0$.

authors. The evolution reduces to the monotone variation of ω_{eq} and Ω_{eq} according to the formulas:

$$\frac{d\omega_{eq}}{dt} = \frac{\sigma}{2\varepsilon^2}(1 - 5\cos^2 i_{eq}),$$

$$\frac{d\Omega_{eq}}{dt} = \frac{\sigma}{\varepsilon^2}\cos i_{eq}.$$

Here ω_{eq} and Ω_{eq} are angles counted from the equatorial plane, and,

$$\sigma = \frac{3}{2}\left(\frac{a_0}{a}\right)^2 \frac{\sqrt{\mu}}{a^{3/2}} c_{20}, \qquad i_{eq} = \text{const.}, \qquad \varepsilon = \text{const}.$$

(3) $\sin I = 0$ – the equatorial plane of the planet coincides with the orbital plane of the perturbing body.

(4) $\cos I = 0$, $\sin i = 0$ – the equatorial plane of the planet is orthogonal to the orbital plane of the perturbing body; the satellite moves in the orbital plane of the perturbing body.

(5) $\cos I = 0$, $\cos i = 0$, $\sin \Omega = 0$ – the equatorial plane of the planet is orthogonal to the orbital plane of the perturbing body; the orbital plane of satellite coincides with the equatorial plane.

(6) $\cos i = 0$, $\cos \Omega = 0$ – the orbital plane of the satellite is orthogonal to the intersection line of the planes of the perturbing body and the equator.

(7) $\varepsilon = 1$ $(e = 0)$ – circular orbits of the satellite for the arbitrary parameters β, I.

(a) In case 4 the evolution reduces to the monotone variation of the longitude of the pericentre of the orbit:

$$\lambda = \Omega + \Delta\omega, \qquad \Delta = \operatorname{sgn} \cos i,$$
$$d\lambda/dn = \Delta\left(\varepsilon_0^{1/2}/5 + \tfrac{1}{2}\beta/\varepsilon^2\right).$$

Here $\varepsilon = \varepsilon_0 = \text{const.}$, $i = i_0 = \text{const.}$

(b) In Figures 2a–d the qualitative pictures of the behaviour of the trajectories for the special case, $\sin I = 0$ and $\cos i = 0$, are given for different values of parameter β.

The equations of the evolution in this case coincide with the equations for case 6. That is why the qualitative pictures of the behaviour of solutions, represented in Figure 2, describe also case 6.

The equations of the evolution for case 5 also coincide with the same equations if one should replace β by $-\frac{1}{2}\beta$ therein.

That is why one can also use the pictures in Figure 2 for the description of case 5.

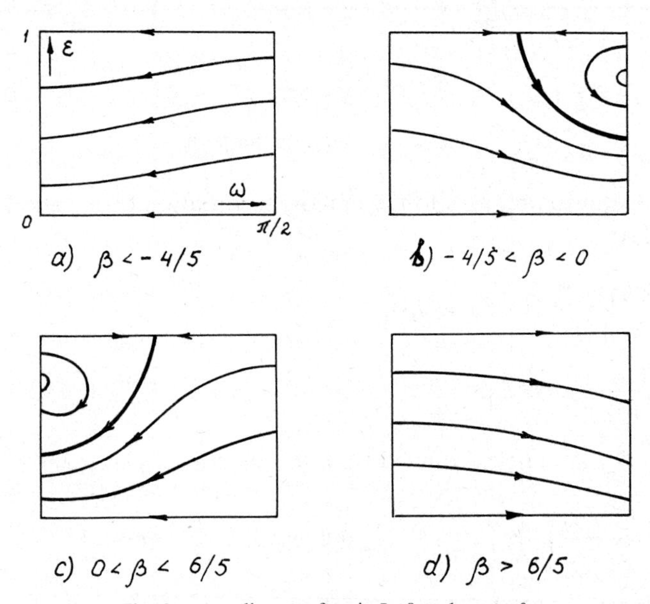

Fig. 2. ε–ω diagram for $\sin I = 0$ and $\cos i = 0$.

In this connection one should only to replace β by -2β in the legends to the pictures.

Integrable case 5 was used by Lidov (1962) for the proof of stability by the eccentricity of the orbits of Uranus' satellites in spite of closeness of their inclination to the ecliptic to 90°.

(c) For the practically important and interesting case 7 the reader is referred to Sekiguchi (1961) and Allan and Cook (1964).

(d) The most complicated and interesting case 3 has been investigated by Kozai (1963) in connection with the problem of evolution of the orbits of the Moon's artifical satellites.

The integration of the problem in this case is based on an additional integral: the projection of the vector of the angular momentum on the axis of symmetry is constant. This integral may be written in the form:

$$\varepsilon \cos^2 i = c_1 = \text{const}.$$

Using a computer Kozai (1963) constructed the field of integral curves in the plane of variables, similar to ε, ω.

In our work this problem is studied completely for all the values c_1 and β. In this connection certain new cases, qualitatively different from those by Kozai, were revealed.

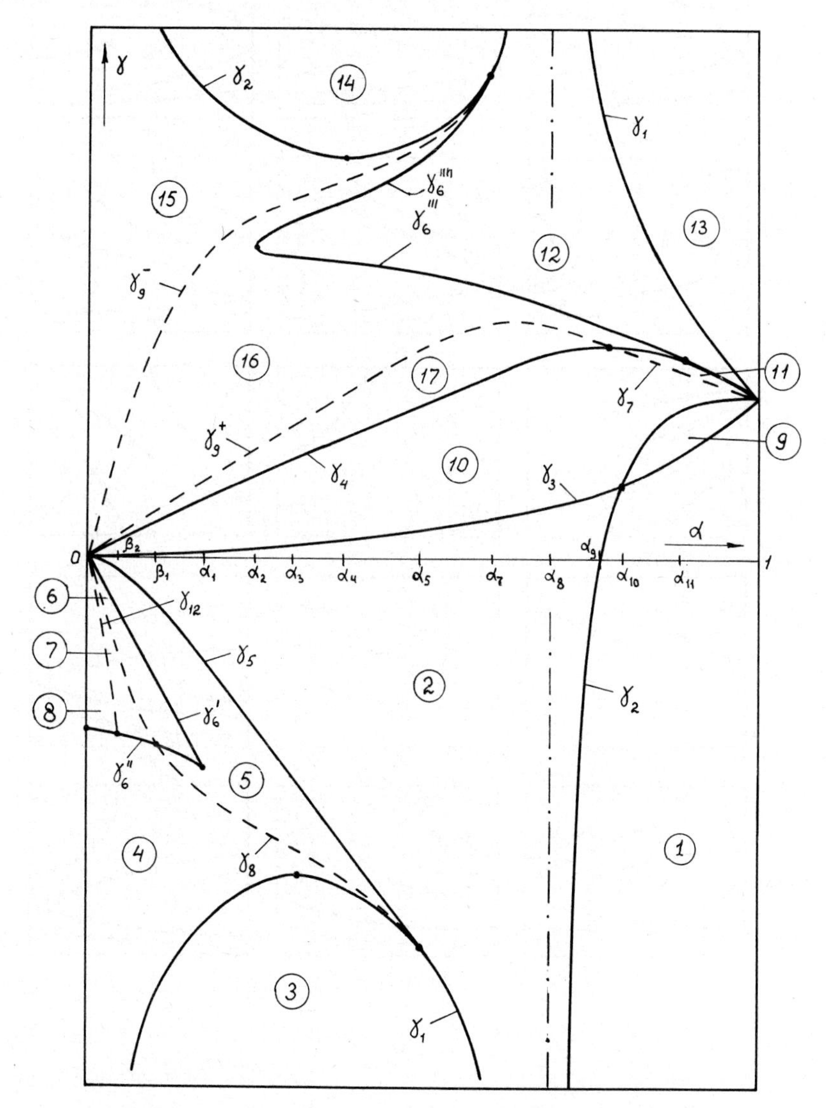

Fig. 3. Condition of appearance (or disappearance) of singularities is expressed by solid curves, and dashed curves define boundaries on which qualitative change takes place for $\sin I = 0$.

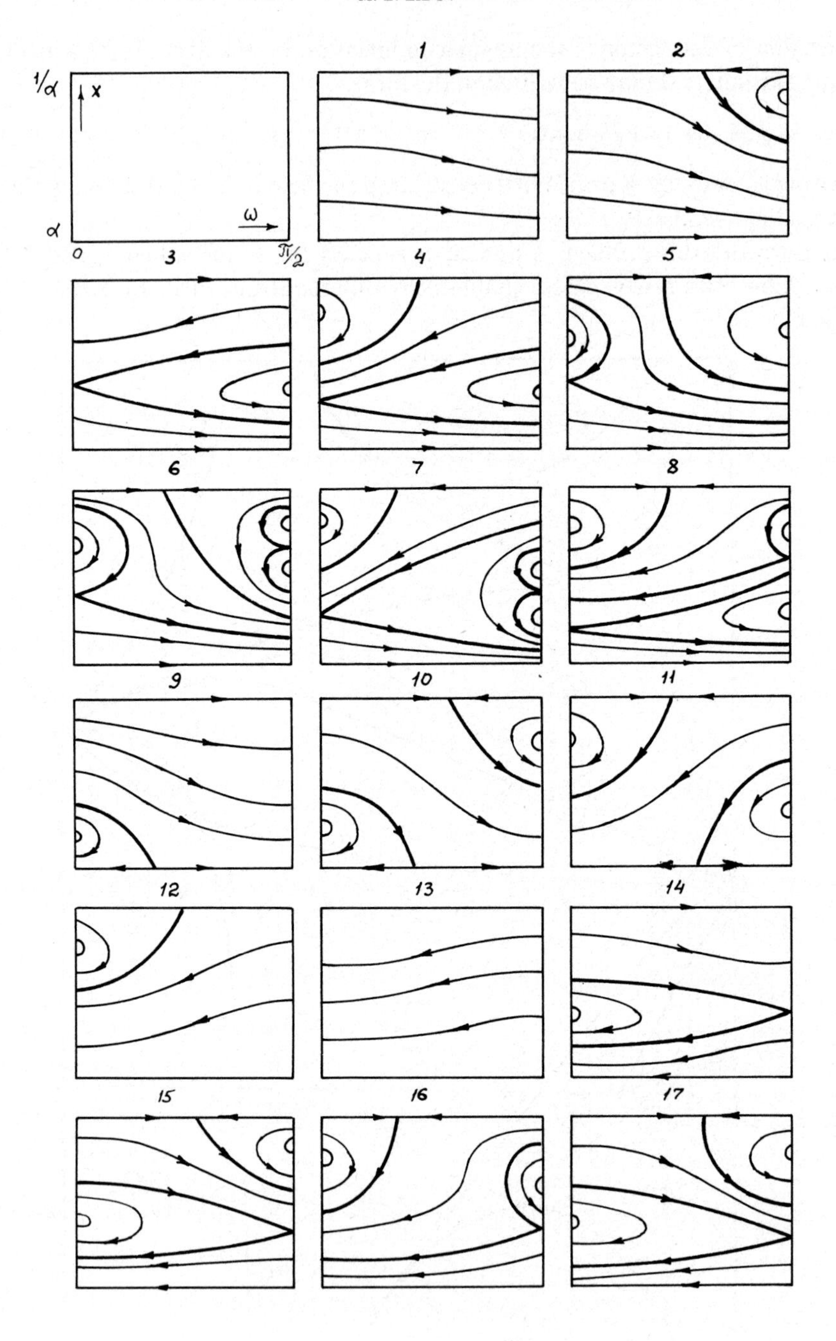

Fig. 4. ε–ω diagrams for $\sin I = 0$.

In this report we shall give only certain results. A more detailed treatment is found in Lidov and Jarskaya (1973).

The Investigation of Case 3 ($\sin I=0$)

Instead of parameters c_1 and β we shall use the parameters: $\gamma=\beta c_1^{-5/4}$ and $\alpha=c_1^{1/4}$. For the elliptical orbits $0\leqslant\alpha\leqslant 1$.

Instead of variable ε it turned out to be more convenient to introduce the variable $x=\varepsilon^{1/2}/\alpha$. It is not difficult to show that, for elliptical orbits of satellites, when α is fixed, x could turn into the values from the interval: $\alpha\leqslant x\leqslant\alpha^{-1}$. In this investigation we studied the whole range of possible values $-\infty<\gamma<+\infty$, although in most applications we could confine ourselves to the case $\gamma\leqslant 0$ ($c_{20}\leqslant 0$).

In Figure 3 the strip, $0\leqslant\alpha\leqslant 1$, $-\infty<\gamma<+\infty$, is divided by the solid curves $\gamma_1(\alpha)$, $\gamma_2(\alpha)$, $\gamma_3(\alpha)$, $\gamma_4(\alpha)$, $\gamma_5(\alpha)$, $\gamma_6'(\alpha),\ldots,\gamma_6''''(\alpha)$ and by the dashed ones $\gamma_7(\alpha)$, $\gamma_8(\alpha)$, $\gamma_9^+(\alpha)$, $\gamma_9^-(\alpha)$, $\gamma_{12}(\alpha)$ in 17 areas. The equations of these curves are given in Lidov and Jarskaya (1973). The construction of these curves requires a cumbersome analysis. The first group of these curves is defined by the condition of appearance (disappearance) of the singular points in the rectangle,

$$P:\{\alpha\leqslant x\leqslant\alpha^{-1},\quad 0\leqslant\omega\leqslant\tfrac{1}{2}\pi\}.$$

The second group of curves (the dashed ones) define the boundaries in which the qualitative change in the behaviour of separatrixrs takes place.

In Figure 4 (1–17) the qualitative pictures of the behaviour of the integral curves are given. The number of 17 pictures corresponds to the areas of parameters shown in Figure 3. In Kozai (1963) there are examples qualitatively coinciding with the cases 2, 3, 4, 5 in Figure 4.

The cases 1, 6, 7, 8, which also take place for $\gamma<0$ evidently, were not revealed. Let us note that the most complicated cases 6–8 (in the rectagle P 6 singular points are

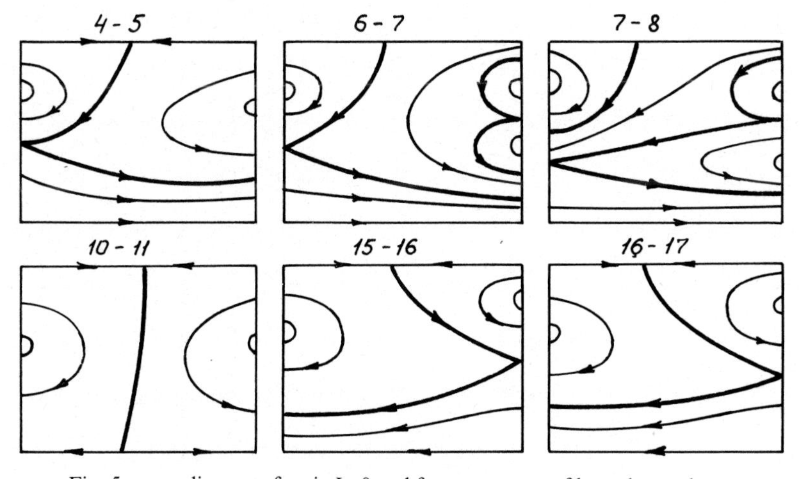

Fig. 5. ε–ω diagrams for $\sin I=0$ and for parameters of boundary values.

present), if the eccentricity of the orbit is not too close to 1, could be realized only when the inclination of the satellite's orbit is close to 90°.

In Figure 5 analogous pictures are given for certain boundary values of the parameters of the problem.

For example, case 4–5 is realized if the parameters α, γ belong to the boundary, separating areas 4 and 5 represented in Figure 3.

In these cases the separatrix, going from one of the singular points of the saddle type, is at the same time the separatrix of another singular point of the same type.

The existence of many stable singular points gives reason to believe that this investigation could turn out to be useful for the practical selection of the special orbits of satellites.

References

Allan, R. R. and Cook, G. E.: 1964, *Proc. Roy. Soc.* **A208**, 97.
Kozai, Y.: 1962, *Astron. J.* **67**, 99.
Kozai, Y.: 1963, *Publ. Astron. Soc. Japan* **15**, 213.
Lidov, L. M.: 1961, *Iskus. sputniky Zemly* **8**, Moscow.
Lidov, L. M.: 1962, in M. Roy (ed.), *Dynamics of Satellites*, Springer Verlag, Berlin, p. 168.
Lidov, L. M. and Jarskaya, M. V.: 1973, in press.
Sekiguchi, N.: 1961, *Publ. Astron. Soc. Japan* **19**, 207.

THE GLOBAL SOLUTION IN THE PROBLEM OF THE CRITICAL INCLINATION*

B. GARFINKEL
Department of Astronomy, Yale University, New Haven, Conn., U.S.A.

The problem of the critical inclination in the artificial satellite theory can be regarded as a problem of resonance arising from the near-commensurability 1:1 between the anomalistic and the draconitic frequencies of the motion, n_1 and n_2. Let us define the main problem by the potential function

$$V = -1/r + J_2 P_2(\sin\theta)/r^3 + J_4 P_4(\sin\theta)/r^5, \tag{1}$$

with

$$J_2 \ll 1, \qquad J_4 = O(J_2^2). \tag{2}$$

The *classical* solution of this problem, expanded in powers of the small parameter J_2, carries the critical divisor

$$\Delta = 5\cos^2 i - 1, \tag{3}$$

as illustrated in Garfinkel (1959), Brouwer (1959), Kozai (1959), Vinti (1963), and Aksnes (1967). This divisor vanishes for the critical inclination

$$i_* = \tan^{-1} 2 = 63.^\circ 4, \tag{4}$$

which corresponds to exact commensurability $n_1 = n_2$, and the solution fails in the neighborhood of i_*.

That the classical singularity at $\Delta = 0$ can be removed by means of the Bohlin-type expansion (1889) in powers of $\sqrt{J_2}$ is illustrated in Hori (1960) and Garfinkel (1960). They showed that the elimination of short-periodic terms from the Hamiltonian reduces the problem to the ideal resonance form as defined in Garfinkel (1966).

Although the literature of the critical inclination is quite extensive, no complete global solution of $O(\sqrt{J_2})$, valid for all inclinations, is presently available. The recent publication of a first-order solution of the ideal resonance problem has opened the way for the construction of such a theory. The present paper furnishes the algorithm for the calculation of the long-periodic and secular terms in the Delaunay variables G, g, h and l that are at least of the first order in $\sqrt{J_2}$. The coefficients of the algorithm are expressed directly in terms of the Delaunay elements L, G, H and the orbital elements a, e, i of the satellite. As a check, it is shown that our *global* solution, valid for all inclinations, includes asymptotically the classical solution with $5\cos^2 i - 1$ as a divisor. The short-periodic terms, absent in our solution, can be taken from any

* Published in full in *Celes. Mech.* **8** (1973), 25.

Y. Kozai (ed.), The Stability of the Solar System and of Small Stellar Systems, 125–127.

classical artificial satellite theory, such as Brouwer (1959). The solution is subject to the normality condition which can be written

$$eG^2/(1+\tfrac{45}{4}e^2)\geqslant O\left[|\tfrac{1}{5}(J_2+J_4/J_2|^{1/4}\right]. \tag{5}$$

The paper is divided into three parts. Part A summarizes the first-order algorithm of the ideal resonance problem, defined by the Hamiltonian

$$F=B(y)+2\mu^2 A(y)\sin^2 x_1, \quad \mu\ll 1. \tag{6}$$

Here y is the momentum-vector y_k, x_1 is the *critical argument*, and x_k for $k>1$ are the ignorable coordinates, which have been eliminated from F in the process of reduction to the ideal form. Part B applies the algorithm of Part A to the problem of the critical inclination. Part C, largely historical, puts the work into a proper perspective in relation to the contributions of Poincaré (1893), E. W. Brown (1931), Hori (1960), Hagihara (1961), Kozai (1961), Iszak (1962), Aoki (1963), and Jupp (1968).

References

Aksnes, K.: 1967, *Astrophys. Norv.* **10**, 69.
Aoki, S.: 1963, *Astron. J.* **68**, 355.
Bohlin, K. P.: 1889, 'Über eine neue Annäherungmethode in der Störungtheorie', *Ak. Handl. Bihang* **14** (afd. 1, Stockholm).
Brouwer, D.: 1959, *Astron. J.* **64**, 378.
Brown, E. W. and Shook, C. A.: 1933, *Planetary Theory*, Dover, Ch. VIII.
Garfinkel, B.: 1959, *Astron. J.* **64**, 353.
Garfinkel, B.: 1960, *Astron. J.* **65**, 624.
Garfinkel, B.: 1964, *Astron. J.* **69**, 453.
Garfinkel, B.: 1966, *Astron. J.* **71**, 657 (Paper I).
Garfinkel, B.: 1972a, *Celest. Mech.* **5**, 189 (Paper III).
Garfinkel, B.: 1972b, *Celest. Mech.* **5**, 451 (Paper IV).
Garfinkel, B.: 1972c, *Celest. Mech.* **6**, 151 (Paper V).
Garfinkel, B.: 1972d, *Celest. Mech.* **7**, 205 (Paper VI).
Garfinkel, B., Jupp, A. H., and Williams, C. A.: 1971, *Astron. J.* **76**, 157 (Paper II).
Hagihara, Y.: 1961, *Smithsonian Contributions to Astrophys.* **5**, 39.
Hagihara, Y.: 1972, *Celestial Mechanics*, MIT Press II, Part I, p. 430.
Hori, G.: 1960, *Astron. J.* **65**, 291.
Iszak, I.: 1962, *Smithsonian Astrophys. Obs.*, Special Report No. 90.
Kozai, Y.: 1961, *Smithsonian Contributions to Astrophys.* **5**, 53.
Message, P. J., Hori, G., and Garfinkel, B.: 1962, *Observatory*, No. 929, 168.
Poincaré, H.: 1893, *Méthodes Nouvelles de la Méchanique Céleste*, Volume II, p. 352, Dover Press.
Tisserand, F. F.: 1889–96, *Mécanique Céleste* **IV**, p. 444.
Vinti, J. P.: 1963, *J. Res. Nat. Bur. Sd.* **67B**, 191.
von Zeipel, H.: 1916, *Arkiv. Mat. Astron. Fys.* **II**.

DISCUSSION

A. H. Jupp: Would you like to comment on my conjecture that your regularizing function would have to be modified if you were to formulate your solution in terms of initial values rather than in terms of the mean values?

B. Garfinkel: The elements used in my theory differ from the initial values only by quantities of the second-order. Therefore, no modification would be required in a first-order theory. As far as the second-order theory is concerned, some of the coefficients may have to be modified, rather than the regularizing function itself.

T. Inoue: Do you agree with me if I say that the perturbations depend on the choice of zeroth approximation and, therefore, if one chooses a proper intermediary orbit in the case of the ideal resonance problem one will obtain a solution free from any singularity?

B. Garfinkel: I do not agree. In my paper in *Astron. J.* (1964) it was shown that all classical solutions carry a singularity at the critical inclination, regardless of the choice of reference orbit or the choice of the coordinate system.

RETROGRADE SATELLITES IN THE CIRCULAR PLANE RESTRICTED THREE-BODY PROBLEM

D. BENEST
Observatoire, Université de Nice, Le Mont-Gros, 06 Nice, France

Characteristics and stability of simple-periodic retrograde satellites of the lighter body are presented for Hill's case and for all values of the mass ratio $m_2/(m_1+m_2)$ between 0 and 0.5.

Nonperiodic orbits are presented for some values of the ratio near a critical one, and a new family of double-periodic orbits is found to take an important part in the stability on nonperiodic orbits, so that they counterbalance the instability of simple-periodic family.

DISCUSSION

T. Inoue: Après avoir calculé beaucoup d'orbites contenant des orbites stables est-ce que vous avez trouvé des conditions analytiques pour que les orbites soient stables?

D. Benest: Je ne l'ai pas encore cherché, mais ce serait certainement très intéressant, si elles existent.

Y. Kozai (ed.), The Stability of the Solar System and of Small Stellar Systems, 129.

A NOTE ON A SEPARATION OF EQUATIONS OF VARIATION OF THE ELLIPTIC RESTRICTED THREE-BODY PROBLEM INTO HILL'S EQUATIONS

V. MATAS

Astronomical Institute of the Czechoslovak Academy of Sciences, Budečská 6, Vinohrady, Praha 2, Czechoslovakia

Abstract. The equations of variation of the three-dimensional elliptic restricted three-body problem corresponding to the equilibrium solutions (the libration points) have been separated into three Hill's equations. As regards the equation 'corresponding' to the motion of the infinitesemal body in the z-axis (perpendicular to the plane of motion of the primaries), the matter is trivial one since the initial equation – as known – reads $d^2z/dv^2+(A_i+e\cos v)/(1+e\cos v)=0$ (e, $0<e<1$, and v are the eccentricity and the true anomaly of the relative motion of the primaries) with $A_i>1$ for the straight-line libration points L_i ($i=1, 2, 3$) and $A_i=1$ for the triangular libration points L_i, $i=4, 5$. As concerns the remaining two components, x and y, of the motion of the infinitesimal body (x, y and z are the Nechvíle's variables), in the case of the straight-line libration points, L_1, L_2 and L_3, the corresponding equations of variation have been transformed and separated into two further – mutually independent – Hill's equations without any limitation. In the case of the equilateral triangle libration points, L_4 and L_5, the separation has been found only when the eccentricity e and the dimensionless mass μ, $0<\mu\leqq\frac{1}{2}$, of the 'minor' primary satisfy the additional conditions:

$$\tfrac{1}{9}\{5+e^2-[8(e^2+2)(1-e^2)]^{1/2}\}\neq 3\mu(1-\mu)< \\ <\tfrac{1}{9}\{5+e^2+[8(e^2+2)(1-e^2)]^{1/2}\}.$$

Let us write the latter two Hill's equations obtained in the form

$$d^2y_k^*/dv^2+I_k(v)\,y_k^*=0,\quad k=1,2,$$

where I_k, $k=1, 2$, are 2π-periodic even functions of the true anomaly v. The functions I_k, $k=1, 2$, are real functions in the case of the straight-line libration points, L_1, L_2 and L_3, without a limitation but in the case of the triangular libration points, L_4 and L_5, they are real only if

$$3\mu(1-\mu)<\tfrac{1}{9}\{5+e^2-[8(e^2+2)(1-e^2)]^{1/2}\}.$$

Provided

$$\tfrac{1}{9}\{5+e^2-[8(e^2+2)(1-e^2)]^{1/2}\}<3\mu(1-\mu)< \\ <\tfrac{1}{9}\{5+e^2+[8(e^2+2)(1-e^2)]^{1/2}\}$$

the functions I_k, $k=1, 2$, are complex-valued functions of the real variable v.

Y. Kozai (ed.), The Stability of the Solar System and of Small Stellar Systems, 131.

COMET SCHWASSMANN-WACHMANN 3 (1930 VI)

N. A. BELYAEV and S. D. SHAPOREV
Institute of Theoretical Astronomy, Leningrad, U.S.S.R.

The definitive orbit of comet 1930 VI basing on 190 observations has been derived. Investigation of the comet orbit evolution for the time span of 200 yr indicated that the 1882 approach (t_{appr} = 1882 Oct. 22.559; Δ_{min} = 0.0057 AU) of the comet to Jupiter resulted in considerable element transformations. The perihelion and aphelion distances reduced from 4.1 AU to 1.2 AU and from 12.2 AU to 5.2 AU, respectively, whereas the period of revolution also underwent a decrease from 23.3 yr to 5.8 yr. The next return of the comet to the perihelion is to be expected in 1974 (T = 1974 March, 17.475).

Y. Kozai (ed.), The Stability of the Solar System and of Small Stellar Systems, 133.

THE MOTION OF COMET WESTPHAL IN 1852–1974

L. M. BELOUS

Institute of Theoretical Astronomy, Leningrad, U.S.S.R.

Two apparitions of Comet Westphal in 1852 and 1913 have been linked on the basis of 101 observations, the mean errors being equal to $\pm 2''55$. The perturbations from nine major planets and the possible influence of nongravitational forces have been taken into account. The elements and the ephemeride for 1974 have been computed.

Y. Kozai (ed.), The Stability of the Solar System and of Small Stellar Systems, 135.

THE MOTION OF COMET ENCKE-BACKLUND OVER 1901–1970

N. A. BOKHAN and Yu. A. CHERNETENKO
Institute of Theoretical Astronomy, Leningrad, U.S.S.R.

The investigation of the 22 apparitions of Comet Encke-Backlund for the time interval 1901–70 was carried out. The integration of equations of motion was performed using the technique devised by Kazimirchak-Polonskaya with automatic step choice, taking into account the perturbations from nine planets (Mercury-Pluto) as well as nongravitational effects.

The observations made before and after the perihelion passage are presented with sufficient accuracy. The existence of a continuous secular acceleration in the cometary motion is confirmed and the secular variation of the eccentricity is shown to be a function of time.

Y. Kozai (ed.), The Stability of the Solar System and of Small Stellar Systems, 137.

AN ITERATIVE METHOD OF GENERAL PLANETARY THEORY

V. A. BRUMBERG
Institute of Theoretical Astronomy, Leningrad, U.S.S.R.

Abstract. This paper deals with an iterative version of the general planetary theory. Just as in Airy's Lunar method the series in powers of planetary masses are replaced here by the iterations to achieve improved approximations for the coefficients of planetary inequalities. The right-hand members of the equations of motion are calculated in closed formulas, and no expansion in powers of small corrections to the planetary coordinates is needed. For the N-planet case this method requires the performance of the analytical operations on a computer with power series of $4N$ polynomial variables, the coefficients being the exponential series of $N-1$ angular arguments. To obtain numerical series of planetary motion one has to solve the secular system using Birkhoff's normalization or the Taylor series in powers of time. A slight modification of the method in the resonant case makes it valid for the treatment of the main problem of the Galilean satellites of Jupiter.

1. Introduction

This paper deals with an iterative version of the general planetary theory proposed by the author in 1969 (Brumberg, 1970). The general method is based on the ideas by Hill (non-Keplerian intermediary), von Zeipel (separation of slow and fast variables) and Birkhoff (reduction of dynamical system to normal form). As in all existing planetary theories the actual performance of this method involves the expansions in powers of planetary masses. Such a method was used in the second-order linear theory of secular perturbations (Brumberg and Egorova, 1971) and in the general first-order theory (Brumberg and Chapront, 1974).

Expansions in powers of planetary masses are convenient in construction of first-order theory permitting us to consider separately the perturbations in motion of planet i due to planet j. In second-order theory in addition to these single perturbations one has to deal with double perturbations caused by the combined action of planets j and k. In third-order theory triple perturbations arise from the combined influence of planets j, k and l, etc. Expressions of these perturbations become rather cumbersome and expansions in powers of planetary masses lose their efficiency. That is why these expansions are not used here. Dependence of masses is taken into account numerically by means of iterations. The only criterion of smallness of any term is its numerical value. This approach is similar to the Airy method used in the modern theory of the Moon. It should be noted that iterative versions of some classic methods of planetary perturbations have been already successfully elaborated (Broucke, 1969; Seidelmann, 1970). In the version suggested here we apply a new algorithm for computation of the right members. This algorithm does not demand expansions in powers of unknown variables.

Construction of the general planetary theory by this iterative method calls for a system of literal manipulation on computer with power series of $4N$ polynomial variables, the coefficients being the exponential series of $N-1$ angular arguments

Y. Kozai (ed.), The Stability of the Solar System and of Small Stellar Systems, 139–155.

(to be more precise, we have N arguments but the sum of indices is always zero). Solution of this problem for N=9 is possible on modern computers.

2. Intermediary

The general theory has been given in detail by Brumberg and Chapront (1974). We give here only the computational algorithm of the iterative version.

If p_i and w_i are related with heliocentric coordinates x_i, y_i and z_i by means of

$$x_i+\sqrt{-1}\, y_i=a_i(1-p_i)\exp\sqrt{-1}\,\lambda_i, \qquad z_i=a_iw_i,$$
$$\lambda_i=n_it+\varepsilon_i, \qquad n_i^2a_i^3=k^2(m_0+m_i),$$

then the equations of motion take the form

$$\begin{aligned}&\ddot{p}_i+2\sqrt{-1}\, n_i\dot{p}_i-\tfrac{3}{2}n_i^2(p_i+q_i)=n_i^2P_i,\\ &\ddot{w}_i+n_i^2w_i=n_i^2W_i.\end{aligned} \tag{1}$$

Here

$$\begin{aligned}P_i=&-1-\tfrac{1}{2}p_i-\tfrac{3}{2}q_i+(1-p_i)\left(\frac{a_i}{r_i}\right)^3+\mu\sum_{j=1}^{N}{}^{(i)}\kappa_{ij}\left\{\zeta_{ij}^{-1}\left(\frac{a_i}{a_j}\right)^2(1-p_j)\left(\frac{a_j}{r_j}\right)^3+\right.\\ &\left.+\left(\frac{a_i}{a_{ij}}\right)^2\left[\frac{a_i}{a_{ij}}(1-p_i)-\frac{a_j}{a_{ij}}(1-p_j)\,\zeta_{ij}^{-1}\right]\left(\frac{a_{ij}}{\Delta_{ij}}\right)^3\right\},\end{aligned} \tag{2}$$

$$\begin{aligned}W_i=&\,w_i-w_i\left(\frac{a_i}{r_i}\right)^3-\mu\sum_{j=1}^{N}{}^{(i)}\kappa_{ij}\left[\left(\frac{a_i}{a_j}\right)^2w_j\left(\frac{a_j}{r_j}\right)^3+\right.\\ &\left.+\left(\frac{a_i}{a_{ij}}\right)^2\left(\frac{a_i}{a_{ij}}w_i-\frac{a_j}{a_{ij}}w_j\right)\left(\frac{a_{ij}}{\Delta_{ij}}\right)^3\right],\end{aligned} \tag{3}$$

where

$$\begin{aligned}&\mu\kappa_{ij}=\frac{m_j}{m_0+m_i}, \qquad \zeta_{ij}=\exp\sqrt{-1}\,(\lambda_i-\lambda_j), \qquad a_{ij}=\max(a_i,a_j),\\ &\left(\frac{r_i}{a_i}\right)^2=(1-p_i)(1-q_i)+w_i^2,\end{aligned} \tag{4}$$

$$\begin{aligned}\left(\frac{\Delta_{ij}}{a_{ij}}\right)^2=&\left[\frac{a_i}{a_{ij}}(1-p_i)-\frac{a_j}{a_{ij}}(1-p_j)\,\zeta_{ij}^{-1}\right]\left[\frac{a_i}{a_{ij}}(1-q_i)-\frac{a_j}{a_{ij}}(1-q_j)\,\zeta_{ij}\right]+\\ &+\left(\frac{a_i}{a_{ij}}w_i-\frac{a_j}{a_{ij}}w_j\right)^2,\end{aligned} \tag{5}$$

and $q_i=\bar{p}_i$. The form of expressions (2) and (3) is simpler for $a_{ij}=a_i$, but the choice made here is more suitable for computation.

We take as intermediary the particular solution of Equations (1), $p_i=p_i^{(0)}$ and $w_i=0$, containing $2N$ arbitrary constants (mean motions n_i and mean longitudes at the epoch ε_i). The planetary masses m_i and mean motions n_i are supposed to have fixed values which lead by Kepler's third law to definite values of the semimajor axes a_i. The intermediate solution can be represented in the exponential series form

$$p_i^{(0)} = \sum p_\gamma^{(i)} \exp\sqrt{-1}\,(\gamma\lambda), \tag{6}$$

where

$$(\gamma\lambda) = \gamma_1\lambda_1 + \gamma_2\lambda_2 + \cdots + \gamma_N\lambda_N, \tag{7}$$

and

$$\gamma_1 + \gamma_2 + \cdots + \gamma_N = 0. \tag{8}$$

Summation in (6) is extended over all integer values of $\gamma_1, \ldots, \gamma_N$. Coefficients $p_\gamma^{(i)}$ have real values which are determined by iterations. To do this we substitute expressions (6) known at some stage of the approximation process into right members (2). The pertinent series P_i are of the same form (6)–(8) with coefficients $P_\gamma^{(i)}$. Improved values of the coefficients of (6) are then obtained as

$$p_0^{(i)} = -\tfrac{1}{3} P_0^{(i)},$$
$$p_\gamma^{(i)} = n_i^2 \frac{\left[(\gamma n)^2 - 2n_i(\gamma n) + \tfrac{3}{2}n_i^2\right] P_\gamma^{(i)} - \tfrac{3}{2} n_i^2 P_{-\gamma}^{(i)}}{(\gamma n)^2 \left[n_i^2 - (\gamma n)^2\right]}, \tag{9}$$

where (γn) means the dot product of $\gamma = (\gamma_1, \ldots, \gamma_N)$ and $n = (n_1, \ldots, n_N)$. The initial values of this process are $p_i^{(0)} = 0$. Having found, with use of (6) the exponential series,

$$Z_i(k, l) = (1 - p_i)^k (1 - q_i)^l, \tag{10}$$

and

$$Z_{ij}(k, l) = \left[\frac{a_i}{a_{ij}}(1 - p_i) - \frac{a_j}{a_{ij}}(1 - p_j)\,\zeta_{ij}^{-1}\right]^k \left[\frac{a_i}{a_{ij}}(1 - q_i) - \frac{a_j}{a_{ij}}(1 - q_j)\,\zeta_{ij}\right]^l, \tag{11}$$

for $k = 1$ and $l = 3$, we have in calculating (2) to convert them into exponential series,

$$T = Z^{-N/2}, \tag{12}$$

with $N = 1$. We meet further the value $N = 2$. This problem can be solved by well-known methods. The initial value of T corresponds to the unperturbed circular orbits ($p_i = 0$), that is, for (10) $T_0 = 1$ and for (11),

$$T_0 = \gamma^{(ij)} \begin{Bmatrix} 0, & -\tfrac{1}{2}kN, & -\tfrac{1}{2}lN, & 0, & a_j/a_i \\ 0, & -\tfrac{1}{2}lN, & -\tfrac{1}{2}kN, & \tfrac{1}{2}(k-l)\,N, & a_i/a_j \end{Bmatrix}$$

with function $\gamma^{(ij)}(n, x, y, v, \alpha)$ determined in Brumberg (1970).

3. Equations of Motion

Putting $p_i = p_i^{(0)} + \delta p_i$ and collecting in (2) and (3) the linear terms we obtain

$$P_i = P_i^{(0)} - \sum_{j=1}^{N} (K_{ij}\delta p_j + L_{ij}\delta q_j) + P_i^*, \tag{13}$$

$$W_i = -\sum_{j=1}^{N} M_{ij} w_j + W_i^*. \tag{14}$$

The unknowns δp_i and w_i must obey the equations

$$\delta\ddot{p}_i + 2\sqrt{-1}\, n_i \delta\dot{p}_i + n_i^2 \sum_{j=1}^{N} [(-\tfrac{3}{2}\delta_{ij} + K_{ij})\, \delta p_j + (-\tfrac{3}{2}\delta_{ij} + L_{ij})\, \delta q_j] = n_i^2 P_i^*, \tag{15a}$$

$$\ddot{w}_i + n_i^2 \sum_{j=1}^{N} (\delta_{ij} + M_{ij})\, w_j = n_i^2 W_i^*. \tag{15b}$$

Assuming (6) to be known one can find functions depending on intermediary as follows:

$$A_i = (1-p_i)\left(\frac{a_i}{r_i}\right)^5, \qquad B_i = \left(\frac{a_i}{r_i}\right)^3, \qquad C_i = (1-p_i)\left(\frac{a_i}{r_i}\right)^3,$$

$$D_i = (1-q_i)\left(\frac{a_i}{r_i}\right)^2, \qquad E_i = \left(\frac{a_i}{r_i}\right)^2, \qquad F_i = (1-p_i)^2\left(\frac{a_i}{r_i}\right)^5,$$

$$A_{ij} = \left[\frac{a_i}{a_{ij}}(1-p_i) - \frac{a_j}{a_{ij}}(1-p_j)\,\zeta_{ij}^{-1}\right]\left(\frac{a_{ij}}{\Delta_{ij}}\right)^5, \qquad B_{ij} = \left(\frac{a_{ij}}{\Delta_{ij}}\right)^3,$$

$$C_{ij} = \left[\frac{a_i}{a_{ij}}(1-p_i) - \frac{a_j}{a_{ij}}(1-p_j)\,\zeta_{ij}^{-1}\right]\left(\frac{a_{ij}}{\Delta_{ij}}\right)^3,$$

$$D_{ij} = \left[\frac{a_i}{a_{ij}}(1-q_i) - \frac{a_j}{a_{ij}}(1-q_j)\,\zeta_{ij}\right]\left(\frac{a_{ij}}{\Delta_{ij}}\right)^2, \qquad E_{ij} = \left(\frac{a_{ij}}{\Delta_{ij}}\right)^2,$$

$$F_{ij} = \left[\frac{a_i}{a_{ij}}(1-p_i) - \frac{a_j}{a_{ij}}(1-p_j)\,\zeta_{ij}^{-1}\right]^2\left(\frac{a_{ij}}{\Delta_{ij}}\right)^5.$$

Determination of these functions in form (6)–(8) calls for conversion (12) with (10) and (11). The corresponding values of integers k, l and N are

	A	B	C	D	E	F
k	3	3	1	1	1	1
l	5	3	3	0	1	5
N	1	1	1	2	2	1.

One may use the check relations

$$A_i = C_i E_i, \qquad B_i = C_i D_i, \qquad A_i = D_i F_i,$$

and the similar expressions for two-indices functions.

With the aid of these functions the coefficients for the left members of (15) can be obtained in form (6)–(8) as follows:

$$K_{ii}=\tfrac{1}{2}\left[1-B_i-\mu\sum_{j=1}^{N}{}^{(i)}\kappa_{ij}\left(\frac{a_i}{a_{ij}}\right)^3 B_{ij}\right],$$

$$K_{ij}=\tfrac{1}{2}\mu\kappa_{ij}\left[\left(\frac{a_i}{a_{ij}}\right)^2\frac{a_j}{a_{ij}}B_{ij}-\left(\frac{a_i}{a_j}\right)^2 B_j\right]\zeta_{ij}^{-1},\quad (i\neq j)$$

$$L_{ii}=\tfrac{3}{2}\left[1-F_i-\mu\sum_{j=1}^{N}{}^{(i)}\kappa_{ij}\left(\frac{a_i}{a_{ij}}\right)^3 F_{ij}\right], \tag{16}$$

$$L_{ij}=\tfrac{3}{2}\mu\kappa_{ij}\left[\left(\frac{a_i}{a_{ij}}\right)^2\frac{a_j}{a_{ij}}\zeta_{ij}F_{ij}-\left(\frac{a_i}{a_j}\right)^2\zeta_{ij}^{-1}F_j\right],\quad (i\neq j)$$

$$M_{ii}=-2K_{ii},\qquad M_{ij}=-2K_{ij}\zeta_{ij}\quad (i\neq j).$$

Put, further,

$$S_i=D_i\delta p_i+\bar{D}_i\delta q_i-E_i(\delta p_i\delta q_i+w_i^2), \tag{17}$$

$$S_{ij}=D_{ij}\left(\frac{a_i}{a_{ij}}\delta p_i-\frac{a_j}{a_{ij}}\zeta_{ij}^{-1}\delta p_j\right)+\bar{D}_{ij}\left(\frac{a_i}{a_{ij}}\delta q_i-\frac{a_j}{a_{ij}}\zeta_{ij}\delta q_j\right)-$$
$$-E_{ij}\left[\left(\frac{a_i}{a_{ij}}\delta p_i-\frac{a_j}{a_{ij}}\zeta_{ij}^{-1}\delta p_j\right)\left(\frac{a_i}{a_{ij}}\delta q_i-\frac{a_j}{a_{ij}}\zeta_{ij}\delta q_j\right)+\left(\frac{a_i}{a_{ij}}w_i-\frac{a_j}{a_{ij}}w_j\right)^2\right] \tag{18}$$

and let T_i and T_{ij} be the series of the same kind

$$T_i=\sum_{k=2}^{\infty}\frac{(3/2)_k}{(1)_k}S_i^k,\qquad T_{ij}=\sum_{k=2}^{\infty}\frac{(3/2)_k}{(1)_k}S_{ij}^k, \tag{19}$$

In virtue of relations

$$\left(\frac{r_i}{a_i}\right)^2=\frac{1-S_i}{E_i},\qquad \left(\frac{a_i}{r_i}\right)^3=B_i(1+\tfrac{3}{2}S_i+T_i),$$

$$\left(\frac{\Delta_{ij}}{a_{ij}}\right)^2=\frac{1-S_{ij}}{E_{ij}},\qquad \left(\frac{a_{ij}}{\Delta_{ij}}\right)^3=B_{ij}(1+\tfrac{3}{2}S_{ij}+T_{ij}),$$

there results

$$P_i^*=-\tfrac{3}{2}A_i(\delta p_i\delta q_i+w_i^2)-\tfrac{3}{2}B_iS_i\delta p_i+(C_i-B_i\delta p_i)\,T_i+$$
$$+\mu\sum_{j=1}^{N}{}^{(i)}\kappa_{ij}\left\{-\tfrac{3}{2}A_{ij}\left[\left(\frac{a_i}{a_{ij}}\delta p_i-\frac{a_j}{a_{ij}}\zeta_{ij}^{-1}\delta p_j\right)\left(\frac{a_i}{a_{ij}}\delta q_i-\frac{a_j}{a_{ij}}\zeta_{ij}\delta q_j\right)+\right.\right.$$
$$\left.+\left(\frac{a_i}{a_{ij}}w_i-\frac{a_j}{a_{ij}}w_j\right)^2\right]-\tfrac{3}{2}B_{ij}S_{ij}\left(\frac{a_i}{a_{ij}}\delta p_i-\frac{a_j}{a_{ij}}\zeta_{ij}^{-1}\delta p_j\right)+$$

$$+\left[C_{ij}-B_{ij}\left(\frac{a_i}{a_{ij}}\,\delta p_i-\frac{a_j}{a_{ij}}\,\zeta_{ij}^{-1}\delta p_j\right)\right]T_{ij}+\zeta_{ij}^{-1}\left(\frac{a_i}{a_j}\right)^2\times$$

$$\left.\times\left[-\tfrac{3}{2}A_j(\delta p_j\delta q_j+w_j^2)-\tfrac{3}{2}B_jS_j\delta p_j+(C_j-B_j\delta p_j)\,T_j\right]\right\}, \tag{20}$$

$$W_i^*=-B_i(\tfrac{3}{2}S_i+T_i)\,w_i-\mu\sum_{j=1}^{N}{}^{(i)}\,\kappa_{ij}\left[B_{ij}(\tfrac{3}{2}S_{ij}+T_{ij})\left(\frac{a_i}{a_{ij}}\,w_i-\frac{a_j}{a_{ij}}\,w_j\right)+\right.$$

$$\left.+\left(\frac{a_i}{a_j}\right)^2 B_j(\tfrac{3}{2}S_j+T_j)\,w_j\right]. \tag{21}$$

These expressions allow us to calculate the right members without expansions in powers of δp_i, δq_i and w_i.

4. Linear Inequalities

According to the general theory the solution of (15) can be presented in the form

$$\delta p_i=\sum p_{pqrs}^{(i)}\prod_{j=1}^{N} a_j^{p_j}\bar{a}_j^{q_j}b_j^{r_j}\bar{b}_j^{s_j}, \tag{22}$$

$$w_i=\sum w_{pqrs}^{(i)}\prod_{j=1}^{N} a_j^{p_j}\bar{a}_j^{q_j}b_j^{r_j}\bar{b}_j^{s_j}, \tag{23}$$

where summation goes over all nonnegative values of indices p_j, q_j, r_j and s_j, with positive sum. The sum of all r_j and s_j is even in (22) and odd in (23). Coefficients $p_{pqrs}^{(i)}$ and $w_{pqrs}^{(i)}$ are series of form (6)–(8). The variables a_j and b_j (the same designation a_j for the unknown variables and semimajor axes should not lead to misunderstanding) are related with slow variables α_j, β_j by simple expressions,

$$a_j=\alpha_j\exp\sqrt{-1}\,\lambda_j,\qquad b_j=\beta_j\exp\sqrt{-1}\,\lambda_j. \tag{24}$$

The unknowns α_j and β_j represent a sort of eccentricity and obliquity variables, respectively. For the two-body problem they can be expressed in terms of the usual Keplerian elements (Brumberg and Chapront, 1974). The secular system determining these variables is treated in Section 6.

Determination of the linear terms of (22) and (23) is performed in the following way. Let $Q=\|Q_{\kappa\nu}\|$ ($\kappa,\nu=1,2,\ldots,6$) be a block matrix, each block representing linear combination of matrices $K=\|K_{ij}\|$, $L=\|L_{ij}\|$ and $M=\|M_{ij}\|$,

$$\begin{aligned}
&Q_{11}=K-L-\bar{K}+\bar{L},\qquad Q_{12}=-\tfrac{2}{3}(K+L+\bar{K}+\bar{L}),\\
&Q_{13}=\tfrac{1}{2}(-K+3L+3\bar{K}-\bar{L}),\qquad Q_{21}=\tfrac{3}{2}(-K+L-\bar{K}+\bar{L}),\\
&Q_{22}=K+L-\bar{K}-\bar{L},\qquad Q_{23}=\tfrac{3}{4}(K-3L+3\bar{K}-\bar{L}),\\
&Q_{3\nu}=\tfrac{1}{2}Q_{1\nu}+\tfrac{2}{3}Q_{2\nu}\quad(\nu=1,2,3),\qquad Q_{43}=-\tfrac{1}{2}Q_{13}+\tfrac{2}{3}Q_{23},\\
&Q_{55}=-Q_{65}=\tfrac{1}{2}M.
\end{aligned} \tag{25}$$

Eight blocks of Q are defined by conjugate relations,

$$\begin{array}{llll} Q_{14}=\bar{Q}_{13}, & Q_{24}=-\bar{Q}_{23}, & Q_{41}=\bar{Q}_{31}, & Q_{42}=-\bar{Q}_{32}, \\ Q_{34}=-\bar{Q}_{43}, & Q_{44}=-\bar{Q}_{33}, & Q_{56}=-\bar{Q}_{65}, & Q_{66}=-\bar{Q}_{55}. \end{array} \tag{26}$$

Remaining 16 blocks are zero matrices (for values $\kappa=1, 2, 3, 4$, $\nu=5, 6$ and $\kappa=5, 6$, $\nu=1, 2, 3, 4$).

Hence, each element $Q_{\kappa\nu}[i, j]$ ($\kappa, \nu=1, 2,\ldots, 6$; $i, j=1, 2,\ldots, N$) of Q can be found in form (6)–(8), that is,

$$Q_{\kappa\nu}[i, j]=\sum Q_{\gamma}^{(\kappa\nu ij)} \exp\sqrt{-1}(\gamma\lambda). \tag{27}$$

Matrices G and S needed for calculation of the linear inequalities are of the same form as Q. Relations (26) remain valid for the blocks of G. The blocks of S satisfy similar relations but with the change of signs,

$$\begin{array}{llll} S_{14}=-\bar{S}_{13}, & S_{24}=\bar{S}_{23}, & S_{41}=-\bar{S}_{31}, & S_{42}=\bar{S}_{32}, \\ S_{34}=\bar{S}_{43}, & S_{44}=\bar{S}_{33}, & S_{56}=\bar{S}_{65}, & S_{66}=\bar{S}_{55}. \end{array} \tag{28}$$

Therefore, each of these matrices is completely defined by 12 blocks enumerated in (25), and it is sufficient to give expressions only for these blocks. G and S satisfy equations

$$G=Q(E+S)-\mathcal{N}^{-1}S\mathcal{N}G^*, \tag{29}$$

$$G=G^*+\tilde{G}, \tag{30}$$

$$\dot{S}+\sqrt{-1}(S\mathcal{N}P-\mathcal{N}PS)=\sqrt{-1}\,\mathcal{N}\tilde{G}, \tag{31}$$

where $\mathcal{N}=\mathrm{diag}\,(n_1, n_2,\ldots, n_N)$ and $P=\|P_{\kappa\nu}\|$ being a constant block matrix with non-zero blocks,

$$P_{12}=P_{33}=P_{55}=E, \qquad P_{44}=P_{66}=-E.$$

E denotes the unit matrix. G and S, like Q, have the factor of planetary masses and can be found from (29)–(31) by iterations. Let us assume that we know the approximate values of elements G and S in the form of (27). Then (29) yields more accurate series for G:

$$G_{\kappa\nu}[i, j]=Q_{\kappa\nu}[i, j]+\sum_{k=1}^{N}\sum_{\lambda=1}^{6}\left(Q_{\kappa\lambda}[i, k]\, S_{\lambda\nu}[k, j)]-\right.$$
$$\left.-\frac{n_k}{n_i}\, S_{\kappa\lambda}[i, k]\, G^*_{\lambda\nu}[k, j]\right). \tag{32}$$

We collect in G^* the terms with coefficients $G_{\gamma}^{(\kappa\nu ij)}$ such that

$$\begin{array}{ll} \gamma_k=0, \quad \text{for} & (\kappa, \nu)=(1, 1), (1, 2), (2, 1), (2, 2), \\ \gamma_k=\delta_{ki}-\delta_{kj}, \quad \text{for} & (\kappa, \nu)=(3, 3), (5, 5) \quad (k=1, 2,\ldots, N). \end{array} \tag{33}$$

Remaining terms are included into $\tilde{G}$. From (31) we obtain the improved series for S:

$$S_\gamma^{(21ij)}=\frac{n_i}{(\gamma n)}G_\gamma^{(21ij)},\qquad S_\gamma^{(22ij)}=\frac{n_i}{(\gamma n)}\left(G_\gamma^{(22ij)}-\frac{n_j}{n_i}S_\gamma^{(21ij)}\right),$$

$$S_\gamma^{(11ij)}=\frac{n_i}{(\gamma n)}(G_\gamma^{(11ij)}+S_\gamma^{(21ij)}),\qquad S_\gamma^{(12ij)}=\frac{n_i}{(\gamma n)}\left(G_\gamma^{(12ij)}-\frac{n_j}{n_i}S_\gamma^{(11ij)}+S_\gamma^{(22ij)}\right),$$

$$S_\gamma^{(23ij)}=\frac{n_i}{(\gamma n)+n_j}G_\gamma^{(23ij)},\qquad S_\gamma^{(13ij)}=\frac{n_i}{(\gamma n)+n_j}(G_\gamma^{(13ij)}+S_\gamma^{(23ij)}),$$

$$S_\gamma^{(31ij)}=\frac{n_i}{(\gamma n)-n_i}G_\gamma^{(31ij)},\qquad S_\gamma^{(32ij)}=\frac{n_i}{(\gamma n)-n_i}\left(G_\gamma^{(32ij)}-\frac{n_j}{n_i}S_\gamma^{(31ij)}\right),$$

$$S_\gamma^{(\kappa\nu ij)}=\frac{n_i}{(\gamma n)+n_j-n_i}G_\gamma^{(\kappa\nu ij)}\quad(\kappa,\nu)=(3,3),(5,5),$$

$$S_\gamma^{(\kappa\nu ij)}=\frac{n_i}{(\gamma n)+n_j+n_i}G_\gamma^{(\kappa\nu ij)}\quad(\kappa,\nu)=(4,3),(6,5).$$

The zero values of divisors correspond to critical terms (33) which are absent in $\tilde{G}$. The associated terms of S are taken as zero. The initial values of this iteration process are

$$G=Q,\qquad S=0.$$

Having found G and S we compute the linear terms in the coordinates with the aid of

$$\delta p_i^{(1)}=\sum_{j=1}^{N}[(-\tfrac{1}{2}\delta_{ij}+c_{ij})\,a_j+(\tfrac{3}{2}\delta_{ij}+d_{ij})\,\bar{a}_j],\tag{34}$$

$$w_i^{(1)}=\sum_{j=1}^{N}[(\delta_{ij}+f_{ij})\,b_j+(\delta_{ij}+\bar{f}_{ij})\,\bar{b}_j],\tag{35}$$

where c_{ij}, d_{ij} and f_{ij} represent elements of matrices,

$$c=S_{13}-\tfrac{2}{3}S_{23}-\tfrac{1}{2}S_{33}+\tfrac{3}{2}S_{43},\tag{36}$$

$$d=-\bar{S}_{13}-\tfrac{2}{3}\bar{S}_{23}+\tfrac{3}{2}\bar{S}_{33}-\tfrac{1}{2}\bar{S}_{43},\tag{37}$$

$$f=S_{55}+S_{65},\tag{38}$$

and hence are available in form (6)–(8). The components of velocities can be obtained, if necessary, from

$$\delta\dot{p}_i^{(1)}=\sqrt{-1}\;n_i\sum_{j=1}^{N}[(-\tfrac{1}{2}\delta_{ij}+c'_{ij})\,a_j+(-\tfrac{3}{2}\delta_{ij}+d'_{ij})\,\bar{a}_j],$$

$$\dot{w}_i^{(1)}=\sqrt{-1}\;n_i\sum_{j=1}^{N}[(\delta_{ij}+f'_{ij})\,b_j-(\delta_{ij}+\bar{f}'_{ij})\,\bar{b}_j],$$

where

$$c' = S_{23} - \tfrac{1}{2}S_{33} - \tfrac{3}{2}S_{43},$$
$$d' = \bar{S}_{23} - \tfrac{3}{2}\bar{S}_{33} - \tfrac{1}{2}\bar{S}_{43},$$
$$f' = S_{55} - S_{65}.$$

This completes the computation of inequalities linear in eccentricities and inclinations. Just as in the case of intermediary we have to deal here only with operations on exponential series of form (6)–(8).

5. Nonlinear Inequalities

In accordance with the general theory the nonlinear problem reduces to the determination of vectors U and Γ from the equations,

$$U = R + Q\Gamma - \mathcal{N}^{-1}\Gamma_y \mathcal{N} G^* Y - \mathcal{N}^{-1}(S + \Gamma_y)\,\mathcal{N} U^*, \tag{39}$$

$$U = U^* + \tilde{U}, \tag{40}$$

$$\Gamma_t + \sqrt{-1}(\Gamma_y \mathcal{N} P Y - \mathcal{N} P \Gamma) = \sqrt{-1}\,\mathcal{N}\tilde{U}. \tag{41}$$

Vectors U and Γ consist of six subvectors U_κ and $\Gamma_\kappa (\kappa = 1, 2, \ldots, 6)$ respectively. Herein

$$\bar{U}_1 = U_1, \qquad \bar{U}_2 = -U_2, \qquad U_4 = -\bar{U}_3, \qquad U_6 = -\bar{U}_5, \tag{42}$$

$$\bar{\Gamma}_1 = -\Gamma_1, \qquad \bar{\Gamma}_2 = \Gamma_2, \qquad \Gamma_4 = \bar{\Gamma}_3, \qquad \Gamma_6 = \bar{\Gamma}_5. \tag{43}$$

Each subvector has N components. Vectors of variables Y and right members R have the same properties as Γ and U, respectively. Their components are

$$\begin{aligned} &Y_1[i] = Y_2[i] = 0, \qquad Y_3[i] = a_i, \qquad Y_5[i] = b_i, \\ &R_1[i] = -P_i^* - \bar{P}_i^*, \qquad R_2[i] = \tfrac{3}{2}(P_i^* - \bar{P}_i^*), \\ &R_3[i] = \tfrac{1}{2}P_i^* - \tfrac{3}{2}\bar{P}_i^*, \qquad R_5[i] = -\tfrac{1}{2}W_i^*. \end{aligned} \tag{44}$$

Rewriting (39) in scalar form we obtain

$$U_\kappa[i] = R_\kappa[i] + \sum_{j=1}^{N}\sum_{\nu=1}^{6} Q_{\kappa\nu}[i,j]\,\Gamma_\nu[j] - \sum_{j=1}^{N}\sum_{\nu=3}^{6} \frac{n_j}{n_i} \times$$
$$\times \left\{\left(S_{\kappa\nu}[i,j] + \frac{\partial\Gamma_\kappa[i]}{\partial Y_\nu[j]}\right) U_\nu^*[j] + \frac{\partial\Gamma_\kappa[i]}{\partial Y_\nu[j]} \sum_{k=1}^{N} G_{\nu\nu}^*[j,k]\, Y_\nu[k]\right\}. \tag{45}$$

In virtue of (42) and (43) for determination of U and Γ only subvectors with numbers $\kappa = 1, 2, 3, 5$ are needed.

The terms of the first order in eccentricities and inclinations found in the previous section serve as initial values in determining the second-order terms, etc. In general, let expansions (22) and (23) be known up to order $m-1$ inclusively ($m=2, 3, \ldots$). Then the terms of order m can be found in the following way:

Expressions (17) and (18) give S_i and S_{ij} up to order $m-1$. Restricting summation in (19) to value $k=m$ we find T_i and T_{ij} up to terms of order m. With obtained expressions of S_i, S_{ij}, T_i and T_{ij} the terms of order m in the right members of the equations of motion result from (20) and (21). Let superscript m denote the terms of order m in the appropriate function. With $P_i^{*(m)}$ and $W_i^{*(m)}$ we immediately obtain from (44) the components $R_\kappa^{(m)}[i]$. Then the corresponding terms of U and Γ can be found by iterations. Let us assume that we have the approximate values for coefficients $U_{pqrs\gamma}^{(\kappa i)}$ and $\Gamma_{pqrs\gamma}^{(\kappa i)}$ of series

$$U_\kappa^{(m)}[i]=\sum U_{pqrs}^{(\kappa i)} \prod_{j=1}^{N} a_j^{p_j}\bar{a}_j^{q_j}b_j^{r_j}\bar{b}_j^{s_j}, \qquad U_{pqrs}^{(\kappa i)}=\sum U_{pqrs\gamma}^{(\kappa i)} \exp\sqrt{-1}(\gamma\lambda),$$

and similar ones for $\Gamma_\kappa^{(m)}[i]$. Indices γ are again related here by (8). The sum of all indices p_j, q_j, r_j and s_j is equal to m. With respect to all b_j and $\bar{b}_j$ the series for $\kappa=1, 2, 3$ are even, that for $\kappa=5$ is odd. The improved values for coefficients $U_{pqrs\gamma}^{(\kappa i)}$ result from (45) or in more detail:

$$\begin{aligned} U_\kappa^{(m)}[i]=R_\kappa^{(m)}[i]&+\sum_{j=1}^{N}\sum_{\nu=1}^{6} Q_{\kappa\nu}[i,j]\,\Gamma_\nu^{(m)}[j]-\sum_{j=1}^{N}\frac{n_j}{n_i}\times \\ &\times\left\{\sum_{\nu=3}^{6} S_{\kappa\nu}[i,j]\,U_\nu^{*(m)}[j]+\sum_{k=1}^{N}\left(G_{33}^*[j,k]\,a_k\frac{\partial}{\partial a_j}-\right.\right. \\ &-\bar{G}_{33}^*[j,k]\,\bar{a}_k\frac{\partial}{\partial \bar{a}_j}+G_{55}^*[j,k]\,b_k\frac{\partial}{\partial b_j}-\left.\bar{G}_{55}^*[j,k]\,\bar{b}_k\frac{\partial}{\partial \bar{b}_j}\right)\times \\ &\times\Gamma_\kappa^{(m)}[i]+\sum_{l=1}^{E[(m-2)/2]}\left(U_3^{*(2l+1)}[j]\frac{\partial}{\partial a_j}-\bar{U}_3^{*(2l+1)}[j]\frac{\partial}{\partial \bar{a}_j}+\right. \\ &\left.\left.+U_5^{*(2l+1)}[j]\frac{\partial}{\partial b_j}-\bar{U}_5^{*(2l+1)}[j]\frac{\partial}{\partial \bar{b}_j}\right)\Gamma_\kappa^{(m-2l)}[i]\right\}. \end{aligned} \tag{46}$$

Functions $U_\kappa^*[i]$ ($\kappa=3, 5$) represent the critical terms of $U_\kappa[i]$, that is, the terms with odd values of m ($m=3, 5, \ldots$), and the following relations for indices hold:

$$\sum_{j=1}^{N}(p_j-q_j+r_j-s_j)=1, \qquad \gamma_j=\delta_{ij}-p_j+q_j-r_j+s_j. \tag{47}$$

For even values of m ($m=2, 4, \ldots$) the critical terms satisfying relations

$$\sum_{j=1}^{N}(p_j-q_j+r_j-s_j)=0, \qquad \gamma_j=-p_j+q_j-r_j+s_j \tag{48}$$

might have generally occurred in subvectors U_1 and U_2. But $U_1^*=0$ if the critical terms in Γ_2 are chosen appropriately. Relation $U_2^*=0$ holds true by itself and serves as a reliable check of calculations. Then the improved values for coefficients $\Gamma_{pqrs\gamma}^{(\kappa i)}$

are determined by

$$\Gamma^{(2i)}_{pqrs\gamma} = \frac{n_i}{((p-q+r-s+\gamma)\,n)}\,U^{(2i)}_{pqrs\gamma},$$

$$\Gamma^{(1i)}_{pqrs\gamma} = \frac{n_i}{((p-q+r-s+\gamma)\,n)}\,(U^{(1i)}_{pqrs\gamma} + \Gamma^{(2i)}_{pqrs\gamma}); \tag{49}$$

$$\Gamma^{(\kappa i)}_{pqrs\gamma} = \frac{n_i}{((p-q+r-s+\gamma)\,n) - n_i}\,U^{(\kappa i)}_{pqrs\gamma} \quad (\kappa = 3, 5).$$

For critical values (47) and (48), which lead to zero divisors, the coefficients are chosen as follows:

$$\Gamma^{(2i)}_{pqrs\gamma} = -U^{(1i)}_{pqrs\gamma}, \qquad \Gamma^{(\kappa i)}_{pqrs\gamma} = 0 \quad (\kappa = 1, 3, 5). \tag{50}$$

The process of determining U and Γ is repeated until the desired accuracy is achieved. The initial conditions in the first application of (46) are zero values for $\Gamma^{(m)}$ and $U^{*(m)}$.

The terms of order m in δp_i and w_i and, if necessary, in $\delta\dot{p}_i$ and $\dot{w}_i$ can be found from expressions

$$\delta p_i^{(m)} = \Gamma_1^{(m)}[i] - \tfrac{2}{3}\Gamma_2^{(m)}[i] - \tfrac{1}{2}\Gamma_3^{(m)}[i] + \tfrac{3}{2}\bar{\Gamma}_3^{(m)}[i], \tag{51}$$

$$w_i^{(m)} = \Gamma_5^{(m)}[i] + \bar{\Gamma}_5^{(m)}[i], \tag{52}$$

$$\delta\dot{p}_i^{(m)} = \sqrt{-1}\; n_i(\Gamma_2^{(m)}[i] - \tfrac{1}{2}\Gamma_3^{(m)}[i] - \tfrac{3}{2}\bar{\Gamma}_3^{(m)}[i]),$$

$$\dot{w}_i^{(m)} = \sqrt{-1}\; n_i(\Gamma_5^{(m)}[i] - \bar{\Gamma}_5^{(m)}[i]).$$

Thus, expansions (22) and (23) become available up to order m and one can proceed to the next order.

Calculation of the nonlinear terms described here calls for performing literal operations on computer. At present, several systems of operations on Poisson series are known, for instance Rom (1970, 1971), Broucke and Carthwaite (1969), Jefferys (1970, 1972), Kovalevsky (1971), Cherniack (1973). Such a system may be applied without much effort to the general planetary theory. Basic series (22) and (23) are arranged in increasing order of power indices. The coefficients of these power series are exponential series of form (6)–(8). They are arranged in increasing order of exponential indices or in decreasing order of magnitude of numerical coefficients. Having approximate estimates for absolute values of a_j and b_j it is useful to evaluate the absolute value of the whole power term. This permits us to judge the accuracy needed in calculating the corresponding exponential series. It is evident that not all possible combinations of power indices occur actually in (22) and (23). As in any planetary theory the main difficulties arise here in calculating the resonant terms due to the close commensurabilities of mean motions. Existence of such terms allow us to calculate the right members with an accuracy far greater than the accuracy of the final series for the coordinates.

6. Secular System

Equations for slow variables α_j and β_j have the form

$$\begin{aligned} \dot{\alpha} &= \sqrt{-1}\, \mathcal{N}\left[A\alpha + \Phi(\alpha, \bar{\alpha}, \beta, \bar{\beta})\right], \\ \dot{\beta} &= \sqrt{-1}\, \mathcal{N}\left[B\beta + \Psi(\alpha, \bar{\alpha}, \beta, \bar{\beta})\right]. \end{aligned} \tag{53}$$

Matrices A and B and vectors Φ and Ψ are derived from G^*_{33}, G^*_{55}, U^*_3 and U^*_5, respectively. There results

$$\begin{aligned} &A[i, i] = G^*_{33}[i, i], && B[i, i] = G^*_{55}[i, i], \\ &A[i, j] = G^*_{33}[i, j]\, \zeta_{ij}^{-1}, && B[i, j] = G^*_{55}[i, j]\, \zeta_{ij}^{-1}, \quad (i \neq j), \end{aligned} \tag{54}$$

and

$$\begin{aligned} \Phi_i &= \sum\nolimits^* U^{(3i)}_{pqrs\gamma} \prod_{j=1}^{N} \alpha_j^{p_j} \bar{\alpha}_j^{q_j} \beta_j^{r_j} \bar{\beta}_j^{s_j}, \\ \Psi_i &= \sum\nolimits^* U^{(5i)}_{pqrs\gamma} \prod_{j=1}^{N} \alpha_j^{p_j} \bar{\alpha}_j^{q_j} \beta_j^{r_j} \bar{\beta}_j^{s_j}. \end{aligned} \tag{55}$$

In (55) an asterisk denotes summation over critical values (47).

Solution of the autonomous system (53) can be performed by further application of Birkhoff's normalization. Iterations with respect to planetary masses are not needed anymore. In this solution α_i and β_i are presented by series of odd power forms relative to u_j, $\bar{u}_j$, v_j and $\bar{v}_j$, where

$$u_j = \xi_j \exp\sqrt{-1}\,(c_j t + \tau_j), \qquad v_j = \eta_j \exp\sqrt{-1}\,(g_j t + \chi_j), \tag{56}$$

ξ_j, η_j, τ_j and χ_j being real constants. Frequencies c_j and g_j are series in powers of ξ_k^2 and η_k^2. Substitution of these expressions for α_i and β_i into (24), (22) and (23) would lead to the planetary theory in trigonometric form. But such a form is not suitable for the practical aims. It is more preferable to compute α_i and β_i for some interval of time and to substitute these constant numerical values into (22) and (23). Then the coordinates and velocities of the planets will be expressed by exponential series relative to the mean longitudes. For the next interval of time new numerical values of α_i and β_i must be used. In such numerical computation, instead of Birkhoff's normalization, it is possible to apply a polynomial representation of α_i and β_i in powers of time. For this purpose we may substitute into (53) the power series

$$\alpha_i = \sum_{k=0}^{\infty} \alpha_k^{(i)} t^k, \qquad \beta_i = \sum_{k=0}^{\infty} \beta_k^{(i)} t^k \tag{57}$$

with complex coefficients. Coefficients $\alpha_0^{(i)}$ and $\beta_0^{(i)}$ are arbitrary constants. The recurrence relations,

$$\begin{aligned} (k+1)\,\alpha_{k+1}^{(i)} &= \sqrt{-1}\, n_i \left(\sum_{j=1}^{N} A_{ij} \alpha_k^{(j)} + \Phi_k^{(i)} \right), \\ (k+1)\,\beta_{k+1}^{(i)} &= \sqrt{-1}\, n_i \left(\sum_{j=1}^{N} B_{ij} \beta_k^{(j)} + \Psi_k^{(i)} \right), \end{aligned} \tag{58}$$

allow us to obtain all subsequent coefficients. $\Phi_k^{(i)}$ and $\Psi_k^{(i)}$ denote here the coefficients of t^k in substituting (57) into Φ_i and Ψ_i. It is to be noted that in treating a similar problem Krasinsky and Pius (1971) have determined the polynomial coefficients by interpolating the results of numerical integration.

7. Application to the Main Problem of the Galilean Satellites of Jupiter

The method described here may be extended into the case of commensurability of mean motions and applied to the main problem of the Galilean satellites of Jupiter. In spite of several classical theories this problem can still not be regarded as exhausted. Among recent investigations one may note Marsden (1966), Ferraz-Mello (1966), Sagnier (1972). The method of the last-mentioned paper is very close to that of the general planetary theory. Therefore, we confine ourselves to some general remarks.

By a main problem is meant here the study of the motion of the Galilean satellites ($N=4$) in the gravitational field of Jupiter considered as a point-mass. Only perturbations due to the mutual attraction of the satellites are taken into account. All other perturbations such as those caused by the Sun or the asphericity of Jupiter are of less importance and can be neglected in the main problem. The mean motions of the satellites are supposed to satisfy one resonant relation,

$$(\gamma^* n)=0, \tag{59}$$

where (8) remains valid:

$$\gamma_1^*=1, \qquad \gamma_2^*=-3, \qquad \gamma_3^*=2, \qquad \gamma_4^*=0. \tag{60}$$

Approximate commensurabilities $n_1-2n_2\approx 0$, $n_2-2n_3\approx 0$ are not regarded as resonance (like the commensurabilities $2n_5-5n_6\approx 0$, $n_7-2n_8\approx 0$ in the planetary theory).

The intermediate solution $p_i^{(0)}$ is presented in the old form (6)–(8). By substituting these series into the right members $P_i^{(0)}$ we separate the resonant terms,

$$P_i^{(0)*}=\sum{}^* P_\gamma^{(i)} \exp\sqrt{-1}(\gamma\lambda),$$

where an asterisk denotes summation only over the resonant values

$$\gamma_j=K\gamma_j^* \quad (K=\pm 1, \pm 2, \ldots). \tag{61}$$

Let $\tilde{P}_i^{(0)}$ stand for the nonresonant terms of the right members. Then

$$P_i^{(0)}=P_i^{(0)*}+\tilde{P}_i^{(0)}. \tag{62}$$

Coefficients $p_\gamma^{(i)}$ can be obtained again by (9) where $P_\gamma^{(i)}$ represent now coefficients of the expansions for functions $\tilde{P}_i^{(0)}$. As usually, coefficients $p_\gamma^{(i)}$ for the resonant values (61) may be set equal to zero.

The variables δp_i and w_i are defined by Equations (15) but the right member $n_i^2 P_i^*$ of Equation (15a) is to be replaced now by $n_i^2(P_i^*+P_i^{(0)*})$. As a result of this change the basic matrix equation equivalent to system (15) takes the form

$$\dot{X}=\sqrt{-1}\,\mathcal{N}\left[(P+Q)\,X+R(X,t)+R^{(0)}(t)\right], \tag{63}$$

where $R^{(0)}(t)$ is determined by (44) with substitution $P_i^*=P_i^{(0)*}$ and $W_i^*=0$. Unlike the nonresonant case one has to include into Birkhoff's transformation a term $\Gamma^{(0)}(t)$, depending on time only. Thus we find that the transformation

$$X=(E+S)\,Y+\Gamma(Y,t)+\Gamma^{(0)}(t) \tag{64}$$

changes (63) into

$$\dot{Y}=\sqrt{-1}\,\mathcal{N}\left[HY+F(Y,t)+F^{(0)}(t)\right], \tag{65}$$

where

$$\sqrt{-1}\,\mathcal{N}H=(E+S)^{-1}\left[\sqrt{-1}\,\mathcal{N}(P+Q)\,(E+S)-\dot{S}\right], \tag{66}$$

$$\sqrt{-1}\,\mathcal{N}F^{(0)}=(E+S)^{-1}\left[\sqrt{-1}\,\mathcal{N}R^{(0)}+\sqrt{-1}\,\mathcal{N}(P+Q)\,\Gamma^{(0)}-\dot{\Gamma}^{(0)}\right], \tag{67}$$

$$\sqrt{-1}\,\mathcal{N}F=(E+S+\Gamma_y)^{-1}\left[\sqrt{-1}\,\mathcal{N}R+\sqrt{-1}\,\mathcal{N}(P+Q)\,\Gamma-\right.$$
$$\left.-\sqrt{-1}\,\Gamma_y\mathcal{N}(HY+F^{(0)})-\Gamma_t\right]. \tag{68}$$

Expressions (63)–(68) underlie the theory. Their only distinction from the corresponding expressions of the planetary theory (Brumberg and Chapront, 1974) consists in the presence of the terms $R^{(0)}$, $\Gamma^{(0)}$ and $F^{(0)}$. To bring both theories closer together impose the relation $F^{(0)}=0$. Then the presence of $R^{(0)}$ has three consequences:

(a) In constructing the linear theory the critical terms in $G_{\kappa\nu}[i,j]$ will be determined, instead of (33), by the following relations:

$$\begin{aligned}
&\gamma_k=K\gamma_k^* && \text{for}\quad (\kappa,\nu)=(1,1),(1,2),(2,1),(2,2),\\
&\gamma_k=\delta_{ki}-\delta_{kj}+K\gamma_k^* && \text{for}\quad (\kappa,\nu)=(3,3),(5,5)\\
&(k=1,2,\ldots,N;\quad K=0,\pm1,\pm2,\ldots).
\end{aligned} \tag{69}$$

Zero value for K corresponds to (33). The expressions for the matrices A and B take the form

$$\begin{aligned}
A[i,j]&=\sum_{K=-\infty}^{\infty} G_\gamma^{(33ij)}\exp\sqrt{-1}\,K(\gamma^*\lambda),\\
B[i,j]&=\sum_{K=-\infty}^{\infty} G_\gamma^{(55ij)}\exp\sqrt{-1}\,K(\gamma^*\lambda),
\end{aligned} \tag{70}$$

where the value of K defines the resonant set of indices,

$$\gamma_k=\delta_{ki}-\delta_{kj}+K\gamma_k^*.$$

(b) In the nonlinear theory the resonant parts of functions $U_\kappa[i]$ will include terms with indices satisfying the relations

$$\sum_{j=1}^{N} (p_j - q_j + r_j - s_j) = 0, \qquad \gamma_j = -p_j + q_j - r_j + s_j + K\gamma_j^*, \qquad K = 0, \pm 1, \pm 2, \ldots, \tag{71}$$

for $\kappa = 1, 2$ and

$$\sum_{j=1}^{N} (p_j - q_j + r_j - s_j) = 1, \qquad \gamma_j = \delta_{ij} - p_j + q_j - r_j + s_j + K\gamma_j^*, \qquad K = 0, \pm 1, \pm 2, \ldots, \tag{72}$$

for $\kappa = 3, 5$. These expressions generalize (48) and (47), respectively. Therefore, nonlinear terms of the secular system will be

$$\Phi_i = \sum_{p,q,r,s}^{*} \sum_{K=-\infty}^{\infty} U_{pqrs\gamma}^{(3i)} \exp\sqrt{-1}\, K(\gamma^* \lambda) \prod_{j=1}^{N} \alpha_j^{p_j} \bar{\alpha}_j^{q_j} \beta_j^{r_j} \bar{\beta}_j^{s_j}, \tag{73}$$

$$\Psi_i = \sum_{p,q,r,s}^{*} \sum_{K=-\infty}^{\infty} U_{pqrs\gamma}^{(5i)} \exp\sqrt{-1}\, K(\gamma^* \lambda) \prod_{j=1}^{N} \alpha_j^{p_j} \bar{\alpha}_j^{q_j} \beta_j^{r_j} \bar{\beta}_j^{s_j}, \tag{74}$$

where indices p, q, r, s and γ are related by (72).

(c) Additional terms caused by the resonance and described by vector $\Gamma^{(0)}$ satisfy the equation

$$\dot{\Gamma}^{(0)} - \sqrt{-1}\, \mathcal{N} P \Gamma^{(0)} = \sqrt{-1}\, \mathcal{N} U^{(0)}, \tag{75}$$

where $U^{(0)}(t)$ is determined by an iterative expression:

$$U^{(0)} = R^{(0)} + Q\Gamma^{(0)}. \tag{76}$$

There results

$$\begin{aligned} \dot{\Gamma}_1^{(0)} - \sqrt{-1}\, \mathcal{N} \Gamma_2^{(0)} &= \sqrt{-1}\, \mathcal{N} U_1^{(0)}, \\ \dot{\Gamma}_2^{(0)} &= \sqrt{-1}\, \mathcal{N} U_2^{(0)}, \\ \dot{\Gamma}_3^{(0)} - \sqrt{-1}\, \mathcal{N} \Gamma_3^{(0)} &= \sqrt{-1}\, \mathcal{N} U_3^{(0)}. \end{aligned} \tag{77}$$

Additional corrections $\delta p_i^{(0)}$ caused by $\Gamma^{(0)}(t)$ can be obtained by (51) with $m = 0$.

System (77) provides a complement to the secular system. If we neglect the libration of the satellites and assume the strict fulfilment of (55) and

$$(\gamma^* \lambda) = 180°, \tag{78}$$

then the right members $P_i^{(0)*}$ will be real constants,

$$P_i^{(0)*} = \sum_{K=-\infty}^{\infty}{}^{(0)} (-1)^K P_\gamma^{(i)}, \qquad \gamma_j = K\gamma_j^*, \tag{79}$$

and up to the terms of order μ^2 inclusively we obtain

$$\Gamma_1^{(0)}[i]=0, \qquad \Gamma_2^{(0)}[i]=2P_i^{(0)*}, \qquad \Gamma_3^{(0)}[i]=P_i^{(0)*}. \tag{80}$$

Subsequent approximations for $\Gamma_\kappa^{(0)}[i]$ are determined by (49) and (50) replacing Γ and U by $\Gamma^{(0)}$ and $U^{(0)}$ and putting $p=q=r=s=0$. Critical terms cannot arise in $U_3^{(0)}$ and in $U_1^{(0)}$, they vanish in virtue of the choice of the appropriate terms in $\Gamma_2^{(0)}$ and their absence in $U_2^{(0)}$ is again the check relation of the method. The secular system will be autonomous in this case and may be solved by the methods of the previous section.

Neglect of the libration which so far has not been discovered from observations signifies the choice of the particular solution with two missing arbitrary constants. Approximate consideration of the libration may be performed in the following way.

From (44) we have

$$R_1^{(0)}[i]=\sum_{k=1}^{\infty} a_k^{(i)} \cos k(\gamma^*\lambda), \qquad R_2^{(0)}[i]=\sqrt{-1}\sum_{k=1}^{\infty} b_k^{(i)} \sin k(\gamma^*\lambda), \tag{81}$$

where

$$a_k^{(i)}=-2(P_\gamma^{(i)}+P_{-\gamma}^{(i)}), \qquad b_k^{(i)}=3(P_\gamma^{(i)}-P_{-\gamma}^{(i)}), \qquad \gamma_j=k\gamma_j^*. \tag{82}$$

Supposing that (59) and (78) are close to strict fulfilment and differentiating the first Equation (77) there results:

$$\ddot{\Gamma}_1^{(0)}+\mathcal{N}^2 U_2^{(0)}\approx 0, \tag{83}$$

or up to the terms of order μ^2 inclusively:

$$\ddot{\Gamma}_1^{(0)}[i]+\sqrt{-1}\, n_i^2 \sum_{k=1}^{\infty} b_k^{(i)} \sin k(\gamma^*\lambda)=0. \tag{84}$$

In the resonant case the mean longitides must have the resonant corrections $\delta\lambda_i$. Therefore,

$$\lambda_i=\lambda_i^{(0)}+\delta\lambda_i. \tag{85}$$

These corrections can be related with $\Gamma_1^{(0)}$ by imposing

$$\Gamma_1^{(0)}[i]=-\sqrt{-1}\,\delta\lambda_i. \tag{86}$$

Substituting (85) and (86) into (84) and denoting $\theta=(\gamma^*\delta\lambda)$ we obtain

$$\delta\ddot{\lambda}_i+n_i^2 \sum_{k=1}^{\infty} (-1)^{k+1}\, b_k^{(i)} \sin k\theta=0, \tag{87}$$

from which after multiplication by γ_i^* and summation follows the equation of libration

$$\ddot{\theta}+\sum_{k=1}^{\infty} (-1)^{k+1}\left(\sum_{i=1}^{N} \gamma_i^* n_i^2 b_k^{(i)}\right) \sin k\theta=0. \tag{88}$$

8. Conclusion

Experience with constructing the first-order theory revealed the practical efficiency of the method. Yet it became evident that the immediate continuation of the work by calculating the terms of the second and higher orders would involve the serious technical difficulties. Iterative version proposed here presents a trial to avoid these difficulties. Realization of the method on modern computers seems to merit attention.

References

Broucke, R.: 1969, *Celes. Mech.* **1**, 110.
Broucke, R. and Garthwaite, K.: 1969, *Celes. Mech.* **1**, 271.
Brumberg, V. A.: 1970, in G. E. O. Giacaglia (ed.), *Periodic Orbits, Stability and Resonances*, D. Reidel, Dordrecht, The Netherlands, p. 410.
Brumberg, V. A. and Chapront, J.: 1974, *Celes. Mech.* **8**, 335.
Brumberg, V. A. and Egorova, A. V.: 1971, *Observations of Artificial Celestial Bodies* **62**, 42 (in Russian); erratum: *Bull. Inst. Theor. Astron.* **13** (1972) 336 (in Russian).
Cherniack, J. R.: 1973, *Celes. Mech.* **7**, 107.
Ferraz-Mello, S.: 1966, *Bull. Astron.* (3) **1**, 287.
Jefferys, W. H.: 1970, *Celes. Mech.* **2**, 474.
Jefferys, W. H.: 1972, *Celes. Mech.* **6**, 117.
Kovalevsky, J.: 1971, *Calculs de mécanique céleste sur ordinateur*, São José dos Campos, Brazil.
Krasinsky, G. A. and Pius, L. Yu.: 1971, *Observations of Artificial Celestial Bodies* **62**, 93 (in Russian).
Marsden, B. G.: 1966, The Motions of the Galilean Satellites of Jupiter, Dissertation, Yale University.
Rom, A.: 1970, *Celes. Mech.* **1**, 301.
Rom, A.: 1971, *Celes. Mech.* **3**, 331.
Sagnier, J. L.: 1972, *Séminaires du Bureau des Longitudes*, 11ème Année, No. 9.
Seidelmann, P. K.: 1970, *Celes. Mech.* **2**, 134.

ON THE CALCULATION OF SECULAR PERTURBATIONS IN THE CASE OF CLOSE COMMENSURABILITY

N. I. LOBKOVA and M. S. PETROVSKAYA
Institute of Theoretical Astronomy, Leningrad, U.S.S.R.

Abstract. In a previous paper the authors derived expansions of the derivatives of the disturbing function for the general case including the orbits close to intersection. The present paper deals especially with the case of close commensurability of the mean motions. A new variable v is introduced characterizing the deviation of the mean anomalies from the exact commensurability, and is considered further as an unknown quantity. In the equations of motion the short-period terms are eliminated. The form of expansions of the right-hand sides is chosen basing on the same principles as in the general case. The factors are separated, corresponding to the poles in the case of circular intersecting orbits. For rapidity of calculation the summation in powers of the major semi-axes ratio is made the inner one.

1. Expansion with Respect to the Negative Powers of the Mutual Distances

For studying the case of close commensurability in the planetary motion von Zeipel's method and its various modifications are generally used. The canonical transformations performed in the methods entail the Hamiltonian expansions in trigonometric series, the elimination of particular arguments and finding the mean and extreme values of the Hamiltonian. All these operations are usually carried out numerically (Giacaglia, 1968; Hori and Giacaglia, 1968; Giacaglia and Nacozy, 1969), since the analytical treatment is complicated by the problem of convergence of the series. The well-known expansions of the disturbing function are valid, provided the ratio of the major axes and other parameters are sufficiently small (Kozai, 1962, 1968). In Petrovskaya (1972) an analytical expansion of the perturbation function is given for the general case, including the intersecting orbits. When resonance problems are considered the possibility offers to improve the convergence of the series. One can take advantage of the fact that the mean anomalies, M_1 and M_2, of the planets are fairly commensurable and so, the distance Δ between the bodies is approximately a periodic function of the time. While the convergence of the two-argument expansion representing $\Delta^{-\gamma}(M_1, M_2)$ depends on the minimum distance between the planetary orbits, the validity of the single-argument expansion of $\Delta^{-\gamma}$ in the case of exact commensurability is dependent on the minimum actual distance between the planets.

Considering the case of close commensurability we put

$$M_j = M_j^0 + \int_{t_0}^{t} N_j \, dt, \quad j = 1, 2,$$

N_j being the mean motions of the planets, and

$$N_2 - (p/q)\, N_1 = v, \tag{1}$$

where p and q are integers, and v is a small quantity.

Y. Kozai (ed.), The Stability of the Solar System and of Small Stellar Systems, 157–163.

Introduce, instead of M_1 and M_2, new variables z and y as follows:

$$z=\frac{1}{q}\int_{t_0}^{t} N_1\, dt, \qquad y=\int_{t_0}^{t} \nu\, dt.$$

Hence M_1 and M_2 are expressed as

$$M_1=qz+M_1^0, \qquad M_2=pz+M_2^0+y. \tag{2}$$

If N_1 and N_2 are strictly commensurable, that is $\nu=0$, we have $y=0$. Thus, y is a slowly changing variable.

In the Lagrange's equations the disturbing function after substituting (2) becomes a function of y and z, the equations for those being

$$dy/dt=\nu, \tag{3a}$$

$$dz/dt=N_1/q. \tag{3b}$$

In the restricted three-body problem one of the two N_j is a constant. Let it be N_2. Then, after finding y from Equation (3a), M_1 will be

$$M_1=(q/p)\,(M_2-M_2^0-y)+M_1^0.$$

We put a: the major semi-axis, e: the eccentricity, i: the inclination, Ω: the longitude of the node, ω: the argument of the perihelion, M^0: the mean anomaly at the epoch.

The mutual distance of the planets $\Delta(a, e, i, \Omega, \omega, M^0, z, y)$, where $a=a_1, a_2,\ldots$, in the case of rigorous commensurability for the undisturbed motion, is

$$\Delta_0=\Delta(a^0, e^0, i^0, \Omega^0, \omega^0, M(t_0), z, 0),$$

being the function of a single variable z and of 12 constants.

Consider a Taylor's expansion,

$$\Delta^{-1}=\Delta_0^{-1}\sum_{\kappa=0}^{\infty}\frac{(\frac{1}{2})_\kappa}{(1)_\kappa}\left(1-\frac{\Delta^2}{\Delta_0^2}\right)^{\kappa}. \tag{4}$$

It is a series in powers of the deviation of Δ^2 from Δ_0^2, which is of order of $\sigma=\max\{\mu, |\nu|\}$, where μ is the disturbing mass, and ν is given by (1). If, for example, $|\nu|\leqslant\mu^{1/2}$, then the terms of series (4) are of the order $\mu^{\kappa/2}$. When $|\nu|>\mu^{1/2}$ there is no need to treat the problem as one of resonance, since in this case the perturbations may be derived by the usual procedure of successive approximations and the solution of the differential equations would be developed in powers of $\mu^{1/2}$.

Thus, we assume that $|\nu|\leqslant\mu^{1/2}$. Then, if the perturbations up to the second order with respect to μ have to be calculated, one should substitute into Lagrange's equations instead of Δ^{-1} the sum of terms of expansion (4) for $\kappa=1, 2, 3, 4$. The zero term disappears after differentiation with respect to the elements. Hence, if one starts from (4) the problem of developing the principal part of the disturbing function is reduced to the expansion of several positive powers of the two-argument function $\Delta^2(z, y)$, and several negative powers of the single-argument function $\Delta_0^2(z)$, all the Keplerian elements in the latter being fixed.

One has to take notice of the following characteristics of expansion (4). It involves no negative powers of the two-argument distrubed function $\Delta(z, y)$ while the positive powers of $\Delta^2(z, y)$ are rapidly convergent series which can be easily differentiated with respect to Keplerian elements. The functions $\Delta_0^{-2\kappa-1}(z)$ are single-argument series which are not differentiated into Langrange's equations, their coefficients being evaluated once with any precision. It is convenient to use expansion (4) when the equations are solved with the short-period terms being eliminated since all the operations performed by numerical integration or by finding out the stationary solutions etc. are applied to the positive powers of Δ^2, which depends analytically on the elements.

We confine ourselves to the evaluation of the perturbations up to the second order. As in (4) the same powers of Δ^2/Δ_0^2 appear several times, the following form of Δ^{-1} is preferable:

$$\Delta^{-1}=\sum_{s=0}^{4} \lambda_s \frac{\Delta^{2s}}{\Delta_0^{2s+1}},$$
$$\lambda_s=\frac{(-1)^s}{(1)_s}\left(\tfrac{1}{2}\right)_s \sum_{q=0}^{4-s} \frac{\left(\frac{1}{2}+s\right)_q}{(1)_q}. \tag{5}$$

2. Expansion of Δ^{2s}

Consider

$$\Delta^{2s}=(\tau_1^2+\tau_2^2-2\tau_1\tau_2 \cos H)^s,$$

where τ_1, τ_2 and H are the distances of Sun-asteroid, Sun-Jupiter and the elongation of the asteroid from Jupiter.

As it is known (Sack, 1964), the following relations hold:

$$\Delta^{2s}=\sum_{\kappa=0}^{s} R_{2s,\kappa}(\tau_1, \tau_2)\, P_\kappa(\cos H),$$

$$R_{2s,\kappa}(\tau_1, \tau_2)=\tau_2^{2s} \frac{(-s)_\kappa}{\left(\frac{1}{2}\right)_\kappa} \left(\frac{\tau_1}{\tau_2}\right)^\kappa F\left(\kappa-s,\, -\tfrac{1}{2}-s,\, \kappa+\tfrac{3}{2};\, \frac{\tau_1^2}{\tau_2^2}\right),$$

or

$$\Delta^{2s}=\sum_{\kappa=0}^{s} \frac{(-s)_\kappa}{\left(\frac{1}{2}\right)_\kappa} P_\kappa(\cos H) \sum_{n=0}^{s-\kappa} \frac{(\kappa-s)_n\left(-\frac{1}{2}-s\right)_n}{\left(\kappa+\frac{3}{2}\right)_n (1)_n}\, \tau_1^{2n+\kappa}\tau_2^{2s-\kappa-2n}. \tag{6}$$

The Legendre's polynominals $P_\kappa(\cos H)$ can be presented by the formula (Brumberg, 1971),

$$P_\kappa(\cos H)=\sum_{l=0}^{\kappa} \sum_{l_1=0}^{\kappa} \sum_{j=0}^{\kappa} (2-\delta_{j,0}) \frac{(1)_{\kappa-j}}{(1)_{\kappa+j}} F_{\kappa jl}(i_1)\times$$
$$\times F_{\kappa jl_1}(i_2) \cos\left[(\kappa-2l)\, u_1-(\kappa-2l_1)\, u_2+j(\Omega_1-\Omega_2)\right], \tag{7}$$

where u_1 and u_2 are the arguments of latitude of the planets and

$$F_{\kappa jl}(i)=\lambda_{\kappa jl}(\sin(\tfrac{1}{2}i))^{|\kappa-j-2l|}(\cos(\tfrac{1}{2}i))^{|\kappa+j-2l|}\times$$
$$\times F(-m, 2\kappa-m+1, 1+|\kappa-j-2l|; \sin^2(\tfrac{1}{2}i).$$
$$m=\kappa-\tfrac{1}{2}(|\kappa-j-2l|+|\kappa+j-2l|),$$

λ are some numerical coefficients, $\delta_{0,0}=1$, $\delta_{j,0}=0$ if $j\neq 0$, F is Gauss' hypergeometric function.

After substituting (7) into (6) Δ^{2s} can be readily presented as a series in the multiples of M_1 and M_2:

$$\Delta^{2s}=\sum_{q_1}\sum_{q_2}\sum_{q_3, q_4, q_5} A_{\{q\}}\cos[q_1M_1+q_2M_2+q_3\omega_1+q_4\omega_2+q_5(\Omega_1-\Omega_2)]. \quad (9)$$

The absolute values of the integers q_3, q_4 and q_5 do not exceed 4. $A_{\{q\}}$ are the functions of the remaining elements a, e, i:

$$A_{\{q\}}=C(a_1, a_2)\, X(e_1)\, X(e_2)\, F(i_1)\, F(i_2),$$

C and F being the polynominals of the order $2s$ and X being Hansen's coefficients. The limits in the sums with respect to q_1 and q_2 depend on the rapidity of diminishing Hansen's coefficients, in other words, on the values of eccentricities.

Expression (9) may be presented as a series with respect to all the variables involved, so that it would be a polynomial in a_1, a_2, $\sin(\tfrac{1}{2}i_1)$ and $\sin(\tfrac{1}{2}i_2)$, a power series in e_1 and e_2, a trigonometric one with respect to M_1 and M_2 and a trigonometric polynomial in ω_1, ω_2, and $\Omega_1-\Omega_2$. This series will be obtained readily as soon as any of well-known expansions of Hansen's coefficients in powers of eccentricities is used. In such a form function (9) and, therefore, (5) can be easily differentiated with respect to the Keplerian elements.

Substituting (2) into (9) it follows:

$$\Delta^{2s}=\sum_{n=-n_2}^{n_2} B_n^{(2s)}\, e^{inz}, \qquad n_1>0, \quad n_2>0, \quad (10)$$

$B_n^{(2s)}$ being the function of Keplerian elements and y. The number of terms taken into consideration, as was mentioned above, depends on the values of the eccentricities.

3. Expansion of $\Delta_0^{-\gamma}$, $\gamma=1, 3, 5,\ldots$

Now we would like to express $\Delta_0^{-\gamma}$ as a Fourier series in z with numerical coefficients. The first step is to develop $\Delta_0^{-\gamma}$ in multiples of M_1 and M_2 by any device of expansion. Taking into account that in Δ_0 M_j are given by (2) under condition $y=0$ one is to construct the expansion

$$\Delta_0^{-\gamma}=\sum_{n=-\infty}^{\infty} C_n^{(\gamma)}\, e^{inz}, \quad \gamma=1, 3,\ldots. \quad (11)$$

In order to calculate the coefficients $C_n^{(\gamma)}$ an intermediate expansion of $\Delta^{-\gamma}$ may be applied which was derived by Petrovskaya (1972). From this expansion $C_n^{(\gamma)}$ are defined as

$$C_n^{(\gamma)} = \sum_{k=0}^{\infty} \sum_{m=0}^{\infty} \frac{(D)_\kappa}{(1)_\kappa} \Phi_\kappa^{(\gamma)}(\beta)\, I_{n\kappa m}^{(\gamma)}, \tag{12}$$

$$I_{n\kappa m}^{(\gamma)} = \frac{1}{2\pi} \int_0^{2\pi} f_{\kappa,m}^{(\gamma)}(z)\, e^{-inz}\, dz, \tag{13}$$

$$f_{\kappa,m}^{(\gamma)}(z) = (\tau_1^0 + \tau_2^0)^{-\gamma} x^\kappa \cos mH^0, \tag{14}$$

$$D = \frac{\partial}{\partial \ln \beta}, \qquad \beta = \frac{a_1^0 a_2^0 (1-e_1^0)(1-e_2^0)}{[a_1^0(1+e_1^0) + a_2^0(1+e_2^0)]^2},$$

$$\Phi_m^{(\gamma)}(\beta) = (2 - \delta_{m,0}) \frac{(\gamma/2)_m}{(1)_m} \beta^m F(m + \tfrac{1}{2}\gamma, m + \tfrac{1}{2}\gamma, 2m+1; 4\beta),$$

$$x = 1 - \beta(\tau_1^0 + \tau_2^0)^2 / \tau_1^0 \tau_2^0.$$

As to the integrals $I_{n\kappa m}^{(\gamma)}$ they may be evaluated either by mechanical quadratures methods or by developing them in an analytical way with respect to Keplerian elements. In order to obtain an analytical expansion the functions $f_{\kappa,m}^{(\gamma)}(z)$ have to be presented as a series in multiples of M_1 and M_2 and then of z, taking into consideration that

$$M_1 = qz + M_1(t_0), \qquad M_2 = pz + M_2(t_0).$$

Finally, one can develop with respect to z each function of $(\tau^0 + \tau_2^0)^{-\gamma}$, x and $\cos H^0$ in (14) and the series for (14) follows from the multiplications.

To calculate the functions $(D)_\kappa\, \Phi_m^{(\gamma)}(\beta)$ in (12) recurrent formulas were obtained earlier. These functions can be also evaluated without application of recurrent relations which sometimes may result in less accurate calculations. The following formula is free from that deficiency:

$$\frac{(D)_\kappa}{(1)_\kappa} \Phi_m^{(\gamma)}(\beta) = (1 - 4\beta)^{1/2 - \gamma/2 - \kappa} (m)_\kappa \times \sum_{i=0}^{\kappa} A_{mi\kappa}\, \Phi_{mi}^{(\gamma)}, \tag{15}$$

$$\Phi_{mi}^{(\gamma)} = (2 - \delta_{m,0}) \frac{(\gamma/2)_m}{(1)_m} \beta^m F(m + 1 - \tfrac{1}{2}\gamma, m + \tfrac{1}{2}, 2m + 1 + i; 4\beta),$$

$$A_{mi\kappa} = \frac{(4\beta)^i (m + \frac{1}{2}\gamma)_i (m + \frac{1}{2})_i (1 - 4\beta)^{\kappa - i}}{(1)_i (m)_i (2m+1)_i (1)_{\kappa - i}}.$$

Formulas (11)–(15) provide the expansion of $\Delta_0^{-\gamma}$.

The coefficients of the series (11) might be calculated by evaluating the integrals

$$C_n^{(\gamma)} = \frac{1}{2\pi} \int_0^{2\pi} \Delta_0^{-\gamma} e^{-inz}\, dz,$$

with the aid of the quadrature formulas. Though this way may lead to difficulties when the minimum distance between the orbits is small ($\Delta \to 0$). In our procedure we suggest to use the harmonic analysis method only for calculating the functions (13), after the singularity $(1-4\beta)^{1/2-\gamma/2-\kappa}$ at $a_1^0 \to a_2^0$, $e_1^0 \to 0$, $e_2^0 \to 0$ is removed from (11) by the formula (15). The absolute value of the integrand $f_{\kappa,m}^{(\gamma)}$ in (13) is less than 1.

4. Averaged Equations of Motion

For simplicity we consider the restricted three-body problem. The formulas (5), (10) and (11) provide the expansion of Δ^{-1} with respect to z. The constant term in Δ^{-1} is

$$[\Delta^{-1}] = \sum_{s=0}^{4} \lambda_s \sum_{n=-n_1}^{n_2} B_n^{(2s)} C_{-n}^{(2s+1)}, \tag{16}$$

where $C_{-n}^{(2s+1)}$ are numerical coefficients, while $B_n^{(2s)}$ are functions of the disturbed Keplerian elements. The limits n_1 and n_2 are defined by the number of terms retained in Δ^{2s}.

After substituting (16) into Lagrange's equations the secular and long-period perturbations will be determined by the set

$$\mathrm{d}\varepsilon/\mathrm{d}t = \mu f(\varepsilon, y), \qquad \mathrm{d}y/\mathrm{d}t = N_2 - (p/q)\, N_1 = \nu, \tag{17}$$

ε being any Keplerian element, $N_2 = \text{const}$, N_1 being a function of a_1, $f(\varepsilon, y)$ being derivatives of the quantity (16).

After solving (17), M_1 is provided by the relations

$$M_1 = (q/p)\,(M_2 - M_2^0 - y) + M_1^0, \qquad M_2 = N_2(t - t_0).$$

We denote

$$\nu = \nu_0(1 + \nu_1),$$

where ν_0 is the value of ν for $\mu = 0$. It is easily seen that, if ν_0 is less than $\mu^{1/2}$, by putting $\tau = \nu_0(t - t_0)$ and $y = \tau + \zeta$ one obtains equations

$$\mathrm{d}\varepsilon/\mathrm{d}\tau = \lambda f(\varepsilon, \tau + \zeta), \qquad \mathrm{d}\zeta/\mathrm{d}\tau = \nu_1, \tag{18}$$

where $\lambda = O(\mu^{1/2})$.

After averaging with respect to τ the equations provide the purely secular perturbations with respect to time, which are power series in λ. The periodic terms in the solution can be found by the usual procedure of von Zeipel.

In the case which is of particular interest for us the order of ν is more than $\mu^{1/2}$ and we return to set (17). If it is solved by numerical integration then at any step of approximation the coefficients $C_{-n}^{(2s+1)}$ in (16) corresponding to the expansion (11) remain unchanged. They may be evaluated once with any precision. The coefficients $B_n^{(2s)}$ which appear in (16) from (10) are literal series with respect to all the Keplerian elements, including the major semi-axes.

Equations (17) may be applied for finding out periodic solutions and those which are close to them.

This consideration remains valid in the case of intersecting resonant oribts of asteroids and in the Neptune–Pluto case.

References

Brumberg, V. A., Evdokimova, L. S., and Kochina, N. G.: 1971, *Celes. Mech.* **3**, 197.

Giacaglia, G. E. O.: 1968, *Smithsonian Astrophys. Obs.*, Special Report No. 278.

Giacaglia, G. E. O. and Nacozy, P. E.: 1969, in G. E. O. Giacaglia (ed.), *Periodic Orbits, Stability and Resonances*, D. Reidel, Dordrecht, The Netherlands, p. 96.

Hori, G. and Giacaglia, G. E. O.: 1968, *Research in Celestial Mechanics and Differential Equations* (University of Paulo) **1**, 4.

Kozai, Y.: 1962, *Astron. J.* **67**, 591.

Kozai, Y.: 1968, in G. E. O. Giacaglia (ed.), *Periodic Orbits, Stability and Resonances*, D. Reidel, Dordrecht, The Netherlands, p. 451.

Petrovskaya, M. S.: 1972, *Celes. Mech.* **6**, 328.

Sack, R. A.: 1964, *J. Math. Phys.* **5**, 245.

LONG PERIOD TERMS IN THE SOLAR SYSTEM

P. BRETAGNON
Bureau des Longitudes, Paris, France

L'ensemble de ce travail fera l'objet d'une publication dans *Astronomy and Astrophysics.*

Nous avons étudié les variations à longues périodes des éléments des huit planètes du système solaire. Pour cela, nous avons calculé la solution de Lagrange puis introduit les termes d'ordre 4 à longues périodes de la fonction perturbatrice ainsi que la contribution, au premier ordre des masses, des termes à courtes périodes. Nous nous sommes, de plus, tout particulièrement attachés au problème de la détermination des constantes d'intégration. En effet, nous avons obtenu une grande modification des constantes d'intégration en particulier pour les constantes relatives aux planètes Mercure, Vénus, Terre et Mars. Ceci provient de l'importance, pour ces planètes, des termes d'ordre 3. La contribution des termes à courtes périodes apporte surtout une modification des fréquences relatives aux planètes Jupiter, Saturne, Uranus et Neptune.

Ce travail montre donc l'importance relative des différentes contributions: il est, par exemple, inutile d'introduire les termes d'ordre 5 à longues périodes si on n'a pas tenu compte des termes à courtes périodes.

Y. Kozai (ed.), The Stability of the Solar System and of Small Stellar Systems, 165.

ON THE THEORY OF THE GALILEAN SATELLITES OF JUPITER

S. FERRAZ-MELLO

Aeronautics Institute of Technology, Astronomical Observatory, 12200 São José dos Campos, Brazil

Abstract. In this communication the main equations for the variables: radius vector, longitude, P and Q (variables built from Laplace's perihelium first integral) are given in closed form. These equations are used for deriving the equations of a second-order theory. At this order, the equations for P and Q, are separated and they are integrodifferential linear equations. The equations for the radius vector and for the longitudes, give, after integration, perturbations which are purely trigonometric. The solution shows the features observed in the motion of Jupiter's Galilean satellites. The results are discussed, and extended to include the space variables.

1. Introduction

As it has been pointed out by Kovalevsky (1962) the tables of the four great satellites of Jupiter (Sampson, 1910) do not allow, nowadays, a precise prediction of the phenomena involving these bodies. The errors of such predictions are of the order of one minute, and, in many occasions they have gone up to several minutes.

Since the nominal time unity of Sampson's tables is the UT, it may be expected that a shift of the time scale be among the main sources of errors. Indeed, the analysis of all photographic observations made in the past 40 years, carried out by Rodrigues (1970) and by Ferraz-Mello and Paula (1973), has shown that this shift is increasing two times faster than the difference ET – UT. The best modern observations, made by D. Pascu at McCormick, allow the following estimate for the shift of the Sampson's tables time scale (ST), for the mean epoch 1968.2:

$$\mathrm{ST} - \mathrm{UT} = 1.0 \pm 0.9 \text{ min.}$$

The standard deviation is much larger than should be expected from the observations themselves. Indeed, after correcting the time scale, the standard deviation for the $(O - C)$ of the mutual distances is 0″.2, while its expected value for the focal length of the telescope employed is 0″.1. So, these deviations are almost entirely due to tables' errors.

If the evolution of precise measurements in the Solar System from radar astronomy, and the increasing need of better ephemeris for astrodynamics are considered, it is clear that a better theory will be necessary in the near future. So, efforts have been made, mainly at the Bureau des Longitudes (Paris), to derive a new theory for the Galilean satellites of Jupiter.

This theory must take into account the main features of the Galilean system:

(a) The ratio of the semimajor axis (0.2 to 0.6) and the masses of the satellites with respect to the primary (10^{-4}) are characteristics of a planetary problem with strong interactions.

Y. Kozai (ed.), The Stability of the Solar System and of Small Stellar Systems, 167–184.

(b) The periods of the satellites are very short and prevent the use of the classical methods of planetary theory, as well as purely numerical theories.

(c) The standard deviations of the best observations already made (0″.03), when translated in terms of the Jovicentric longitudes of the satellites, give rise to very high values (50″, 34″, 21″ and 13″, respectively).

These features allow us to characterize the problem of the motion of Jupiter's Galilean satellites as a problem of research of absolute orbits, i.e., orbits of low precision, but, which remain valid for very long time intervals. The first attempts to construct absolute planetary theories are due to H. Gyldén and, after him, to G. W. Hill (see Brouwer, 1959). The most important results obtained hereto are those by Brumberg (1970) and by Sagnier (1973a, b), the latter in intimate relation with Jupiter's satellites; in connection with these works the researches by Krasinsky (1968, 1969) providing a very strong mathematical tool for the integration of a certain kind of systems of linear differential equations with periodic coefficients and providing an existence theorem for quasi-periodic solutions of the first kind (*première sore*) in the planar N-body problem, should also be mentioned.

This paper deals with the problem of the construction of second-order absolute orbits for the Galilean system of satellites. The method is the same already used for deriving the equations of a first-order theory (Ferraz-Mello, 1966; Hagihara, 1972). The main ideas for deriving the second-order theory are those shortly described in Ferraz-Mello (1969a, b).

2. The Equations

Let a Jovicentric system of moving axes be considered. Following a suggestion by De Sitter (1918) the angular velocity of this frame is taken so that the mean motions of the three inner satellites are exactly commensurable. Such a choice is possible since the absolute mean motions of these satellites are such that

$$n_1 - 3n_2 + 2n_3 = 0. \tag{1}$$

If ν_1, ν_2 and ν_3 are the Eulerian mean motions of these satellites, it follows that

$$n_i = \nu_i + N, \tag{2}$$

where N is the angular velocity of rotation of the equatorial axes, and so

$$n_1 - 2n_2 = \nu_1 - 2\nu_2 - N = 0^\circ 739\,507\,42 \text{ days}^{-1},$$
$$n_2 - 2n_3 = \nu_2 - 2\nu_3 - N = 0^\circ 739\,507\,42 \text{ days}^{-1},$$

i.e.,

$$N = -0^\circ 739\,507\,42 \text{ days}^{-1}.$$

For the plane variables, let Hill's normalized variables

$$u_j = (x_j + iy_j)/a_j, \qquad s_j = (x_j - iy_j)/a_j \tag{3}$$

be introduced. The normalization factors a_j are the mean distances from the satellites to the planet. The heights over the fundamental plane are also normalized by this factor:

$$Z_j = z_j / a_j.$$

Let also a new independent variable,

$$\zeta = \exp i v_3 t, \tag{4}$$

and the operator

$$D = \zeta \, \mathrm{d}/\mathrm{d}\zeta, \tag{5}$$

be introduced.

In the computations it is wise to take into account that the motions will not depart too much from coplanar circular uniform motions, whose angular velocities are the observed mean motions and whose radii are the mean distances a_j. The zeroth-order solution for each satellite is given by

$$u_j^0 = \sigma_j \zeta^{g_j}, \qquad s_j^0 = \sigma_j^* \zeta^{-g_j}, \quad (\sigma_j \sigma_j^* = 1), \tag{6}$$

where

$$g_j = v_j / v_3, \tag{7}$$

and $\sigma_j = \exp i\theta_{0j}$ gives the position of the jth satellite at the time origin. Let then the variables U_j and S_j be introduced through

$$\begin{aligned} u_j &= \sigma_j \zeta^{g_j} (1 + U_j), \\ s_j &= \sigma_j^* \zeta^{-g_j} (1 + S_j). \end{aligned} \tag{8}$$

The equations of motion are, then,

$$\begin{aligned} (D + \kappa_j)^2 \, U_j + \kappa_j^2 &= \lambda_j (a_j / r_j)^3 \, (1 + U_j) + \mathcal{R}_j, \\ (D - \kappa_j)^2 \, S_j + \kappa_j^2 &= \lambda_j (a_j / r_j)^3 \, (1 + S_j) \mathcal{T}_j, \\ D^2 Z_j &= \lambda_j (a_j / r_j)^3 \, Z_j + \mathcal{V}_j, \end{aligned} \tag{9}$$

where

$$\lambda_j = G m_0 (1 + m_j) / v_3^2 a_j^3, \tag{10}$$

m_j are the masses of the satellites with respect to the mass of Jupiter (m_0), G is the constant of gravitation, r_j are the vector radii of the satellites,

$$\kappa_j = g_j + m, \tag{11}$$

and

$$m = N / v_3 = -0.01448391. \tag{12}$$

$\mathcal{R}_j$, $\mathcal{T}_j$ and $\mathcal{V}_j$ are the disturbing forces for these variables. If only the mutual interactions are considered, then

$$\begin{aligned}
\mathscr{R}_j &= \sum_{i \neq j} \frac{Gm_i \sigma_j^* \zeta^{-g_j}}{v_3^2} \left\{ \frac{u_j - \alpha_{ij} u_i}{r_{ji}^3} + \frac{\alpha_{ij} u_i}{r_i^3} \right\}, \\
\mathscr{T}_j &= \sum_{i \neq j} \frac{Gm_i \sigma_j \zeta^{g_j}}{v_3^2} \left\{ \frac{s_j - \alpha_{ij} s_i}{r_{ji}^3} + \frac{\alpha_{ij} s_i}{r_i^3} \right\}, \\
\mathscr{V}_j &= \sum_{i \neq j} \frac{Gm_i}{v_3^2} \left\{ \frac{Z_j - \alpha_{ij} Z_i}{r_{ji}^3} + \frac{\alpha_{ij} Z_i}{r_i^3} \right\},
\end{aligned} \tag{13}$$

where r_{ji} are the mutual distances, and

$$\alpha_{ij} = a_i / a_j. \tag{14}$$

It must be observed that the planar equations are conjugated throughout the transformation,

$$U_j \to S_j, \qquad S_j \to U_j, \qquad t \to -t \quad (\zeta \to \zeta^{-1}, D \to -D).$$

This property is very useful for checking the calculations throughout this work, and will be used in this paper in order to avoid duplicated derivations.

3. The Area Integral and Its Application

The integral of the areas in the two-body problem may be used to introduce a new pair of variables for each satellite, and these variables are intimately related to the proper oscillations perpendicular to the orbit. Let this integral be considered in the form

$$\mathbf{c} = \mathbf{r} \times \mathbf{v}, \tag{15}$$

where $\mathbf{r}$ is the relative position of the second body and $\mathbf{v}$ its relative velocity in an inertial frame. The vector $\mathbf{c}$ is a constant vector directed along the positive normal to the orbital plane.

Let now $\mathbf{k}$ be the unit vector of the z-axis of the Eulerian frame defined in Section 2, and K_0 the projection of $\mathbf{c}$ along this axis. It follows that

$$K_0 = (\mathbf{r} \times \mathbf{v}) \cdot \mathbf{k},$$

or

$$K_0 = (\mathbf{r} \times \mathbf{v}_e) \cdot \mathbf{k} + m v_3 [\mathbf{r} \times (\mathbf{k} \times \mathbf{r})] \cdot \mathbf{k},$$

where

$$\mathbf{v}_e = \mathbf{v} - m v_3 \mathbf{k} \times \mathbf{r}$$

is the velocity vector with respect to the Eulerian system of reference. With the normalized Hill's variables, it writes

$$K_0 = v_3 a^2 (\kappa + \tfrac{1}{2} K_2), \tag{16}$$

where

$$K_2=(1+S)\cdot DU-(1+U)\cdot DS+2\kappa(1+U)(1+S)-2\kappa. \tag{17}$$

We have now the following proposition: K_2 is a second-order quantity with respect to the eccentricity. Indeed, the above calculations were made in the frame of the problem of the two bodies where, to the first order, we must have the classical result $K_1=\kappa\nu_3 a^2$ $(=na^2)$.

The integral of the areas is then used to define two new variables. Let K and H be respectively defined by

$$\begin{aligned} K&=\frac{\sigma^*\zeta^{-\kappa}}{K_1}(c_x+ic_y),\\ H&=\frac{\sigma\zeta^{\kappa}}{K_1}(c_x-ic_y), \end{aligned} \tag{18}$$

where c_x and c_y are the projections of $\mathbf{c}$ over inertial axes in the fundamental plane of reference. Easy calculations lead to

$$K=\frac{ia^2\zeta^{-\kappa}}{K_1}\{-\dot{Z}\zeta^{\kappa}(1+U)+Z[\zeta^{\kappa}(1+U)]^{\cdot}\},$$

$$H=\frac{ia^2\zeta^{\kappa}}{K_1}\{-\dot{Z}\zeta^{-\kappa}(1+S)+Z[\zeta^{-\kappa}(1+S)]^{\cdot}\},$$

and, by introducing the operator D, to

$$\begin{aligned} \kappa K&=(1+U)\cdot DZ-Z[DU+\kappa(1+U)],\\ \kappa H&=-(1+S)\cdot DZ+Z[DS-\kappa(1+S)]. \end{aligned} \tag{19}$$

The pair of variables introduced in this way will serve to describe the proper oscillations along the z-axis and it is close to the Poincaré's variables $I\ \exp -i(l-\Omega)$ and $I\ \exp i(l-\Omega)$. They will be introduced in the system as follows: We have for each satellite one equation like

$$D^2Z=\psi(Z).$$

We normalize this equation by introducing $W=DZ$. Then we have

$$DW=\psi(Z), \qquad DZ=W,$$

which are transformed, by introducing the relations

$$K=K(Z,W), \qquad H=H(Z,W),$$

into two equations for the new variables.

4. Laplace's First Integral and Its Application

For each satellite, a second couple of variables will be introduced through functional relations which allow us to transform the remaining equations to Weierstrass' normal form. Laplace's first integral for the two-body problem is the best suitable to suggest the functional relations to be introduced. This integral gives the invariance of the apsidal line and of the eccentricity in this problem. It writes

$$\mathbf{p} = -\mathbf{r}/r - (G\mu)^{-1}(\mathbf{r}\times\mathbf{v})\times\mathbf{v}, \tag{20}$$

where μ is the sum of the masses and $\mathbf{p}$ is a constant vector directed to the pericenter.

Decomposing $\mathbf{p}$ along a couple of inertial axes in the fundamental plane of reference, we write

$$P = (p_x + ip_y)\,\sigma^*\zeta^{-\kappa},$$
$$Q = (p_x - ip_y)\,\sigma\zeta^{\kappa},$$

then, to the third order in the orbital eccentricity,

$$\begin{aligned} P &= -\frac{a}{r}(1+U) + \frac{1}{\kappa}\left(1+\frac{K_2}{2\kappa}\right)[DU + \kappa(1+U)] - \frac{1}{\kappa}K\cdot DZ, \\ Q &= -\frac{a}{r}(1+S) - \frac{1}{\kappa}\left(1+\frac{K_2}{2\kappa}\right)[DS - \kappa(1+S)] + \frac{1}{\kappa}H\cdot DZ, \end{aligned} \tag{21}$$

where K, H and K_2 are that defined in Section 3. It should be emphasized that all computations in the derivation of P and Q are made in the frame of the two-body problem. Indeed, the aim of these derivations is only to give rise to the functional relations we want. So, for example, Kepler's third law: $G = \kappa^2 v_3^2 a^3$ may be used without any constraint.

In order to solve these equations with respect to DU and DS let them be used to derive a new equation for K_2 where DU and DS are replaced by P and Q. From Equations (21), it follows:

$$\begin{aligned} (1+S)P + (1+U)Q = &-2\frac{a}{r}(1+U)(1+S) + \\ &+\frac{1}{\kappa}\left(1+\frac{K_2}{2\kappa}\right)[(1+S)DU - (1+U)DS + 2\kappa(1+U)(1+S)] - \\ &-\frac{1}{\kappa}DZ[(1+S)K - (1+U)H] + 0(\text{4th}); \end{aligned}$$

and then, taking into account that

$$(1+S)\cdot DU - (1+U)\cdot DS + 2\kappa(1+U)(1+S) = 2\kappa + K_2,$$

and that K_2 is a second-order quantity in the problem of the two bodies, and solving with respect to K_2, we have

$$K_2 = -A + \tfrac{1}{2}DZ[(1+S)K - (1+U)H] + 0(\text{4th}), \tag{22}$$

where

$$A=\kappa\left[1-\frac{a}{r}(1+U)(1+S)\right]-\tfrac{1}{2}\kappa[(1+S)P+(1+U)Q]. \tag{23}$$

It follows, to the third order,

$$\begin{aligned} DU &= \kappa P+\kappa\left(\frac{a}{r}-1\right)(1+U)+\tfrac{1}{2}A(1+U)+W_2(1+U),\\ DS &= -\kappa Q-\kappa\left(\frac{a}{r}-1\right)(1+S)-\tfrac{1}{2}A(1+S)-W_2^*(1+S), \end{aligned} \tag{24}$$

where

$$\begin{aligned} W_2 &= -\tfrac{1}{4}DZ[(1+S)K-(1+U)H]+(1+S)K\cdot DZ,\\ W_2^* &= -\tfrac{1}{4}DZ[(1+S)K-(1+U)H]-(1+U)H\cdot DZ. \end{aligned} \tag{25}$$

Equations (24) will be taken as defining our new parameters P and Q. They will be introduced in the system as follows: We have for each satellite one pair of equations like

$$D^2U=\psi_1(U,S), \qquad D^2S=\psi_2(U,S).$$

We normalize these equations by introducing the functional relations

$$DU=\varphi_1(U,S,P,Q), \qquad DS=\varphi_2(U,S,P,Q).$$

From their derivatives and the original equations, we have

$$DP=\varphi_3(U,S,P,Q), \qquad DQ=\varphi_4(U,S,P,Q).$$

And this completes the transformation of our system of equations in a system of first-order equations.

The reasons for the choice of the functional relations given by Equations (24) may be discussed. Since this choice is arbitrary it would be possible in this theory to use the same functional relations already used in the first-order theory. The new choice corresponds to have P and Q close to the Poincaré's variables $e\exp-i(l-\varpi)$ and $e\exp i(l-\varpi)$ to the second order in the elliptical parameters; this fact is of an utmost importance for it warrants linear equations for P and Q in the second-order theory. We would also ask about the possibility of introducing a third variable by taking the space component of p, which conjugates itself with Z in the same way as P and Q conjugate themselves with U and S. The difficulty for such a procedure, from an algebraic point of view, lies in the fact that the space component of $\mathbf{p}$ is a third-order quantity. On the other hand such possibility would not lead to easy interpretations of the variables as describing proper oscillations.

5. The Central Problem

We call central problem in the theory of the four great satellites of Jupiter the restricted problem in which the satellites and the planet lie in a fixed plane under the action of their mutual attractions, disregarding the effects arising from their shapes or from external bodies. This restricted problem has the main difficulties of the general problem and its solution shows the main features of the observed motions. Indeed, aside the characteristics already discussed in Section 1, some others may be considered:

(a) *The quasi-resonances.* The sidereal mean motions of the satellites are such that $n_1 - 2n_2$ and $n_2 - 2n_3$ are small and will give rise to small divisors in the integration step. For example, in Laplace's theory (Tisserand, 1896), the longitude of Io has the inequality

$$\delta v_1 = \frac{m_2 n_1 f(\alpha_{12})}{n_1 - 2n_2 + g(J_2)} \sin(2l_1 - 2l_2). \tag{26}$$

Since its period is close to the period of the satellite it is named induced equation of the centre, and its half amplitude is the forced eccentricity. For the four satellites we have respectively,

Proper eccentricity	Forced eccentricity
0.00001	0.00412
0.00013	0.00943
0.00139	0.00063
0.00736	–

These values show that the first two satellites depart from uniform circular motions more owning to perturbations than they do owning to proper oscillations. In the choice of the criteria for defining the small quantities of the theory, this fact is determinant.

(b) *The proper oscillations.* In reason of the strong interactions we cannot consider each orbit as having its own free oscillation (equation of the centre). The strong interactions do not allow to take as intermediate solutions those arising from separated integrations. The intermediate orbit must arise from the integration of the system formed by the four pairs of variables P_i and Q_i, simultaneously. So we will have four proper oscillations which will be apparent in the orbit of each satellite. The free oscillations in the longitudes of the satellites will have the form

$$\delta v_j = 2 \sum_i M_{ji} \sin(l_j - \varpi_i). \tag{27}$$

The fact that the system oscillates as a whole leads to the necessity of having the P_i and Q_i close to the Poincaré's variables at least to the second order in the elliptical parameters for the sake of having linear equations for these quantities in the second-order theory.

(c) *The libration.* The Galilean resonance, which arises from

$$n_1 - 3n_2 + 2n_3 = 0, \tag{28}$$

will give rise to libration's inequalities in the longitudes of the first three satellites.

The best results indicate for their half amplitudes respectively 8.″7, 24″ and 2.″3. These values must be compared with those giving the standard deviations of the best observations (Section 1), and such comparison allows to disregard a deeper study of this phenomenon when deriving a theory for ephemeris purposes, notwithstanding its very high mathematical interest. The complete modern treatment of this phenomenon has been made by Sagnier (1973b), who succeeded in deriving formal quasi-periodic solutions of the second kind for the central problem including the Galilean libration.

6. The Equations of the Central Problem

The equations of the central problem are those given in Section 2, when restricted to the plane variables U_j and S_j and to the disturbing functions arising from the mutual interactions. The functional relations which are to be introduced are

$$\begin{aligned} DU_j &= \kappa_j P_j + \kappa_j(1/C_j - 1)(1+U_j) + \tfrac{1}{2}A_j(1+U_j), \\ DS_j &= -\kappa_j Q_j - \kappa_j(1/C_j - 1)(1+S_j) - \tfrac{1}{2}A_j(1+S_j), \end{aligned} \tag{29}$$

where

$$C_j = [(1+U_j)(1+S_j)]^{1/2}, \tag{30}$$

and

$$A_j = \kappa_j(1-C_j) - \tfrac{1}{2}\kappa_j[(1+S_j)P_j + (1+U_j)Q_j]. \tag{31}$$

The conjugacy of all equations, already mentioned in Section 1 is preserved; the equations are invariant with respect to the transformation

$$U_j \to S_j, \qquad S_j \to U_j, \qquad P_j \to Q_j, \qquad Q_j \to P_j, \qquad t \to -t. \tag{32}$$

If the technique of utilization of the functional relations already discussed in Section 4 is adopted, we have, after some appropriate differentiations and substitutions, the equations

$$\begin{aligned} DP_j + \kappa_j P_j &= \frac{\lambda_j - \kappa_j^2}{\kappa_j C_j^3}(1+U_j) + \frac{1}{\kappa_j}\mathcal{R}_j - \\ &\quad - L_j - \frac{1}{4\kappa_j}B_j(1+U_j) + \frac{1}{4\kappa_j}\chi_j(1+U_j), \\ DQ_j - \kappa_j Q_j &= -\frac{\lambda_j - \kappa_j^2}{\kappa_j C_j^3}(1+S_j) - \frac{1}{\kappa_j}\mathcal{T}_j + \\ &\quad + L_j^* - \frac{1}{4\kappa_j}B_j(1+S_j) + \frac{1}{4\kappa_j}\chi_j(1+S_j), \end{aligned} \tag{33}$$

where

$$\begin{aligned} L_j &= \tfrac{1}{2}A_j P_j - \tfrac{1}{8}A_j(1+U_j)[Q_j(1+U_j) - P_j(1+S_j)] + \\ &\quad + \frac{A_j}{C_j^3}(C_j^2 - 1)(1+U_j) + \frac{1}{4\kappa_j}A_j^2(1+U_j). \end{aligned} \tag{34}$$

L_j^* is its conjugate through the above defined transformation

$$B_j=(1-\tfrac{1}{2}C_j^2)^{-1}\,\kappa_j[(1+S_j)\,L_j-(1+U_j)\,L_j^*], \tag{35}$$

and

$$\chi_j=(1-\tfrac{1}{2}C_j^2)^{-1}[(1+S_j)\,\mathscr{R}_j-(1+U_j)\,\mathscr{T}_j]. \tag{36}$$

The new equations of the motion are the set formed by Equations (29) and (33) which are normalized, with respect to the variables U_j, S_j, P_j and Q_j.

We must notice that these equations are exact, that is, no approximation has been made in the course of their derivation. The fact that the functional relations (Equations (29)) are themselves approximate relations in the two-body problem, do not matter. Indeed, the two-body problem has been used only to suggest the form of Equations (29), which are exact as they define the P_j and Q_j.

7. First Integrals of the Functional Relations. Poisson Technique

When $P_j=Q_j=0$, Equations (29) may be written

$$\begin{aligned} D(1+U_j)&=\mathscr{A}_j(1+U_j),\\ D(1+S_j)&=-\mathscr{A}_j(1+S_j), \end{aligned} \tag{37}$$

where

$$\mathscr{A}_j=\mathscr{A}_j(U_j,S_j)=\kappa_j(1/C_j-1)+\tfrac{1}{2}\kappa_j-\tfrac{1}{2}\kappa_jC_j. \tag{38}$$

Two first integrals may be obtained. Firstly, from Equations (37), it follows:

$$(1+S_j)\cdot D(1+U_j)+(1+U_j)\cdot D(1+S_j)=0,$$

and then

$$(1+U_j)\,(1+S_j)=C_j^2=\text{const}. \tag{39}$$

On account of the meaning of U_j and S_j it is easily seen that this integral accounts for the circularity of the motion when $P_j=Q_j=0$. From Equations (37) it follows, still, that

$$(1+S_j)\cdot D(1+U_j)-(1+U_j)\cdot D(1+S_j)=2\mathscr{A}_j(1+U_j)\,(1+S_j),$$

or, if we put

$$\xi_j=(1+U_j)/(1+S_j), \tag{40}$$

that

$$D\xi_j/\xi_j=2\mathscr{A}_j. \tag{41}$$

Let it be remarked that, for $P_j=Q_j=0$, $\mathscr{A}_j$ is a constant. Indeed, from its definition, and from Equations (39), it follows

$$\mathscr{A}_j=-\tfrac{1}{2}\kappa_j+\kappa_j/C_j-\tfrac{1}{2}\kappa_jC_j.$$

So, the integration of Equation (41) may be easily performed and leads to

$$\log \xi_j = \log C'_j + 2\mathscr{A}_j \log \zeta ,$$

or

$$\frac{1+U_j}{1+S_j} \zeta^{-2\mathscr{A}_j} = C'_j. \tag{42}$$

This integral is related to the uniformity of the motion when $P_j = Q_j = 0$.

These integrals may be extended to the general case ($P_j \neq 0$ and $Q_j \neq 0$), by means of Poisson's method for the variation of the first integrals (see Kurth, 1959). Let Equations (29) be written completely:

$$\begin{aligned} DU_j &= \mathscr{A}_j(1+U_j) + G_j, \\ DS_j &= -\mathscr{A}_j(1+S_j) - H_j, \end{aligned} \tag{43}$$

where G_j and H_j are the terms which vanish when $P_j = Q_j = 0$, and which are to be treated as perturbations. The variational equations of Poisson, for this system, write (Ferraz-Mello, 1966)

$$\begin{aligned} DC_j &= \frac{\partial C_j}{\partial U_j} G_j - \frac{\partial C_j}{\partial S_j} H_j, \\ DC'_j &= \frac{\partial C'_j}{\partial U_j} G_j - \frac{\partial C'_j}{\partial S_j} H_j. \end{aligned} \tag{44}$$

If the partial derivatives are computed and substituted, it follows, after some algebra:

$$\begin{aligned} DC_j &= \tfrac{1}{2}\kappa_j [P_j(C'_j)^{-1/2} \zeta^{-\mathscr{A}_j} - Q_j(C'_j)^{1/2} \zeta^{\mathscr{A}_j}], \\ DC'_j &= \mathscr{D}_j [P_j(C'_j)^{1/2} \zeta^{-\mathscr{A}_j} + Q_j(C'_j)^{3/2} \zeta^{\mathscr{A}_j}] - \\ &\quad - \kappa_j \mathscr{A}'_j [P_j(C'_j)^{1/2} \zeta^{-\mathscr{A}_j} - Q_j(C'_j)^{3/2} \zeta^{\mathscr{A}_j}] \log \zeta , \end{aligned} \tag{45}$$

where

$$\mathscr{A}'_j = \mathrm{d}\mathscr{A}_j / \mathrm{d}C_j = -\kappa_j (C_j^{-2} + \tfrac{1}{2}), \tag{46}$$

and

$$\mathscr{D}_j = \kappa_j C_j (C_j^{-2} - \tfrac{1}{2}). \tag{47}$$

It is easily seen that any iterative procedure of integration (the P_j and Q_j being assumed as known Fourier's series) will lead to Poisson's secular terms. This fact is well apparent when the equation for DC'_j is modified taking into account that $\mathscr{A}'_j DC_j = D\mathscr{A}_j$:

$$D \log C'_j = \mathscr{D}_j [P_j(C'_j)^{-1/2} \zeta^{-\mathscr{A}_j} + Q_j(C'_j)^{1/2} \zeta^{\mathscr{A}_j}] + 2\mathscr{A}_j - 2D(\mathscr{A}_j \log \zeta). \tag{48}$$

These Poisson terms are of the same kind as those arising in the formulation of

Lagrange's equations of variation of the parameters. They may be avoided by making use of the Tisserand's transformation (Tisserand, 1868). Let the parameter

$$\Gamma_j = \log C'_j + 2\mathscr{A}_j \log \zeta, \tag{49}$$

be introduced instead of C'_j. Equations (45) become:

$$\begin{aligned} DC_j &= \tfrac{1}{2}\kappa_j [P_j/\gamma_j - Q_j\gamma_j], \\ D\Gamma_j &= 2\mathscr{A}_j + \mathscr{D}_j [P_j/\gamma_j + Q_j\gamma_j], \end{aligned} \tag{50}$$

where

$$\gamma_j^2 = \exp \Gamma_j = C'_j \zeta^{2\mathscr{A}_j}. \tag{51}$$

For these equations we can get a formal quasi-periodic solution provided that the Fourier's series in the right hand side are of zero average. These solutions are formal first integrals of the motion.

8. The Integration

In order to integrate these equations, successive approximations may be used. Let be remarked that the radius vector and the longitude of the satellites are given by

$$\begin{aligned} r_j &= a_j C_j, \\ \theta_j &= \theta_{0j} + g_j \nu_3 t - \Gamma_j/2i, \end{aligned} \tag{52}$$

so that the start solution may be

$$C_j = 1, \qquad \Gamma_j = 0 \quad (\gamma_j = 1). \tag{53}$$

We must observe that $\mathscr{A}_j$ is one order lesser than the other terms of the differential Equations (50). So, at each step, C_j must be computed before Γ_j, and it must be taken for getting the new value of $\mathscr{A}_j$ in the computation of Γ_j. The order-one solution is computed thereafter.

We have, first,

$$DC_j = \tfrac{1}{2}\kappa_j (P_j - Q_j),$$

and then,

$$C_j = 1 + \tfrac{1}{2}\kappa_j D^{-1}(P_j - Q_j). \tag{54}$$

The choice of the constant of integration is provided by the fact that at each approximation, if P_j and Q_j are given by Fourier series, then the mean value of C_j must be equal to one. This value of C_j allows to compute the first-order approximation for $\mathscr{A}_j$:

$$\mathscr{A}_j = -\tfrac{3}{4}\kappa_j^2 D^{-1}(P_j - Q_j). \tag{55}$$

Then we have

$$D\Gamma_j = \tfrac{1}{2}\kappa_j (P_j + Q_j) - \tfrac{3}{2}\kappa_j^2 D^{-1}(P_j - Q_j),$$

and

$$\Gamma_j = \tfrac{1}{2}\kappa_j D^{-1}(P_j + Q_j) - \tfrac{3}{2}\kappa_j^2 D^{-2}(P_j - Q_j). \tag{56}$$

Here, the arbitrary constant is chosen such that the mean value of Γ_j be equal to zero.

In both cases the operator D^{-1} has the meaning of the primitive function of the trigonometric function involved, in the ordinary sense, i.e., without integration constant.

The complete integration involves also Equations (33) where the new parameters C_j and Γ_j are to be introduced through

$$(1 + U_j) = C_j\gamma_j; \qquad (1 + S_j) = C_j/\gamma_j. \tag{57}$$

The integration procedure may be chosen among the usual techniques. Nevertheless, this choice must be made with some care. The technique should not generate Poisson's secular terms, and it must allow for an adequate study of the free oscillations (see Section 5). If we wish only a low-order solution, approximations like those above computed may be used for eliminating C_j and γ_j from the equations for P_j and Q_j. The resulting equations are integrodifferential equations.

For example, to get the second-order solutions, it is enough to take the first order approximations given by Equations (54) and (55) for C_j and Γ_j. The resulting equations are, to the second order,

$$\begin{aligned} DP_j + \kappa_j P_j = {} & \\ = \frac{\lambda_j - \kappa_j^2}{\kappa_j} & [1 + \tfrac{1}{4}\kappa_j D^{-1}(P_j + Q_j) - \tfrac{3}{4}\kappa_j^2 D^{-2}(P_j - Q_j) - \kappa_j D^{-1}(P_j - Q_j)] + \\ & + \frac{1}{\kappa_j}\mathscr{R}_j - A_j^I[\tfrac{7}{8}P_j - \tfrac{5}{8}Q_j + \tfrac{7}{8}\kappa_j D^{-1}(P_j - Q_j)] + \\ & + \frac{1}{4\kappa_j}\chi_j[1 + \tfrac{1}{2}\kappa_j D^{-1}(P_j - Q_j) + \tfrac{1}{4}\kappa_j D^{-1}(P_j + Q_j) - \tfrac{3}{4}\kappa_j^2 D^{-2}(P_j - Q_j)], \end{aligned} \tag{58}$$

and their conjugates. In these equations A_j^I represent the first-order part of A_j. From Equations (33) we see that

$$DA_j^I = -\tfrac{1}{2}\chi_j,$$

and then

$$A_j^I = -\tfrac{1}{2}D^{-1}\chi_j, \tag{59}$$

where the constant of integration is taken as zero, since $A_j^I = 0$ for the undisturbed motion.

For the integration Krasinsky's method may be used since the equations are linear. In this order of approximation, Krasinsky's method reduces to the derivation of a transformation of coordinates having the form

$$P_j = P_j^* + b_j + \sum_i [c_{ji}P_i^* + d_{ji}Q_i^* + e_{ji}D^{-1}P_i^* + f_{ji}D^{-1}Q_i^* + g_{ji}D^{-2}P_i^* + h_{ji}D^{-2}Q_i^*],$$
$$Q_j = Q_j^* + b_j' + \sum_i [c_{ji}'Q_i^* + d_{ji}'P_i^* - e_{ji}'D^{-1}Q_i^* - f_{ji}'D^{-1}P_i^* + g_{ji}'D^{-2}Q_i^* + h_{ji}'D^{-2}P_i^*], \tag{60}$$

where $b_j, \ldots, h_{ji}'$ are quasi-periodic functions of the time through ζ, all of them being of the first order. This transformation is built in such a way that the resulting system, which will remain linear, has constant coefficients. It has the same nature as Euler's differential equations. These equations are not homogeneous: they have a constant independent term.

It must be emphasized that a necessary condition for success in getting constant coefficient equations through Equations (60) as defined, is that $m \neq 0$. Indeed all variable coefficients and terms in Equations (58) depend on the time through $\zeta^{(I|g)}$, where $I \in \mathbf{Z}^4$, and $I_1 + I_2 + I_3 + I_4 = 0$; since g_1, g_2 and g_3 are integers and since $m \neq 0$, i.e., $\kappa_j \neq g_j$, the only possibility of having $(I \mid g)$ too close to κ_j is for higher-order resonances involving g_4.

As a final remark to this section let us mention that an exact set of equations could be found instead of Equations (58) if the variables $\mathscr{P}_j = P_j/\gamma_j$ and $\mathscr{Q}_j = Q_j\gamma_j$ were used instead of P_j and Q_j. Nevertheless, this change corresponds to modifying the functional relations introduced in Section 4 and lead to nonlinear second-order terms. So, the main characteristics of the result obtained above – the linearity of the second-order equations – would be lost.

9. The Constant Perturbation. The Libration

After the integration of Equations (58) is performed, the results are to be introduced in Equations (50). If C_j are replaced by the first-order quantities

$$\varepsilon_j = C_j - 1,$$

these equations write

$$D\varepsilon_j = \tfrac{1}{2}\kappa_j(P_j - Q_j) - \tfrac{1}{4}\kappa_j(P_j + Q_j)\,\Gamma_j,$$
$$D\Gamma_j = -\tfrac{3}{2}\kappa_j\varepsilon_j + \kappa_j\varepsilon_j^2 + \tfrac{1}{2}\kappa_j(P_j + Q_j) - \tfrac{3}{2}\kappa_j(P_j + Q_j)\,\varepsilon_j - \tfrac{1}{4}\kappa_j(P_j - Q_j)\,\Gamma_j. \tag{62}$$

P_j and Q_j are, now, known functions of ζ, involving the four circulatory frequencies g_j and the four oscillatory frequencies ϖ_j introduced by the integration of the constant coefficient equations for the P_j^* and Q_j^*.

Once again Krasinsky's method may be used in order to eliminate periodic coefficients. This aim is achieved notwithstanding the fact that this set of equations has the nonlinear term $\kappa_j\varepsilon_j^2$; indeed, the coefficient of ε_j^2 is constant and we are interested only in second-order equations. As in Section 8, Krasinsky's method reduces itself to the derivation of the transformation of coordinates,

$$\varepsilon_j = (1 + \beta_j)\,\varepsilon_j^* + \delta_j + \eta_j\Gamma_j^*,$$
$$\Gamma_j = (1 + \beta_j')\,\Gamma_j^* + \delta_j' + \eta_j'\varepsilon_j^*, \tag{63}$$

where $\beta_j, \ldots, \eta'_j$ are quasi-periodic functions of first order. The transformation is such that the resulting system has constant coefficients:

$$\begin{aligned} D\varepsilon_j^* &= T_2' \Gamma_j^*, \\ D\Gamma_j^* &= T_0 - \tfrac{3}{2}\kappa_j \varepsilon_j^* + T_1 \varepsilon_j^* + \kappa_j \varepsilon_j^{*2}. \end{aligned} \tag{64}$$

The solution of this system to the second order shows a shift for Γ_j^* proportional to T_0. This fact contradicts the working hypothesis after which v_j is already the observed mean motion and this phase shift may not exist. So, the constants T_0 must be made equal to zero, which allows to determinate the normalization factors a_j, i.e., the mean distances from the satellites to the planet (these constant terms gave the so-called constant perturbation). On the other hand, the first-order parts of T_1 and T_2' are proportional to the first-order part of T_0. Equations (64) reduce so, to

$$\begin{aligned} D\varepsilon_j^* &= 0, \\ D\Gamma_j^* &= -\tfrac{3}{2}\kappa_j \varepsilon_j^* + \kappa_j \varepsilon_j^{*2}, \end{aligned} \tag{65}$$

for which the trivial solution, $\varepsilon_j^* = 0$ and $\Gamma_j^* = 0$, is chosen.

This completes the integration of the central problem to the second-order, when libration is disregarded. Indeed in the calculations shown above there was no question to investigate the nature of the constant terms involved. They could be of an essential nature – i.e. true constant terms – or of an accidental nature since $g_1 - 3g_2 + 2g_3 = 0$.

If we suppose that the relation does not exactly hold,

$$g_1 - 3g_2 + 2g_3 = G,$$

Equations (64) must be slightly modified in order to avoid the integration of terms in ζ^G in the computation of the $\beta_j, \ldots, \eta'_j$ for $j = 1, 2, 3$. The resulting system is

$$\begin{aligned} D\varepsilon_j^* &= T_0' + T_2' \Gamma_j^*, \\ D\Gamma_j^* &= T_0 - \tfrac{3}{2}\kappa_j \varepsilon_j^* + T_1 \varepsilon_j^* + T_2 \Gamma_j^* + \kappa_j \varepsilon_j^{*2}, \end{aligned} \tag{66}$$

where T_0, T_1 and T_2' are the same constants as before, plus functions of $(\zeta^{kG} + \zeta^{-kG} - 2)$; T_0' and T_2 are functions of $(\zeta^{kG} - \zeta^{-kG})$. The constants may be eliminated in the same way as before. The discussion of the remaining system will be the subject of a separate paper.

10. The Complete Second-Order Theory

Let us first show that the equations for the space variables are, at the second order, completely independent of the solutions for the planar parameters. Let the technique discussed in Section 3 be adopted. From Equations (19) it follows:

$$\begin{aligned} \kappa_j(DK_j + \kappa_j K_j) &= \quad (1 + U_j) \cdot D^2 Z_j - Z_j[(D + \kappa_j)^2\, U_j + \kappa_j^2], \\ \kappa_j(DH_j - \kappa_j H_j) &= -(1 + S_j) \cdot D^2 Z_j + Z_j[(D - \kappa_j)^2\, S_j + \kappa_j^2]. \end{aligned} \tag{67}$$

When substituting Equations (9) into the right-hand sides of these equations, the

Keplerian parts will disappear and it results,

$$\begin{aligned}\kappa_j(DK_j+\kappa_jK_j)&=\ \ (1+U_j)\,\mathscr{V}_j-Z_j\,\mathscr{R}_j,\\ \kappa_j(DH_j-\kappa_jH_j)&=-(1+S_j)\,\mathscr{V}_j+Z_j\mathscr{T}_j,\end{aligned}\tag{68}$$

It is easily seen that the right-hand sides are of the second order. So the Z_j may be substituted by the first-order approximations,

$$Z_j=-\tfrac{1}{2}(K_j+H_j).\tag{69}$$

So, the second-order equations are

$$\begin{aligned}\kappa_j(DK_j+\kappa_jK_j)&=\ \ \mathscr{V}_j+\tfrac{1}{2}(K_j+H_j)\,\mathscr{R}_j,\\ \kappa_j(DH_j-\kappa_jH_j)&=-\mathscr{V}_j-\tfrac{1}{2}(K_j+H_j)\,\mathscr{T}_j,\end{aligned}\tag{70}$$

where the u_i, s_i, r_i and r_{ji} in $\mathscr{R}_j$, $\mathscr{T}_j$ and $\mathscr{V}_j$ may be taken as for the circular zeroth-order approximation (Equations (6)). So Equations (70) are homogeneous linear with quasi-periodic coefficients, not involving the U_j, S_j, P_j and Q_j. Again Krasinsky's method may be used. The transformation

$$\begin{aligned}H_j&=(1+\varrho_j)\,H_j^*+\tau_jK_j^*,\\ K_j&=(1+\varrho_j')\,K_j^*+\tau_j'H_j^*\end{aligned}\tag{71}$$

may be derived in such a way that the coefficients in the equations for H_j^* and K_j^* are constants. The necessary condition to succeed in getting constant coefficient equations through Equations (71) is the same as for Equations (60) and it is fulfilled.

For the planar variables the same technique employed in solving the central problem is adopted. The functional relations (24) are adopted, but, for the sake of simplicity, all space terms are collected in W_{2j} and W_{2j}^*. So, A_j is taken as for the central problem (Equation (31)), and for W_{2j} and W_{2j}^* we take

$$\begin{aligned}W_{2j}&=\ \ \tfrac{1}{4}(3K_j+H_j)\cdot DZ_j-\tfrac{1}{4}\kappa_jZ_j^2,\\ W_{2j}^*&=-\tfrac{1}{4}(3H_j+K_j)\cdot DZ_j-\tfrac{1}{4}\kappa_jZ_j^2.\end{aligned}\tag{72}$$

The functional relations then write

$$DU_j=\kappa_jP_j+\kappa_j(C_j^{-1}-1)\,(1+U_j)+\tfrac{1}{2}A_j(1+U_j)+W_{2j}(1+U_j),$$

and their conjugates. The resulting equations for the P_j and Q_j are the same as for the central problem (Equations (33)) and they are solved in the same way. The second-order space terms came through

$$W_{3j}=DW_{2j}+2\kappa_jW_{2j}-\tfrac{3}{4}\kappa_j(W_{2j}-W_{2j}^*)+\tfrac{3}{2}\kappa_j^2Z_j^2,$$

and their conjugates. But, it is easily seen, by using Equation (69), and the first-order relations $DK_j=-\kappa_jK_j$, $DH_j=\kappa_jH_j$, $D^2Z_j=\kappa_j^2Z_j$, and

$$DZ_j=\tfrac{1}{2}\kappa_j(K_j-H_j),\tag{74}$$

that W_{3j} is indeed a third-order quantity.

At last, the equations for ε_j and Γ_j must be considered. The new equations for C_j and C'_j are those given by Equations (45) to which we add the space contributions

$$\delta(DC_j) = \frac{\partial C_j}{\partial U_j} W_{2j}(1+U_j) - \frac{\partial C_j}{\partial S_j} W^*_{2j}(1+S_j),$$

$$\delta(DC'_j) = \frac{\partial C'_j}{\partial U_j} W_{2j}(1+U_j) - \frac{\partial C'_j}{\partial S_j} W^*_{2j}(1+S_j),$$

i.e., after some algebra,

$$\delta(DC_j) = \tfrac{1}{4}\kappa_j(K_j+H_j)(K_j-H_j),$$

$$\delta(DC') = \tfrac{1}{4}\kappa_j C'_j(K_j-H_j)^2 - \tfrac{1}{8}\kappa_j C'_j(K_j+H_j)^2 -$$
$$- \mathscr{A}'_j C'_j\left[\tfrac{1}{2}\kappa_j(K_j+H_j)(K_j-H_j)\right]\log\zeta,$$

and for Γ_j, defined by Equation (49), the new equation is Equation (50) with the additive space term,

$$\delta(D\Gamma_j) = \tfrac{1}{4}\kappa_j(K_j-H_j)^2 - \tfrac{1}{8}\kappa_j(K_j+H_j)^2.$$

The integration of these equations is made exactly as it has been discussed in Section 9.

11. Conclusion

Notwithstanding the fact that this theory has been derived with special regard to the problem of the motion of the Galilean satellites of Jupiter, it may be useful in the study of other problems of planetary kind, in which the motions are close to circularity and coplanarity, and, in which, quasi-resonances lead to strong perturbations. Nevertheless, usual resonant problems must be excluded: the Galilean resonance is a too particular kind of resonance and does not involve the same kind of difficulties as, e.g., Hecubian resonance (see Sagnier, 1973b).

The main characteristic of the theory is that it allows to keep the main frequencies fixed from the earlier stages, and so, to have a purely trigonometric solution. Also the distances are to be fixed from the earlier stages; but the observational data for distances are not so good, and these distances are to be modified after computing the constant perturbation (Section 9).

In practice the Laplace coefficients in the development of the disturbing function are to be taken numerically. The algebra of the series may be, then, performed, by using a computer. The work is made iteratively by getting better distances at each step. It must be remarked that this algebra is not too involved and does not require too powerful computers.

Acknowledgements

Mr C. Basta and Miss M. Sato are working on different aspects of this theory; they are to be acknowledged for their valuable suggestions in discussing this paper. I am indebted to Dr J. L. Sagnier for many discussions on the subject and to Dr C. A. B. Borges for helping in the preparation of the manuscript. This research has been supported partly by the Brazilian Council of Research, proc. 3729/73 and by the Research Foundation of São Paulo, procs. 68/864, 69/373 and 71/1264.

References

Brouwer, D.: 1959, *Proc. Symp. Appl. Math.*, Vol. 9, *Orbit Theory*, Am. Math. Soc., Providence, p. 152.

Brumberg, V. A.: 1970, in G. E. O. Giacaglia (ed.), *Periodic Orbits, Stability and Resonances*, D. Reidel, Dordrecht, The Netherlands, p. 410.

De Sitter, W.: 1918, *Ann. Sterrew. Leiden* **12** (1).

Ferraz-Mello, S.: 1966, *Bull. Astron.*, 3e série, **1**, 287.

Ferraz-Mello, S.: 1969a, *Compt. Rend. Acad. Sci. Paris* **268**, 198.

Ferraz-Mello, S.: 1969b, *Compt. Rend. Acad. Sci. Paris* **268**, 985.

Ferraz-Mello, S. and Paula, L. R.: 1973 (to be published).

Hagihara, Y.: 1972, *Celestial Mechanics*, Vol. 2, *Perturbation Theory*, MIT Press, Cambridge, Mass.

Kovalevsky, J.: 1962, *Trans. IAU* **11B**, 455.

Krasinsky, G. A.: 1968, *Soviet Math. Dokl.* **9**, 641.

Krasinsky, G. A.: 1969, *Trudy Inst. Teoret. Astron.* **13**, 105.

Kurth, R.: 1959, *Introduction to the Mechanics of the Solar System*, Pergamon Press, London.

Rodrigues, C. M.: 1970, M. Sc. Thesis, Inst. Tecnol. Aer., São José dos Campos, Brazil.

Sagnier, J. L.: 1973a, *Astron. Astrophys.* **25**, 113.

Sagnier, J. L.: 1973b, *Astron. Astrophys.* (in press).

Sampson, R. A.: 1910, *Tables of the Four Great Satellites of Jupiter*, Wesley, London.

Tisserand, F.: 1868, *J. Math. Pures Appl.*, ser. 2, **13**, 255.

Tisserand, F.: 1896, *Traité de mécanique céleste*, Vol. 4, Gauthier-Villars, Paris.

DISCUSSION

K. Ziolkowski: Did you try to use your theory to other satellite systems or to planets?

S. Ferraz-Mello: Yes, I am trying to use my theory to an asteroid, Hestia.

CONDITIONS FOR ESCAPE AND RETENTION

J. S. GRIFFITH

Lakehead University, Thunder Bay, Ontario, Canada P7B SE1

Abstract. The methods used in deriving conditions for escape, retention and containment for the three-body problem are applied to the n-body case, and similar conditions are obtained. In the n-body problem less stringent conditions are derived, and in the case of retention a further condition is imposed.

1. Introduction

The classification of the types of motion of the general three-body problem as the time becomes infinite is well known (Chazy, 1922), but the determination of the type of motion for any given initial conditions is difficult. Standish (1971, 1972) has given sufficient conditions for the retention or escape of a member of a three-body system. These conditions were strengthened by Griffith and North (1973), using a similar technique to that of Stándish. Yoshida (1972), with an alternative approach, obtained slightly better conditions for escape, but with a much lengthier derivation. In this paper the results and methods of Griffith and North for the three-body problem are recapitulated, expanded and applied to the general N-body problem. Section 2 deals with escape or retention in the three-body problem, Section 3 with a new containment theorem for the three body problem, while Section 4 extends the results to the general N-body problem. All results apply to motion of the escaping body with respect to the barycentre of the remaining bodies.

2. Three-Body Escape or Retention

The three masses are denoted m_a, m_b, m_c with r the distance of mass m_a from mass m_b, ϱ the distance of mass m_c from the centre of mass of m_a and m_b. The case where m_c passes directly between m_a and m_b is avoided by the condition $r \leqslant \varrho_{ac}, \varrho_{bc}$ where $\varrho_{ac}, \varrho_{bc}$ are respectively the distances between m_a and m_c and between m_b and m_c.

The equation of motion for ϱ is

$$\ddot{\varrho} = g_2^2 (p_\theta^2/\varrho^3 \cos^2\phi + p_\phi^2/\varrho^3) + g_2(\partial F/\partial\varrho), \tag{1}$$

where

$$g_2 = M/m_c(m_a + m_b), \qquad M = m_a + m_b + m_c$$

and

$$F = G\left(\frac{m_a m_b}{r} + \frac{m_a m_c}{\varrho_{ac}} + \frac{m_c m_b}{\varrho_{bc}}\right).$$

The previous proofs of Standish depend upon obtaining upper and lower bounds

Y. Kozai (ed.), The Stability of the Solar System and of Small Stellar Systems, 185–200.

for $\ddot{\varrho}$, multiplying by $\dot{\varrho}$ and integrating with respect to time. This procedure will be followed here, with

$$M_a = m_a/(m_a + m_b) \quad \text{and} \quad M_b = 1 - M_a.$$

ESCAPE THEOREM

If at some time, t_0

(i) $\varrho_0 > r_*$ (the maximum separation of m_a, $m_b = G(m_a m_b + m_b m_c + m_c m_a)/|E|$, where E is the total energy),

(ii) $\dot{\varrho}_0 > 0$, and

(iii) $$\dot{\varrho}_0^2/2 > GM\left[\frac{1}{\varrho_0} + \frac{M_a M_b r_*}{\varrho_0^2}\left\{\frac{M_b}{\varrho_0 - M_b r_*} + \frac{M_a}{\varrho_0 - M_a r_*}\right\}\right],$$

then $\varrho \to \infty$ as $t \to \infty$.

Now

$$\ddot{\varrho} \geqslant g_2 \frac{\partial F}{\partial \varrho} = GM \frac{\partial}{\partial \varrho}\left(\frac{M_a}{\varrho_{ac}} + \frac{M_b}{\varrho_{bc}}\right) =$$

$$= GM \frac{\partial}{\partial \varrho}\left[\frac{1}{\varrho} + \frac{M_a M_b}{\varrho} \sum_{i=2}^{\infty} (M_b^{i-1} + (-M_a)^{i-1}) \left(\frac{r}{\varrho}\right)^i P_i(q)\right],$$

where the $P_i(q)$ are Legendre polynomials,

$$= GM \frac{\partial}{\partial \varrho}\left[\frac{1}{\varrho} + \frac{M_a M_b}{\varrho} \sum_{i=2}^{\infty} \left(\left(\frac{M_b r}{\varrho}\right)^i \frac{P_i(q)}{M_b} + \left(\frac{-M_a r}{\varrho}\right)^i \frac{P_i(q)}{M_a}\right)\right] =$$

$$= GM\left[-\frac{1}{\varrho^2} - \frac{M_a M_b}{\varrho} \sum_{i=2}^{\infty} (i+1)\left(\left(\frac{M_b r}{\varrho}\right)^i \left(\frac{P_i(q)}{M_b}\right) + \left(\frac{-M_a r}{\varrho}\right)^i \frac{P_i(q)}{M_a}\right)\right].$$

Our aim is to establish upper and lower bounds to this expression. The sharpest conditions may be found by determining for what value of q this attains a maximum or minimum, but this appears difficult as differentiation with respect to q yields $\sin q = 0$ or $\cos q$ as the root of a 7th-order equation. Only for equal masses ($m_a = m_b$) does the equation have simpler roots. Cruder estimates of the bounds are available, for as $|P_i(q)| \leqslant 1$,

$$\ddot{\varrho} \geqslant GM\left[-\frac{1}{\varrho^2} - \frac{M_a}{\varrho^2} \sum_{i=2}^{\infty} (i+1)\left(\frac{M_b r}{\varrho}\right)^i - \frac{M_b}{\varrho^2} \sum_{i=2}^{\infty} (i+1)\left(\frac{M_a r}{\varrho}\right)^i\right].$$

It is at this point that my approach diverges from that of Standish, for he used the additional approximation that $|M_b^{i-1} + (-M_a)^{i-1}| \leqslant 1$ and hence obtained a simpler, less precise, expression.

Using r_* as the maximum value of r, we have

$$\ddot{\varrho} \geqslant GM\left[-\frac{1}{\varrho^2} - M_a \sum_{i=2}^{\infty} \frac{(i+1)(M_b r_*)^i}{\varrho^{i+2}} - M_b \sum_{i=2}^{\infty} \frac{(i+1)(M_a r_*)^i}{\varrho^{i+2}}\right].$$

For escape, assume $\dot{\varrho}>0$ in some interval of time (t_0, t_1), where $t_1>t_0$, then

$$\dot{\varrho}\ddot{\varrho}\geqslant GM\left[-\frac{\dot{\varrho}}{\varrho^2}-M_a\sum_{i=2}^{\infty}\frac{(i+1)(M_b r_*)^i}{\varrho^{i+2}}\dot{\varrho}-M_b\sum_{i=2}^{\infty}\frac{(i+1)(M_a r_*)^i}{\varrho^{i+2}}\dot{\varrho}\right].$$

Integrating from t_0 to t_1 gives

$$\tfrac{1}{2}\dot{\varrho}_1^2\geqslant GM\left[\frac{1}{\varrho_1}+\frac{M_a(M_b r_*)^2}{\varrho_1^2(\varrho_1-M_b r_*)}+\frac{M_b(M_a r_*)^2}{\varrho_1^2(\varrho_1-M_a r_*)}\right]+K,$$

where

$$K=\tfrac{1}{2}\dot{\varrho}_0^2-GM\left[\frac{1}{\varrho_0}+\frac{M_a(M_b r_*)^2}{\varrho_0^2(\varrho_0-M_b r_*)}+\frac{M_b(M_a r_*)^2}{\varrho_0^2(\varrho_0-M_a r_*)}\right],$$

and $\dot{\varrho}_1$, ϱ_1 are respectively the values of $\dot{\varrho}$, ϱ at time t_1.

For $K\geqslant 0$, then $\dot{\varrho}_1^2>0$ for all finite values of ϱ_1. Then $\dot{\varrho}$ remains positive for all time and $\frac{1}{2}\dot{\varrho}^2\geqslant K$ for all $t>t_0$ or $\varrho\geqslant\sqrt{(2K)}(t-t_0)+\varrho_0$. So $\varrho\to\infty$ as $t\to\infty$ if $K>0$ and escape occurs. Hence the escape theorem is true.

The value for condition (iii) given by Standish differs from this with the difference decreasing asymptotically as ϱ_0^{-4}. For small values of ϱ, this revised value of K is markedly better than that of Standish.

To compare these conditions in detail, we follow Standish with $\alpha=\varrho_0/r_*$, $G=M=$ $=r_*=1$ to obtain

$$GM\left[\frac{1}{\varrho_0}+\frac{M_a M_b r_*^2}{\varrho_0^2}\left\{\frac{M_b}{\varrho_0-M_b r_*}+\frac{M_a}{\varrho_0-M_a r_*}\right\}\right]=$$
$$=\frac{1}{\alpha}+\frac{M_a M_b}{\alpha^2}\left(\frac{M_b}{\alpha-M_b}+\frac{M_a}{\alpha-M_a}\right),$$

compared to Standish's expression $1/\alpha+M_a M_b/\alpha^2(\alpha-1)$.

The numerical comparison is given in Table I, with the upper entry being the preceding expression, the middle entry that of Standish and the lower that of Tevzadze (1962). It is seen that our condition is superior to the others in all cases. Tevzadze used

$$GM\left[\frac{M_b}{\varrho_0-M_a r_*}-\frac{M_a}{\varrho_0-M_b r_*}\right]=\frac{M_b}{\alpha-M_a}+\frac{M_a}{\alpha-M_b}.$$

If $\varrho<r_*$ the preceding analysis will not work. Expansion in powers of ϱ/r is not effective, as the result hinges on the right hand side of the inequality reversing sign on integration and becoming positive, thus enabling us to assert that $K>0$. Without inverse powers of ϱ to integrate, this reversal is not possible. There are also difficulties in expansions in the intermediate region $r<\varrho<r_*$.

For equal masses,

$$g_2\frac{\partial F}{\partial\varrho}=\frac{GM}{2}\frac{\partial}{\partial\varrho}\left(\frac{1}{\varrho_{ac}}+\frac{1}{\varrho_{bc}}\right),$$

TABLE I

Comparison of conditions (iii)

α	$M_a=0.1$	0.3	0.5
1.1	1.2512	1.2779	1.2534
	1.6529	2.6446	2.9752
	1.4000	1.6250	1.6667
1.3	0.8935	0.9515	0.9541
	0.9467	1.1834	1.2623
	1.0000	1.2000	1.2500
1.5	0.7295	0.7717	0.7778
	0.7467	0.8533	0.8889
	0.8095	0.9583	1.0000
2.0	0.5196	0.5375	0.5417
	0.5225	0.5525	0.5625
	0.5646	0.6425	0.6667
3.0	0.3380	0.3430	0.3444
	0.3383	0.3450	0.3472
	0.3580	0.3897	0.4000
5.0	0.20086	0.2019	0.2022
	0.20090	0.2021	0.2025
	0.20806	0.2187	0.2222
10.0	0.100098	0.10022	0.10026
	0.100100	0.10023	0.10028
	0.101898	0.10442	0.10526

where

$$\varrho_{ac}^2=\varrho^2+\tfrac{1}{2}r+\varrho r\cos q,$$
$$\varrho_{bc}^2=\varrho^2+\tfrac{1}{2}r-\varrho r\cos q.$$

Using $(\partial/\partial q)(g_2(\partial F/\partial\varrho))=0$ we find $\sin q=0$ or

$$\frac{2\varrho^2-(\frac{1}{2}r)^2+\frac{1}{2}\varrho r\cos q}{2\varrho^2-(\frac{1}{2}r)^2-\frac{1}{2}\varrho r\cos q}=\frac{(\varrho^2+(\frac{1}{2}r)^2+\varrho r\cos q)^{5/2}}{(\varrho^2+(\frac{1}{2}r)^2-\varrho r\cos q)^{5/2}}.$$

This last equation gives $\cos q=0$ or

$$\begin{aligned}\frac{\gamma^6\cos^6 q}{4}&+\gamma^4\cos^4 q((2\varrho^2-(\tfrac{1}{2}r)^2)^2+5(2\varrho^2-(\tfrac{1}{2})^2)(\varrho^2+(\tfrac{1}{2}r)^2)+\\&+5(\varrho^2+(\tfrac{1}{2}r)^2)^2)+\gamma^2\cos^2 q(10(\varrho^2+(\tfrac{1}{2}r)^2)^2(2\varrho^2-(\tfrac{1}{2}r))^2+\\&+10(\varrho^2+(\tfrac{1}{2}r)^2)^3(2\varrho^2-(\tfrac{1}{2}r)^2)+(\tfrac{5}{2})(\varrho^2+(\tfrac{1}{2}r)^2)^4)+\\&+(2\varrho^2-(\tfrac{1}{2}r)^2)(\varrho^2+(\tfrac{1}{2}r)^2)^5+(2\varrho^2-(\tfrac{1}{2}r)^2)^2(\varrho^2+(\tfrac{1}{2}r)^2)^4=0,\end{aligned}$$

where $\gamma=\varrho r$.

As $\varrho>r/2\sqrt{2}$ the only solutions are $\cos q=\neq 1$, giving motion along the perpendicular to the lines of centres. Using the extreme values of $\cos q=\pm 1$, we have

$$\ddot{\varrho}\geqslant-\frac{GM(\varrho^2+(\frac{1}{2}r)^2)}{(\varrho^2-(\frac{1}{2}r)^2)^2}\geqslant\frac{GM(\varrho^2+r_*^2)}{(\varrho^2-r_*^2)^2}.$$

On multiplication by $\dot{\varrho}$ and integrating we may compare this new value of K (equal masses) with the value of K from the escape theorem with $M_a = M_b = \frac{1}{2}$, which is

$$\tfrac{1}{2}\dot{\varrho}_0^2 - GM\left(\frac{1}{\varrho_0} + \frac{1}{4}\frac{r_*^2}{\varrho_0^2(\varrho_0 - r_*/2)}\right).$$

Retention Theorem

If the mutual distance between the bodies of mass m_a and m_b is bounded by $r_* \geqslant r \geqslant$ $\geqslant r_* m_a m_b/(m_a m_b + m_b m_c + m_c m_a)$ and if at some time t_0

(i) $\varrho_0 > r_*$,

(ii) $\dot{\varrho}_0 > 0$,

(iii)
$$\tfrac{1}{2}\dot{\varrho}_0^2 < \frac{GM}{\varrho_0} - \frac{GMM_aM_b}{\varrho_0^2}\frac{r_*^2}{\varrho_0 - r_*} - \frac{2Gg_2m_c}{r_*}\left\{\frac{m_a}{M_b}\ln(1 + M_b r_*/\varrho_0) + \right.$$
$$\left. + \frac{m_b}{M_a}\ln(1 + M_a r_*/\varrho_0)\right\},$$

then m_c is retained by the system, at least until $\dot{\varrho}$ becomes negative and ϱ becomes less than r_*.

The differential equation for ϱ (Equation (1)) may be written

$$\ddot{\varrho} = |\boldsymbol{\varrho} \wedge \dot{\boldsymbol{\varrho}}|^2/\varrho^3 + g_2(\partial F/\partial \varrho).$$

Standish deduced a time independent upper bound for $|\boldsymbol{\varrho} \wedge \dot{\boldsymbol{\varrho}}|^2$ before multiplication by $\dot{\varrho}$ and integration. One way of refining his conditions is to use

$$\dot{\boldsymbol{\varrho}}^2 = \{(\boldsymbol{\varrho}\dot{\boldsymbol{\varrho}})^2 + (\boldsymbol{\varrho} \wedge \dot{\boldsymbol{\varrho}})^2\}/\varrho^2 = \dot{\varrho}^2 + (\boldsymbol{\varrho} \wedge \dot{\boldsymbol{\varrho}})^2/\varrho^2,$$

so that

$$\frac{1}{\varrho^3}|\boldsymbol{\varrho} \wedge \dot{\boldsymbol{\varrho}}|^2 = \frac{\dot{\boldsymbol{\varrho}}^2 - \dot{\varrho}^2}{\varrho} \leqslant \frac{(E+F)\,2g_2 - \dot{\varrho}^2}{\varrho},$$

where E is the total energy of the system and F the potential.

The difficulty of bounding this expression is the presence of the $m_a m_b/r$ term in F, which may become large if the minimum interparticle distance is small. A set of conditions not yet fully utilized are those contained in the energy integral. From

$$E = \frac{1}{2g_1}\dot{\mathbf{r}}^2 + \frac{1}{2g_2}\dot{\boldsymbol{\varrho}}^2 - F,$$

we have $E + F \geqslant 0$ (here $g_1 = (m_a + m_b)/m_a m_b$), i.e.

$$E \geqslant -F = -G\left(\frac{m_a m_b}{r} + \frac{m_a m_c}{\varrho_{ac}} + \frac{m_b m_c}{\varrho_{bc}}\right) \geqslant -G(m_a m_b + m_a m_c + m_b m_c)/r.$$

If E is negative,

$$r \leqslant r_* = G(m_a m_b + m_a m_c + m_b m_c)/|E|,$$

which was noted by Standish. We have from

$$E \geqslant -G\left(\frac{m_a m_b}{r} + \frac{m_a m_c}{\varrho_{ac}} + \frac{m_b m_c}{\varrho_{bc}}\right)$$

the condition that, if ϱ does become arbitrarily large, $E \geqslant G m_a m_b / r$ or $r \leqslant r_* m_a m_b / (m_a m_b + m_a m_c + m_b m_c)$.

If ϱ is allowed to become arbitrarily large, we require

$$\frac{m_a m_b + m_a m_c + m_b m_c}{r_*} \leqslant \frac{m_a m_b}{r},$$

i.e.

$$r \leqslant \frac{r_* m_a m_b}{m_a m_b + m_a m_c + m_b m_c}.$$

Let us assume that $r_* \geqslant r \geqslant r_* m_a m_b / (m_a m_b + m_a m_c + m_b m_c)$. Thus, to keep the third body within the system, we do not allow the other two bodies to become so close as to allow their lost potential energy to manifest itself in the escape of the third body. With this lower bound on r we return to

$$\ddot{\varrho} = |\boldsymbol{\varrho} \wedge \dot{\boldsymbol{\varrho}}|^2/\varrho^3 + g_2(\partial F/\partial \varrho), \quad \text{and} \quad |\boldsymbol{\varrho} \wedge \dot{\boldsymbol{\varrho}}|^2/\varrho^3 \leqslant 2g_2(E+F)/\varrho,$$

to obtain

$$\ddot{\varrho} \leqslant 2g_2\left\{E + G\left(\frac{(m_a m_b + m_a m_c + m_b m_c)}{r_*} + \frac{m_a m_c}{\varrho - M_b r_*} + \frac{m_b m_c}{\varrho - M_a r_*}\right)\right\}/\varrho + \\ + GM\left[-\frac{1}{\varrho^2} + M_a M_b \sum_{i=2}^{\infty} (i+1)\frac{r_*^i}{\varrho^{i+2}}\right].$$

Assume that $\dot{\varrho} > 0$ for all time $t > t_0$. Then, multiplying the expression for $\ddot{\varrho}$ by $\dot{\varrho}$ and integrating from t_0 to t, gives

$$\tfrac{1}{2}(\dot{\varrho}_1^2 - \dot{\varrho}_0^2) \leqslant 2g_2\left\{E + G\frac{(m_a m_b + m_a m_c + m_b m_c)}{r_*}\right\}\ln\frac{\varrho_1}{\varrho_0} - \\ - \frac{2g_2 m_a m_c G}{M_b r_*}\left[\ln\left(\frac{\varrho - M_b r_*}{\varrho}\right)\right]_0^1 - \\ - \frac{2g_2 m_b m_c G}{M_a r_*}\left[\ln\left(\frac{\varrho - M_a r_*}{\varrho}\right)\right]_0^1 + \left[\frac{GM}{\varrho} - \sum_{i=2}^{\infty} GMM_a \frac{M_b r_*^i}{\varrho^{i+1}}\right]_0^1.$$

Now $E + G(m_a m_b + m_c + m_b m_c)/r_* = 0$ giving

$$\tfrac{1}{2}\dot{\varrho}_1^2 \leqslant \left[\tfrac{1}{2}\dot{\varrho}_0^2 + \frac{2g_2 m_a m_c G}{M_b r_*}\ln\left(\frac{\varrho_0 - M_b r_*}{\varrho_0}\right) + \frac{2g_2 m_b m_c G}{M_a r_*}\ln\left(\frac{\varrho_0 - M_a r_*}{\varrho_0}\right) - \\ - \frac{GM}{\varrho_0} + \frac{GMM_a M_b r_*^2}{\varrho_0^2(\varrho_0 - r_*)}\right] - \frac{2g_2 m_a m_c G}{M_b r_*}\ln\left(\frac{\varrho_1 - M_b r_*}{\varrho_1}\right) -$$

$$-\frac{2g_2 m_b m_c G}{M_a r_*}\ln\left(\frac{\varrho_1 - M_a r_*}{\varrho_1}\right)+\frac{GM}{\varrho_1}-\frac{GMM_aM_b}{\varrho_1^2}\frac{r_*^2}{(\varrho_1 - r_*)},$$

so that

$$\tfrac{1}{2}\dot{\varrho}_1^2 \leqslant K+\frac{GM}{\varrho_1}-\frac{GMM_aM_b}{\varrho_1^2}\frac{r_*^2}{\varrho_1 - r_*}-\frac{2Gm_c g_2}{r_*}\times$$

$$\times\left\{\frac{m_a}{M_b}\ln(1-M_b r_*/\varrho_1)+\frac{m_b}{M_a}\ln(1-M_a r_*/\varrho_1)\right\},$$

where

$$K=\tfrac{1}{2}\dot{\varrho}_0^2-\frac{GM}{\varrho_0}+\frac{GMM_aM_b}{\varrho_0^2}\frac{r_*^2}{\varrho_0 - r_*}+$$

$$+\frac{2g_2 m_c G}{r_*}\left\{\frac{m_a}{M_b}\ln\left(\frac{\varrho_0+M_b r_*}{\varrho_0}\right)+\frac{m_b}{M_a}\ln\left(\frac{\varrho_0+M_a r_*}{\varrho_0}\right)\right\}.$$

Now, if $\dot{\varrho}>0$ for all $t>t_0$, ϱ_1 can be made as large as desired by the proper choice of t, leading to the expression

$$\dot{\varrho}_1^2<0 \quad \text{if} \quad K<0.$$

Thus if $K<0$ it must *not* be the case that $\dot{\varrho}>0$ for all time $t>t_0$.

In the expression for K, the factor $m_c g_2$ is $M/(m_a+m_b)$, so that the undesirable nature of the expression obtained by Standish (which was

$$\frac{\dot{\varrho}_0^2}{2}<\frac{GM}{\varrho_0}\left[1-\frac{M_aM_b r_*^2}{\varrho_0(\varrho_0 - r_*)}\right]-\frac{Q^2}{\varrho_0^2},$$

where

$$Q=\frac{M|\mathbf{L}|}{m_c(m_a+m_b)}+\left\{\frac{2GM^2M_aM_b r_*}{m_c}\times\left[\frac{M_aM_b}{m_c}(m_a+m_b)+\frac{r_*}{\varrho_0}+\frac{M_aM_b r_*}{\varrho_0^2(\varrho_0 - r_*)}\right]\right\}^{1/2},$$

and $\mathbf{L}$ is the total angular momentum) with its inverse dependence on m_c is not present. However, we have replaced this by the requirement that the two remaining bodies are always sufficiently separated, a requirement that may require numerical integration to check.

Note that

$$r_* \geqslant r \geqslant r_* m_a m_b/(m_a m_b + m_a m_c + m_b m_c)$$

gives less variation in the relative positions m_a, m_b if m_c is small than if m_c is large. A large mass needs more energy to escape from the system, and hence the variation in the distance between m_a, m_b can be larger without causing escape.

We are only assured of retention as long as $\dot{\varrho}>0$, and $\dot{\varrho}<0$ together with close passage of m_c to one of the other bodies renders the retention theorem invalid.

For motion of the third mass towards the other two, $\dot{\varrho}_0$ is negative. We expect $\dot{\varrho}$

to increase until, on passage between or close to one or both of the other two masses, $\dot{\varrho}$ changes sign and becomes positive. This change of sign must be accompanied by an instant when $\dot{\varrho}$ is zero. Can we obtain any conditions in the region of close approach? There are difficulties, as the case of collisions presents obvious singularities.

We have from the energy equation

$$E+F\geqslant 0,$$

or

$$E\geqslant -G\left(\frac{m_a m_b}{r}+\frac{m_a m_c}{\varrho_{ac}}+\frac{m_b m_c}{\varrho_{bc}}\right).$$

If we wish to avoid ϱ_{ac}, ϱ_{bc} becoming infinite (i.e. escape), then we take $E<-Gm_a m_b/r_*$ and have

$$Gm_a m_b\left(\frac{1}{r}-\frac{1}{r_*}\right)>-G\left(\frac{m_a m_b}{\varrho_{ac}}+\frac{m_b m_c}{\varrho_{bc}}\right),$$

or

$$\frac{m_a m_c}{\varrho_{ac}}+\frac{m_b m_c}{\varrho_{bc}}>0.$$

For any configuration we know the total energy E, and can assert that if $E<-Gm_a m_b/r_*$, then the system is bound, as the mass m_c cannot escape. This condition is, of course, less stringent than the retention theorem, but the retention theorem needs to be tested for each $\dot{\varrho}>0$, $\varrho>r_*$ occurrence.

Combining the conditions for escape or retention, we find the region of indeterminancy given by

$$\frac{-M_a M_b}{\varrho_0-r_*}-\frac{2\varrho_0^2}{r_*^3}\left\{\frac{M_a}{M_b}\ln\left(1+\frac{M_b r_*}{\varrho_0}\right)+\frac{M_b}{M_a}\ln\left(1+\frac{M_a r_*}{\varrho_0}\right)\right\}\leqslant$$
$$\leqslant\left(\frac{\dot{\varrho}_0^2}{2}-\frac{GM}{\varrho_0}\right)\frac{\varrho_0^2}{GMr_*^2}\leqslant\left(\frac{\varrho_0-2M_a M_b r_*}{(\varrho_0-M_a r_*)(\varrho_0-M_b r_*)}\right)M_a M_b.$$

3. Containment Theorem

The procedure used for examination of the possibility of retention may be used to derive a containment theorem. Let us examine the condition that the mass m_c does not move further than a distance R from the barycentre of m_a, m_b. We require $\dot{\varrho}_1^2<0$ for some $\varrho<R$ and can use Equation (2) of Section 2 to give

$$\tfrac{1}{2}(\dot{\varrho}_1^2-\dot{\varrho}_0^2)\leqslant K+\frac{GM}{R}-\frac{GMM_a M_b}{R^2}\frac{r_*^2}{R-r_*}-\frac{2Gm_c g_2}{r_*}\left\{\frac{m_a}{M_b}\ln(1+M_b r_*/R)+\right.$$
$$\left.+\frac{m_b}{M_a}\ln(1+M_a r_*/R)\right\}.$$

To ensure return within a sphere of radius R we need

$$\tfrac{1}{2}\dot{\varrho}_0^2+K+\frac{GM}{R}-\frac{GM_aM_b}{R^2}\frac{r_*^2}{R-r_*}-\frac{2GM_cg_2}{r_*}\left\{\frac{m_a}{M_b}\ln(1+M_br_*/R)+\right.$$
$$\left.+\frac{m_b}{M_a}\ln(1+M_ar_*/R)\right\}\leqslant 0,$$

or

$$\tfrac{1}{2}\dot{\varrho}_0^2+\frac{GM}{R\varrho_0}(\varrho_0-R)-\frac{GM_aM_br_*^2}{R^2\varrho_0^2(R-r_*)(\varrho_0r_*)}(\varrho_0^3-R-r_*(\varrho_0-R))-$$
$$-\frac{2GM_cg_2}{r_*}\left\{\frac{m_a}{M_b}\ln\left(\frac{R+M_br_*}{\varrho_0+M_br_*}\right)+\frac{m_b}{M_a}\ln\left(\frac{R+M_ar_*}{\varrho_0+M_ar_*}\right)+\right.$$
$$\left.+\left(\frac{m_a^2+m_b^2}{m_am_b}\right)(m_a+m_b)\ln\left(\frac{\varrho_0}{R}\right)\right\}\leqslant 0.$$

CONTAINMENT THEOREM

If the mutual distance between the bodies of mass m_a and m_b is bounded by $G(m_am_b+m_bm_c+m_cm_a)/|E|=r_*\geqslant r\geqslant m_am_br_*/(m_am_b+m_bm_c+m_cm_a)$ and if at some time t_0

(i) $\varrho_0>r_*$,

(ii) $\dot{\varrho}_0>0$,

(iii)

$$\tfrac{1}{2}\dot{\varrho}_0^2\leqslant\frac{GM}{R\varrho_0}(R-\varrho_0)-\frac{GM_aM_br_*^2(R-\varrho_0)}{R^2\varrho_0^2(R-r_*)(\varrho_0-r_*)}(R^2+\varrho_0^2+\varrho_0R-r_*)+$$
$$+\frac{2GM_cg_2}{r_*}\left\{\frac{m_a}{M_b}\ln\left(\frac{R+M_br_*}{\varrho_0+M_br_*}\right)+\frac{m_b}{M_a}\ln\left(\frac{R+M_ar_*}{\varrho_0+M_ar_*}\right)+\right.$$
$$\left.+\frac{(m_a^2+m_b^2)}{m_am_b}(m_a+m_b)\ln\left(\frac{\varrho_0}{R}\right)\right\}.$$

then the mass m_c does not move outside a sphere, centred on the barycentre of m_a, m_b of radius R. Again, this theorem only applies to this particular portion of motion, and m_c may escape from the sphere after another passage near the centre.

4. The *n*-Body Problem

Take $n+1$ bodies, with the possibility of the $(n+1)$th body being captured by or escaping from the n remaining bodies being of interest.

The Newtonian equations of motion relative to a 'Newtonian origin' N are

$$\ddot{\mathbf{r}}_{N,i}=-G\sum_{j=1}^{n+1}m_j\frac{(\mathbf{r}_{N,i}-\mathbf{r}_{N,j})}{|\mathbf{r}_{N,i}-\mathbf{r}_{N,j}|^3},$$

while the barycentre of the n particles has motion given by

$$(M-m_{n+1})\,\ddot{\mathbf{r}}_{N,b}=\sum_{j=1}^{n} m_j\ddot{\mathbf{r}}_{N,j},$$

where $M=\sum_{j=1}^{n+1} m_j$.

The equation of motion of the $(n+1)$th particle with respect to the barycentre is

$$\ddot{\varrho}=\ddot{\mathbf{r}}_{N,n+1}-\ddot{\mathbf{r}}_{N,b}$$
$$=-\sum_{j=1}^{n}\frac{m_j}{M-m_{n+1}}\ddot{\mathbf{r}}_{N,j}.$$

We know that the centre of gravity c of the $(n+1)$ particles has

$$\ddot{\mathbf{r}}_{NC}=0.$$

So, as

$$\mathbf{r}_{NC}=\sum_{j=1}^{n+1} m_j\mathbf{r}_{N,j},$$
$$0=\ddot{\mathbf{r}}_{NC}=\sum_{j=1}^{n+1} m_j\ddot{\mathbf{r}}_{N,j},$$

so that

$$\sum_{j=1}^{n} m_j\ddot{\mathbf{r}}_{N,j}=-m_{n+1}\ddot{\mathbf{r}}_{N,n+1},$$

and

$$\ddot{\boldsymbol{\varrho}}=\frac{m_{n+1}}{M-m_{n+1}}\ddot{\mathbf{r}}_{N,n+1}+\ddot{\mathbf{r}}_{N,n+1}$$
$$=\frac{M}{M-m_{n+1}}\ddot{\mathbf{r}}_{N,n+1},$$

giving

$$\ddot{\varrho}=-\frac{GM}{M-m_{n+1}}\sum_{j=1}^{n}\frac{m_j\mathbf{r}_{j,n+1}}{|\mathbf{r}_{j,n+1}|^3},$$

where $\boldsymbol{\varrho}$ is the radius vector of the $(n+1)$th body with respect to the barycentre of the remaining n bodies, $\mathbf{r}_{j,n+1}$ the vector between the jth body and the $(n+1)$th body, $M=\sum_{j=1}^{n+1} m_j$. This equation allows for the recoil of the cluster in order to keep the barycentre fixed.

Clearly

$$\ddot{\varrho}\geqslant\frac{-GM}{M-m_{n+1}}\sum_{j=1}^{n}\frac{m_j}{r_{j,n+1}^2}\geqslant$$
$$\geqslant\frac{-GM}{M-m_{n+1}}\sum_{j=1}^{n}\frac{m_j}{(\varrho-r_*)^2}=-\frac{GM}{(\varrho-r_*)^2},$$

where r_* is the maximum distance of any of the n bodies from the barycentre.

We follow a similar procedure to that in Section 2 for the escape theorem to find n-body escape theorem.

If at some time, t_0, (i) $\varrho_0 > r_*$, (ii) $\dot{\varrho}_0 > 0$, and (iii) $\frac{1}{2}\dot{\varrho}_0^2 > GM/(\varrho_0 - r_*)$, then $\varrho \to \infty$ as $t \to \infty$ where M is the total mass of the system. Of course, for a spherically symmetric cluster we expect $\frac{1}{2}\dot{\varrho}_0^2 > GM/\varrho_0$, so this result is rather weak in a physical sense, but is a rigorous proof for any distribution of matter and velocities.

Using the α notation, $GM/(\varrho_0 - r_*)$ becomes $1/(\alpha - 1)$, which is clearly a much less stringent condition than that for the three-body problem, unless α is of the order of 10. This loss is caused by the replacement of $(\mathbf{r}_{j,n+1} \cdot \hat{\varrho})$ by $r_{j,n+1}$, so that the angular position of the escaping mass has not been utilized and by the approximation used for $r_{j,n+1}^2$ which, in the three-body case, was expanded in terms of Legendre polynomials. A tighter expression for $r_{j,n+1}^2$ would strengthen this result, which resembles placing the entire mass of the cluster, the minimum distance away $(\varrho - r_*)$.

However, if we know the form of the distribution of the n particles, we can improve on the estimate of $\ddot{\varrho}$. For example, given n bodies constrained to move along a fixed straight line, then for motion of the $(n+1)$th body along this line we cannot readily improve on $\varrho \geqslant -GM/(\varrho - r_*)^2$ and hence readily improve on the original condition (iii), but for motion of the $(n+1)$th body perpendicular to the line, if θ_j is given by $\tan\theta_j = r_{b,j}/\varrho$ then

$$\ddot{\varrho} = -\frac{GM}{M - m_{n+1}} \sum_{j=1}^{n} \frac{m_j \mathbf{r}_{j,n+1}}{|\mathbf{r}_{j,n+1}|^3},$$

and

$$\begin{aligned}\ddot{\varrho} &\geqslant \frac{GM}{M - m_{n+1}} \sum_{j=1}^{n} \frac{m_j \cos\theta_j}{(r_{j,n+1})^2} = \\ &= -\frac{GM}{M - m_{n+1}} \sum_{j=1}^{n} M_j \frac{\varrho}{(\varrho^2 + r_{b,j}^2)^{3/2}} \geqslant \\ &\geqslant -GM \frac{\varrho}{(\varrho^2 + r_*^2)^{3/2}}.\end{aligned}$$

In this case condition (iii) in the escape theorem becomes

$$\tfrac{1}{2}\dot{\varrho}_0^2 > GM/(\varrho_0^2 + r_*^2)^{1/2},$$

which approximates the spherical case if r_* is small, but which for r_* large gives less stringent conditions. If the N bodies remain in three groups of mass M_1, M_2 and M_1 respectively, and can be approximated by three point masses, with M_2 at the barycentre, then

$$\ddot{\varrho} = \frac{-GM}{M - m_{n+1}} \sum_{j=1}^{n} \frac{m_j \mathbf{r}_{j,n+1}}{|\mathbf{r}_{j,n+1}|^3},$$

yields

$$\ddot{\varrho} \geqslant \frac{-GM}{M - m_{n+1}} \left\{ \frac{2M_1 \varrho}{(\varrho^2 + r_*^2)^{3/2}} + \frac{M_2}{\varrho^2} \right\},$$

and condition (iii) is, for motion perpendicular to the line of remaining masses,

$$\tfrac{1}{2}\dot{\varrho}_0^2 > \frac{GM}{M-m_{n+1}}\left\{\frac{2M_1}{(\varrho_0^2+r_*^2)^{1/2}}+\frac{M_2}{\varrho}\right\},$$

which demonstrates how an increase in separation r_* decreases the velocity required for escape.

For the remaining masses constrained to lie in a plane, we would expect escape from the plane to be easier for motion of circularly the $(n+1)$th body out of the plane. With a symmetric distribution in the plane, motion in the plane gives $\frac{1}{2}\dot{\varrho}_0^2 > GM/(\varrho_0 - r_*)$, while motion along the line of symmetry perpendicular to the plane gives

$$\ddot{\varrho} \geqslant \frac{-GM}{M-m_{n+1}}\sum_{j=1}^{n}\frac{m_j\varrho}{(\varrho^2+r_{b,j}^2)^{3/2}},$$

with $\frac{1}{2}\dot{\varrho}_0^2 > GM/(\varrho_0^2 + r_*^2)^{3/2}$.

With mass M_2 at the barycentre and M_1 distributed in a ring of radius r_*, motion perpendicular to the plane gives

$$\ddot{\varrho} \geqslant \frac{-GM}{M-m_{n+1}}\left\{\frac{M_2}{\varrho^2}+\frac{M_1\varrho}{(\varrho^2+r_*^2)^{3/2}}\right\},$$

with

$$\tfrac{1}{2}\dot{\varrho}_0^2 \geqslant \frac{GM}{M-m_{n+1}}\left\{\frac{M_2}{\varrho_0}+\frac{M}{(\varrho_0^2+r_*^2)^{1/2}}\right\}.$$

For containment we need to extend the energy argument to the n-body problem, but unfortunately appear to require additional assumptions.

Clearly

$$E \geqslant -\tfrac{1}{2}\sum_{i=1}^{n+1}\sum_{\substack{j=1\\ j\neq i}}^{n+1}\frac{m_i m_j}{r_{ij}}.$$

Let the maximum and minimum separation of the n particles be r_*, s_* respectively

$$E \geqslant -\tfrac{1}{2}\sum_{i=1}^{n}\sum_{j=1}^{n}\frac{m_i m_j}{s_*} - m_{n+1}\sum_{i=1}^{n}\frac{m_i}{\varrho - r_*}.$$

If ϱ is allowed to become large, $E \geqslant \frac{1}{2}\sum_{i=1}^{n}\sum_{j=1}^{n} m_i m_j/s_*$ and, for E negative and finite

$$0 < s_* \leqslant \frac{\sum_{i=1}^{n}\sum_{j=1}^{n} m_i m_j}{2|E|}.$$

If

$$s_* \geqslant \sum_{i=1}^{n}\sum_{j=1}^{n}\frac{m_i m_j}{|E|},$$

ϱ cannot become infinite. This lower bound on the mutual distance between the n bodies ensures that the cluster remains bound. If there is an upper bound r_* to the mutual distances, we apparently do not have the restriction, found in the three-body case, that r_* necessarily exists for E negative. Escape of more than one body would not be unexpected, however, with these bounds on the distances between the n bodies, we return to the n-body form of Equation (1)

$$\ddot{\varrho} = |\boldsymbol{\varrho} \wedge \dot{\boldsymbol{\varrho}}|^2 \varrho^3 + g_2 \, \partial F/\partial \varrho,$$

where $g_2 = M/m_{n+1}(M - m_{n+1})$ and $M = \sum_{i=1}^{n} m_i$ to find

$$\ddot{\varrho} \leqslant \frac{2g_2(E+F)}{\varrho} + g_2 \frac{\partial F}{\partial \varrho} \leqslant$$

$$\leqslant 2g_2 \left\{ E + \tfrac{1}{2} G \sum_{i=1}^{n} \sum_{j=1}^{n} \frac{m_i m_j}{s_*} + G \sum_{i=1}^{n} \frac{m_{n+1} m_i}{\varrho - r_*} \right\} / \varrho +$$

$$+ \frac{GM}{M - m_{n+1}} \left[\sum_{j=1}^{n} \frac{m_j (\mathbf{r}_{j,n+1} \cdot \hat{\varrho})}{r_{j,n+1}^3} \right] \leqslant$$

$$\leqslant \frac{2g_2}{\varrho} \left\{ E + \frac{G}{2} \frac{1}{s_*} \left(\sum_{i=1}^{n+1} \sum_{j=1}^{n+1} m_i m_j \right) + \sum_{i=1}^{n} \frac{m_{n+1} m_i}{\varrho - r_*} \right\} +$$

$$+ \frac{-GM}{M - m_{n+1}} \left[\sum_{j=1}^{n} \frac{m_j \mathbf{r}_{j,n+1} \cdot \hat{\varrho}}{r_{j,n+1}^3} \right] \leqslant$$

$$\leqslant \frac{2g_2 G}{\varrho} \left\{ \sum_{i=1}^{n} \frac{m_{n+1} m_i}{\varrho - r_*} \right\} - \frac{GM}{M - m_{n+1}} \sum_{j=1}^{n} \frac{m_j}{r_{j,n+1}^2} \leqslant$$

$$\leqslant \frac{2g_2 G}{\varrho} \times \frac{m_{n+1}(M - m_{n+1})}{\varrho - r_*} + \frac{GM}{M - m_{n+1}} \sum_{j=1}^{n} \frac{m_j}{(\varrho - r_*)^2} =$$

$$= \frac{2g_2 G}{\varrho(\varrho - r_*)} m_{n+1}(M - m_{n+1}) - \frac{GM}{(\varrho - r_*)^2}.$$

If $\dot{\varrho} > 0$

$$\dot{\varrho}\ddot{\varrho} \leqslant \frac{2g_2 G m_{n+1}(M - m_{n+1})}{\varrho(\varrho - r_*)} - \frac{GM\dot{\varrho}}{(\varrho - r_*)},$$

$$\tfrac{1}{2}\dot{\varrho}^2 - \tfrac{1}{2}\dot{\varrho}_0^2 \leqslant \frac{-2g_2 G m_{n+1}(M - m_{n+1})}{r_*} \ln \frac{\varrho}{\varrho - r_*} + \frac{GM}{(\varrho - r_*)} +$$

$$+ \frac{2g_2 G m_{n+1}(M - m_{n+1})}{r_*} \ln \frac{\varrho_0}{\varrho_0 - r_*} - \frac{GM}{\varrho_0 - r_*}.$$

Here $\dot{\varrho}^2$ is forced to become zero if

$$\frac{2G g_2 m_{n+1}(M - m_{n+1})}{r_*} \ln\left(\frac{\varrho_0}{\varrho_0 - r_*} \right) - \frac{GM}{\varrho_0 - r_*} + \tfrac{1}{2}\dot{\varrho}_0^2 < 0,$$

or

$$\tfrac{1}{2}\dot{\varrho}_0^2<\frac{GM}{\varrho_0-r_*}+\frac{2GM}{r_*}\ln(1-r_*/\varrho_0).$$

We obtain: n-body retention theorem.

If the mutual distances between the n bodies are bounded by

$$r_*>r>s_*=\sum_{i=1}^{n}\sum_{j=1}^{n}\frac{m_i m_j}{|E|},$$

and if at some time t_0 (i) $\varrho_0>r_*$, (ii) $\dot{\varrho}_0>0$, (iii)

$$\tfrac{1}{2}\dot{\varrho}_0^2<\frac{GM}{\varrho_0-r_*}+\frac{2GM}{r_*}\ln(1-r_*/\varrho_0),$$

then the body of mass m_{n+1} does not escape from the system on this particular passage. Subsequent motion after $\varrho<r_*$, following $\dot{\varrho}<0$, may allow escape, and this theorem needs to be re-applied to every $\varrho>r_*$, $\dot{\varrho}>0$ situation. We have not shown that the rest of the cluster does not 'blow up', but include this restriction in the conditions of the theorem.

If the total energy is negative,

$$\sum_{i=1}^{n+1}\tfrac{1}{2}m_i\mathbf{v}_i^2-\tfrac{1}{2}\sum_{\substack{j=1\\ j\neq i}}^{n+1}\frac{M_i M_j}{r_{ij}}<0,$$

and the cluster cannot totally disintegrate. At least one of the distances r_{ij} must be finite.

As for the n-body escape theorem, the conditions of the retention theorem can be improved for certain specific mass distributions.

The retention theorem technique may be applied to a capture situation.

Given (i) $\varrho_0>r_*$, (ii) $\dot{\varrho}_0<0$, (iii)

$$\tfrac{1}{2}\dot{\varrho}_0^2<\frac{GM}{\varrho_0-r_*}+\frac{2GM}{r_*}\ln(1-r_*/\varrho_0),$$

then the body of mass m_{n+1} will either be permanently captured (if $\varrho>r_*$ for all time) or will pass closer to the barycentre of the n particles than the distance r_*.

With G the centre of gravity of the entire system and B the barycentre of the first n particles we have, from the angular momentum integral

$$\sum_{i=1}^{n+1} m_i\mathbf{r}_{Ni}\wedge\dot{\mathbf{r}}_{Ni}=\text{constant}.$$

Using

$$\sum_{i=1}^{n+1} m_i\mathbf{r}_{Gi}=0,\qquad m_{n+1}=m,$$

$$\sum_{i=1}^{n+1} m_i \mathbf{r}_{NG} = \mathbf{a}t + \mathbf{b}, \quad \mathbf{r}_{B,n+1} = \varrho,$$

$$\sum_{i=1}^{n} m_i \mathbf{r}_{Bi} = 0, \qquad M = \sum_{i=1}^{n} M_i,$$

$$\mathbf{r}_{Bi} = \frac{-\sum_{i=1}^{n} m_i \mathbf{r}_{Ni}}{m},$$

we find

$$\sum_{i=1}^{n+1} m_i \mathbf{r}_{Ni} \wedge \dot{\mathbf{r}}_{Ni} = \sum_{i=1}^{n+1} m_i \mathbf{r}_{Gi} \wedge \dot{\mathbf{r}}_{Gi} =$$
$$= \sum_{i=1}^{n+1} m_i (\mathbf{r}_{GB} + \mathbf{r}_{Bi}) \wedge (\dot{\mathbf{r}}_{GB} + \dot{\mathbf{r}}_{Bi}).$$

Now

$$\mathbf{r}_{Gi} = \mathbf{r}_{GB} + \mathbf{r}_{Bi},$$

$$0 = \sum_{i=1}^{n+1} m_i \mathbf{r}_{Gi} = \sum_{i=1}^{n+1} m_i \mathbf{r}_{GB} + \sum_{i=1}^{n+1} m_i \mathbf{r}_{Bi} =$$
$$= (M+m)\, \mathbf{r}_{GB} + \sum_{i=1}^{n} m_i \mathbf{r}_{Bi} + m\boldsymbol{\varrho} =$$
$$= (M+m)\, \mathbf{r}_{GB} + m\boldsymbol{\varrho},$$

so the angular momentum is

$$\sum_{i=1}^{n+1} m_i \mathbf{r}_{GB} \wedge \dot{\mathbf{r}}_{GB} + m\boldsymbol{\varrho} \wedge \dot{\mathbf{r}}_{GB} + \mathbf{r}_{GB} \wedge m\dot{\boldsymbol{\varrho}} + m\boldsymbol{\varrho} \wedge m\dot{\boldsymbol{\varrho}} + \sum_{i=1}^{n} m_i \mathbf{r}_{Bi} \wedge \dot{\mathbf{r}}_{Bi} =$$
$$= -m\boldsymbol{\varrho} \wedge \dot{\mathbf{r}}_{GB} + m\boldsymbol{\varrho}\dot{\mathbf{r}}_{GB} - \frac{m\boldsymbol{\varrho}}{M+m} \wedge m\dot{\varrho} + m\boldsymbol{\varrho} \wedge \dot{\boldsymbol{\varrho}} + \sum_{i=1}^{n} m_i \mathbf{r}_{Bi} \wedge \dot{\mathbf{r}}_{Bi} =$$
$$= \frac{mM}{M+m} \boldsymbol{\varrho} \wedge \dot{\boldsymbol{\varrho}} + \sum_{i=1}^{n} m_i \mathbf{r}_{Bi} \wedge \dot{\mathbf{r}}_{Bi}.$$

In polar coordinates

$$\varrho^2 \dot{\theta} = -\frac{(M+m)}{mM} \left[\mathbf{a} \cdot \hat{\theta} - \sum_{i=1}^{n} \hat{\theta} \cdot m_i \mathbf{r}_{Bi} \wedge \dot{\mathbf{r}}_{Bi} \right],$$

$$\varrho^2 \cos\theta \dot{\phi} = \frac{(M+m)}{mM} \left[a \cdot \hat{\phi} - \sum_{i=1}^{n} \hat{\phi} \cdot m_i \mathbf{r}_{Bi} \wedge \dot{\mathbf{r}}_{Bi} \right].$$

For $\varrho \to \infty$, $\dot{\theta}$ must $\to 0$ and $\theta \to \frac{1}{2}\pi$ or $\dot{\phi} \to 0$.

In other words the escaping particle must eventually be moving away from the system. Note that the origin is moving – as it is the barycentre of the n particles. Reversing time, we must initially 'fire' particles towards system (from infinity!).

The region of indeterminancy of the n-body problem is given by

$$\frac{GM}{\varrho_0 - r_*} > \frac{GM}{\varrho_0 - r_*} + \frac{2GM}{r_*} \ln(1 - r_*/\varrho_0).$$

The use of velocity of escape of the form $\frac{1}{2}v_\infty^2 = GM/r_*$ is criticized by Kurth (1957) who states that

It is ... usually assumed that a star with this or with a higher energy will actually leave the system.... It is doubtful whether this procedure is reliable. The motion of the star in the course of time also depends, for example, on its initial direction and, to a large extent, on the motions of the remaining stars.

In this paper I derive rigorous conditions for escape, which also show that the usual assumption of a velocity of escape is valid, provided that the remaining cluster is bounded in space for all time.

It is also known (Jacob's criterion of stability) that a gravitating system is unstable if its total energy E is positive. Kurth (p. 65) states

up to the present no satisfactory criterion has been found, for systems with negative total energy, to decide between the two possibilities of periodicity and disintegration. It appears to be one of the most important unsolved problems in the mechanics of stellar system to discover such a criterion ....

Kurth also states that (p. 61) if the mass centre of the system is taken as origin, then $\sum_{i=1}^{n} M_i r_i = 0$ shows that if one particle escapes to infinity, at least another and most probably two will escape to infinity. Hence if a system disintegrates, at least three bodies will, as a rule, escape to infinity. This conclusion is not valid if, as one particle escapes from the cluster, the rest of the cluster en masse 'escapes to infinity', but remains bound together. Such a motion will satisfy the preceding equation, but will from the point of view of an external observer, present the picture of a single particle escaping from a moving system.

Acknowledgements

I thank Drs Myles Standish and Aarseth for their comments, together with those of two anonymous referees. This research was aided by grant NRC A6529 (Canada).

References

Chazy, J.: 1922, *Ann. Sci. École Norm. Sup.* **39**.
Griffith, J. S. and North, R. D.: 1974, *Celes. Mech.* **8**, 473.
Kurth, R.: 1957, *Introduction to the Mechanics of Stellar Systems*, Pergamon, p. 92.
Standish, E. M.: 1971, *Celes. Mech.* **4**, 44.
Standish, E. M.: 1972, *Celes. Mech.* **6**, 352.
Tevzadze, G. A.: 1962, *Izv. Akad. Nauk Armyan SSR* **15**, No. 5, p. 67.
Yoshida, J.: 1972, *Publ. Astron. Soc. Japan* **24**, 391.

DISCUSSION

S. J. Aarseth: It is of interest to note that this type of escape criterion is sharper than is usually required for numerical studies. Thus the computations of three-body systems by Szebehely showed in every case that all escaping particles satisfied both the simple two-body criterion discussed here, when using $r_* = 15$.

INFLUENCE OF THE DYNAMICAL FIGURE OF THE MOON ON ITS ROTATIONAL-TRANSLATIONAL MOTION

G. I. EROSHKIN
Institute of Theoretical Astronomy, Leningrad, U.S.S.R.

Abstract. The influence of the dynamical figure of the Moon on its rotation with respect to its mass centre (the physical libration) is determined by means of the theorem on the angular moment of a rigid body. In the expansion of the Moon's force function in spherical harmonics all the second and the third order harmonics are taken into consideration. For the determination of the Moon's physical libration components a linear system of differential equations of the second order with constant coefficients is constructed.

The integration displays the essential influence of the new terms in the force function expansion. For evaluation of the disturbed elements of the lunar orbit due to the nonsphericity of the Moon's dynamical figure the Lagrange's equations are solved. The disturbing function is taken in an expansion form in powers of the eccentricity of the lunar orbit and of the inclinations of the Moon's equator and its orbit with respect to the ecliptic. The commensurability of the Moon's mean motion and its angular velocity of rotation produces in the major semi-axis of the lunar orbit secular perturbations of the first order.

1. Introduction

The dynamical figure of the Moon implies the geometric figure of a homogeneous rigid body for which the expansion of the force function in spherical harmonics has the same coefficients C_{kj} and S_{kj} as the real Moon.

The expansion of the Moon's force function with respect to spherical harmonics has the form

$$U_m = \kappa \frac{m}{r} \left[1 + \sum_{k=2}^{\infty} \sum_{j=0}^{k} \left(\frac{b}{r}\right)^k (C_{kj} \cos j\lambda + S_{kj} \sin j\lambda)\, P_k^j(\sin\delta) \right],$$

where κ is the gravitational constant, m is the Moon's mass, b is its mean radius, $P_k^j(\sin\delta)$ represent the associated Legendre's polynomials, the coefficients C_{kj} and S_{kj} being calculated in the selenocentric equatorial coordinate system $Oxyz$. The axis Oz is directed along the Moon's rotational axis, the axis Ox coincides with its 'first radius', the plane Oxy is the lunar equator. Selenocentric equatorial coordinates of a point outside the Moon are r, λ, δ, the longitudes being counted from the axis Ox.

From the optical observations the Moon's dynamical figure is found to be a triaxial ellipsoid, its axis of the minimum moment of inertia A being associated with the lunar 'first radius', the one of the maximum moment of inertia C being the Moon's rotational axis. The optical observations make it possible to find the ratios of the moments of inertia of the Moon, α and β, and, therefore, to evaluate the coefficients C_{20} and C_{22} in the series (1). In Goudas (1964) some coefficients C_{kj} for $k=2, 3, 4$ were calculated on the basis of Moon's optical observations assuming that the near and the far sides of the Moon are symmetrical. From the calculations it follows that under such an assumption the coefficient C_{40} has to be not less than C_{20}.

The examination of the Moon's gravity field with the aid of Lunar Artificial Sat-

Y. Kozai (ed.), The Stability of the Solar System and of Small Stellar Systems, 201–207.

ellites (LAS) provides more exact characteristics of the dynamical figure of the Moon. Lately, on the basis of data received from observations of different LAS a number of investigations were carried out for deriving the coefficients C_{kj}, S_{kj} (Akim, 1966; Michael *et al.*, 1970; Lorell, 1970; Michael and Blackshear, 1972, among others). The numerical values of C_{kj} S_{kj} with the same indices in these papers differ from each other appreciably except for C_{20} and C_{22}. Nevertheless the values of C_{k0}, C_{k1} and C_{22} are nearly of the same order for $k>2$. These data allow us to conclude that the dynamical figure of the real Moon differs essentially from the triaxial ellipsoid, for which the values of C_{k0} and C_{k1} would tend to zero with the increase of k.

In previous papers concerning the influence of nonsphericity of the Moon's gravity field on its rotational and translational motions the Moon's dynamical figure was determined by the values of only two coefficients, C_{20} and C_{22}. Now the Moon's dynamical figure is characterized by the first 12 coefficients C_{kj} and S_{kj} ($k=2, 3$; $j=0, \ldots, k$), the values of which are taken from Michael *et al.* (1970):

$$
\begin{array}{ll}
C_{20}=-2.0707\times10^{-4} & \\
C_{21}=-0.4425\times10^{-6} & S_{21}=-0.4573\times10^{-5}\\
C_{22}=0.2242\times10^{-4} & S_{22}=0.2119\times10^{-6}\\
C_{30}=-0.6303\times10^{-5} & \\
C_{31}=0.2437\times10^{-4} & S_{31}=0.2301\times10^{-5}\\
C_{32}=0.5016\times10^{-5} & S_{32}=0.2031\times10^{-5}\\
C_{33}=0.1657\times10^{-5} & S_{33}=-0.6798\times10^{-6}.
\end{array}
$$

The coefficients C_{21}, S_{21} and S_{22} determine the positions of the principal axes of inertia with respect to the coordinate system $Oxyz$, while C_{3j} and S_{3j} characterize the deviation of the Moon's dynamical figure from the triaxial ellipsoid.

2. Physical Libration of the Moon

The problem of the lunar physical libration in the gravity field of the point-Earth is considered. The force function of the mutual attraction of the Moon and the Earth has the form $U=U_m m_0$, where m_0 is the Earth's mass and by r, λ, δ in U_m the Earth's coordinates are implied. The lunar kinetic energy depends both on moments and products of inertia. The orbital motion of the Moon is performed in accordance with Brown's theory.

For the variables ξ, η, τ (Hayn, 1923) the system of the linear differential equations with the constant coefficients is deduced as

$$
\begin{aligned}
\ddot{\eta}+a_{11}\dot{\xi}+a_{12}\eta&=\Sigma_1,\\
\ddot{\xi}-a_{21}\dot{\eta}+a_{22}\xi&=\Sigma_2,\\
\ddot{\tau}+a_{32}\tau&=\Sigma_3.
\end{aligned}
\tag{2}
$$

The equations of this system are similar to ones derived by Hayn and Koziel (Koziel, 1948) except that a_{22} and a_{32} depend now on C_{31} and C_{33} in addition to

C_{20} and C_{22}. Furthermore, the sums of the trigonometric terms in Σ_i have as a factor each of C_{kj}, S_{kj} ($k=2, 3; j=0, ..., k$) and two constant terms in Σ_1 and Σ_2 are connected with kinetic energy, namely with the products of inertia defined by C_{21} and S_{21}.

In Hayn's paper in the equation for τ a resonance phenomenon takes place for the value of the mechanical ellipticity of the Moon, $f=0.662$. Now, existence of this resonance for such a value of f which is called critical, depends on magnitudes of C_{31} and C_{33}.

For the usual variables τ, $\sin I\sigma$, ϱ the main terms in the solution of system (2) are as follows (except for those with coefficients C_{20} and C_{22}):

	τ	$\sin I\sigma$	ϱ
C_{21}	$+8''.51 \cos\omega$	$-360''.53 \cos(g+\omega)$	$+360''.65 \sin(g+\omega)$
S_{21}	$-43''.92 \sin\omega$	$+39''.41 \sin\omega$	$+40''.61 \cos\omega$
S_{21}		$+5824''.51 \sin(g+\omega)$	$+5824''.51 \cos(g+\omega)$
S_{22}	$-971''.82$	$-14''.12$	
C_{32}	$-8''.82 \cos\omega$	$+67''.78 \cos(g+\omega)$	$-67''.80 \sin(g+\omega)$
S_{33}	$+104''.29$		

where g is the lunar mean anomaly and ω is the argument of the perigee of the lunar orbit.

However, according to the optical observations the maximum amplitude of the lunar physical libration must not exceed 100″–120″ (Weimer, 1968). This contradiction seems to be due to very large values of C_{21}, S_{21} and S_{22} accepted here. It is worth noting that these values of C_{21}, S_{21} and S_{22} are the least among those in the studies cited above on the determination of the lunar gravity field.

The terms of the solution dependent on C_{20} and C_{22} are in good agreement with solutions by Hayn and Koziel except for the terms in $\sin I\sigma$ and ϱ with the argument $2g'+2\omega'$ (doubled longitude of the Sun counted from the ascending node of the lunar orbit on the ecliptic). According to Hayn the amplitude of these terms is equal to 3″, but here it is less than 0″.1. A similar discrepancy with Hayn's results was noted by Habibullin (1966).

3. Influence of Nonsphericity of the Moon's Dynamical Figure on Its Translational Motion

Now we deal with the problem of the motion of the mass centre of the rigid Moon disturbed by nonsphericity of its dynamical figure in the gravitational field of the point-Earth. The Moon's rotation with respect to its mass centre is performed in accordance with Cassini's laws, the Moon's physical libration being neglected.

Introduce the ecliptic coordinate system $O_1x_1y_1z_1$ with the origin in the Earth's mass centre the axes of which are parallel to those of $Ox_1y_1z_1$. The dimension and the position of the lunar orbit with respect to $O_1x_1y_1z_1$, as well as the position of the Moon's mass centre on the orbit are defined by six osculating elements as follows:

a and e are the major semi-axis and the eccentricity of the lunar orbit, i represents the inclination of the orbit with respect to the ecliptic, Ω means the longitude of the ascending node of the lunar orbit in ecliptic, ω is the argument of the lunar orbit perihelion and M_0 is the Moon's mean anomaly at the initial epoch. For the determination of the perturbation of these elements the system of Lagrange's equations is constructed, the disturbing function R being of the form

$$R=\kappa \frac{m+m_0}{r} \sum_{k=2}^{\infty} \sum_{j=0}^{k} \left(\frac{b}{r}\right)^k (C_{kj} \cos j\lambda + S_{kj} \sin j\lambda)\, P_k^j(\sin \delta), \tag{3}$$

where r, λ and δ mean the selenocentric equatorial coordinates of the Earth which have to be expressed in terms of the osculating elements of the lunar orbit.

If one considers the problem of calculating the perturbations of the osculating elements of the Earth's selenocentric orbit disturbed by the nonsphericity of the Moon's gravity field, then the disturbing function R_1 would be of the same form (3) but r, λ and δ should be expressed in terms of the osculating elements of the Earth's selenocentric orbit a', e', i', Ω', ω' and M_0'. In this case it is convenient to take R_1 in the form of an expansion in powers of e', i' and ϑ (Brumberg *et al.*, 1971) since those are the small quantities. It is evident that between the osculating elements of the lunar geocentric orbit and those of the Earth's selenocentric orbit there exists a simple relation,

$$a=a', \qquad e=e', \qquad i=i', \qquad \Omega=\Omega', \qquad \omega=\omega'+180^\circ, \qquad M_0=M_0'.$$

Therefore, R can be presented as an expansion in powers of e, i, ϑ which is similar to the one mentioned above,

$$R=\kappa \frac{m+m_0}{r} \sum_{k=2}^{\infty} \sum_{j=0}^{k} \sum_{l=-k}^{+k} \sum_{s=0}^{k} \sum_{q=-\infty}^{+\infty} \left(\frac{b}{a}\right)^k \times$$
$$\times (-1)^h A_{kjl}(\vartheta)\, F_{kls}(i)\, X_{k-2s+q}^{-k-1,\,k-2s}(e)\, (C_{kj} \cos \Delta + S_{kj} \sin \Delta), \tag{4}$$

where

$$h=E\left(\frac{k-j}{2}\right)+E\left(\frac{k-l}{2}\right)+\max\{0, j-l\}+k-l-j,$$
$$\Delta=(k-2s+q)\, M+(k-2s)\, \omega+l(\Omega-\psi)-j\varphi-\nabla_{kj}90^\circ,$$

M is the Moon's mean anomaly,

$$\nabla_{kj}=\begin{cases} 0, & k-j=2p, \\ 1, & k-j=2p+1, \end{cases}$$

where p is integer, A_{kjl}, F_{kls} are hypergeometric functions (Brumberg *et al.*, 1971), $X_{k-2s+q}^{-k-1,\,k-2s}$ being Hansen's coefficients (Brumberg, 1967).

Due to the smallness of C_{kj} and S_{kj}, it is preferable to find out the solution of the Lagrange's equations for the osculating elements with the disturbing function in the form (4) by the method of successive approximations:

$$a=a_0+\delta_1 a+\cdots+\delta_\nu a+\cdots,$$
$$e=e_0+\delta_1 e+\cdots+\delta_\nu e+\cdots,$$
$$\cdots \quad \cdots \quad \cdots \quad \cdots \quad \cdots \quad \cdots,$$

where a_0, $e_0, \ldots$ are the values of the osculating elements of the lunar orbit for the certain epoch, and $\delta_\nu a$, $\delta_\nu e, \ldots$ are the perturbations of the νth order.

In order to obtain the first-order perturbations one should substitute into R, instead of a, $e, \ldots$ their undisturbed values corresponding to some epoch, and instead of φ, ψ and ϑ, their values related to the undisturbed rotation of the spherical body as follows:

$$\varphi=n(t-t_0)+\varphi_0, \qquad \psi=\psi_0, \qquad \vartheta=\vartheta_0,$$

where t_0 is the initial epoch, and φ_0, ψ_0 and ϑ_0 are the values of the angles for $t=t_0$. Then R becomes

$$R=\sum H_{\sin}^{\cos}[(k-2s+q)\,M-j\varphi+P], \tag{5}$$

where H are constant coefficients which depend in linear way on C_{kj} and S_{kj}. H vanishes simultaneously with all C_{kj} and S_{kj}, $M=n(t-t_0)+M_0$ and P are linear combinations of ω_0, Ω_0, ψ_0. Since the Moon's mean motion equals to its mean angular velocity of rotation with respect to the proper mass centre the major semi-axis of the lunar orbit and the Moon's mean motion have the secular perturbations of the first order. Let us recall that here the two-body problem is considered, with the disturbing function being due to nonsphericity of the Moon's dynamical figure. For this case the theorem can be stated on the secular first order perturbations of the major semi-axis of the planetary orbit which is analogous to that by Laplace–Lagrange for the three-body problem.

THEOREM 1. Let the dynamical figure of the planet have no axial symmetry. If the mean planetary motion n_1 with respect to the central body is not commensurable with its mean angular velocity of rotation n_2 about its proper mass centre, then the major semi-axis and the mean planetary motion have no secular perturbations of the first order.

Indeed, in this case the disturbing function is of the form (5), where $M=n_1(t-t_0)+$ $+M_0$ and $\varphi=n_2(t-t_0)+\varphi_0$. Since n_1 and n_2 are not commensurable, $\partial R/\partial M_0$ has a constant part only for $k-2s+q=0$ and $j=0$. But in this case $\partial R/\partial M_0\equiv 0$ and, therefore, the major semi-axis of the planetary orbit has no secular perturbations of the first order.

THEOREM 2. Let the dynamical figure of the planet have the axial symmetry and the planet performs its rotation about this axis. Then the major semi-axis of the planetary orbit has no secular perturbations of the first order even in the case of commensurability between the mean planetary motion n_1 and its mean angular velocity of rotation n_2.

The axial symmetry of the dynamical figure of a planet assumes that its dynamical figure is defined by the values of C_{k0}. All the other coefficients of (1) are identically equal to zero. In this case R reduces to

$$R=\sum H_{\sin}^{\cos}[(k-2s+q)M+P],$$

where H depends lineary on C_{k0} and vanishes simultaneously with all C_{k0}. In $\partial R/\partial M_0$ the terms corresponding to $k-2s+q=0$ do not depend on time, but $\partial R/\partial M_0 \equiv 0$ whatever values n_1 and n_2 have. Thus, the major semi-axis of the planetary orbit has no secular perturbations of the first order even in the case of commensurability of n_1 and n_2.

Since, for the Moon, $n_1=n_2$, the secular perturbations of the lunar orbital elements are produced due to the terms in R, and its derivatives with indices k, j, s, q satisfying the relation $k-2s+q-j=0$. After differentiating R with respect to the osculating elements the Eulerian angles φ, ψ, ϑ are expressed in terms of the orbital elements according to Cassini's laws. The first-order perturbations of the major semi-axis determined by S_{22} and S_{33} are as follows (per Julian century):

$$(1/a_0)\,\delta_1 a=+0''.089-0''.009.$$

The periodic perturbations of the osculating elements of the lunar orbit are determined by the terms of the perturbation function R and its derivatives with indices k, j, s, q satisfying the condition $k-2s+q-j\neq 0$. As was expected the periodic perturbations appeared to be small. The largest of them are the following:

$$\delta_1\tau=+0''.02\sin M,$$
$$\delta_1 M_0=-0''.02\sin M,$$

the amplitude of these perturbations being determined by C_{22}. Since in Section 2 the conclusion was drawn that C_{21}, S_{21}, S_{22}, accepted here exceed their real values, the perturbations due to the corresponding terms in R are too large.

Let us note that the secular perturbations of Ω and π which are determined by C_{20} and C_{22} are in good agreement with the results obtained earlier (Eckert, 1965).

4. Conclusion

The work carried out demonstrates the essential influence of the formerly neglected harmonics of the third order ($k=3$) of the Moon's force function expansion (1) on its rotational and translational motions. The terms of the disturbing function with coefficients C_{21}, S_{21} and S_{22} appeared to influence essentially the Moon's rotational-translational motion. The discrepancy found between the results of the theory of the Moon's libration proposed here and the results of observations allows us to conclude that the values of the coefficients C_{21}, S_{21} and S_{22} are exaggerated, though the values of these coefficients are the least among those in other papers on the determination of the Moon's gravity field. In Lidov and Neishtadt (1973) the same conclusion is made.

The evaluation of the perturbations of the rotational-translational motion corresponding to some other values of C_{kj} and S_{kj} can be readily performed.

References

Akim, E. L.: 1966, *Dokl. Akad. Sci.* **170**, 4, 799 (in Russian).
Brumberg, V. A.: 1967, *Bull. Inst. Theor. Astron.* **11**, 73 (in Russian).
Brumberg, V. A., Evdokimova, L. S., and Kochina, N. G.: 1971, *Celes. Mech.* **3**, 197.
Eckert, W. J.: 1965, *Astron. J.* **70**, 787.
Goudas, C. L.: 1964, *Icarus* **3**, 375.
Habibullin, Sh. T.: 1966, *Trudy KGO* **34**, 3 (in Russian).
Hayn, F.: 1923, *Encykl. math. Wiss.* **6**, 2a, 1020.
Koziel, K.: 1948, *Acta Astron.*, Ser. a, **4**, 61.
Lidov, M. L. and Neishtadt, A. I.: 1973, Preprint IPM N9.
Lorell, J.: 1970, *The Moon* **1**, 190.
Michael, W. N. Jr. and Blackshear, W. Th.: 1972, *The Moon* **3**, 388.
Michael, W. N. Jr., Blackshear, W. Th., and Gapcynski, J. P.: 1970, in B. Morando (ed.), *Dynamics of Satellites*, Springer Verlag, Berlin, p. 42.
Weimer, Th.: 1966, in *La Lune à l'ère spatiale*, Paris.

A COMPARISON OF THE MEAN-VALUE AND INITIAL-VALUE SOLUTIONS OF THE IDEAL RESONANCE PROBLEM WITH AN APPLICATION IN RIGID-BODY MECHANICS

A. H. JUPP
The University of Liverpool, U.K.

Abstract. A solution of the ideal resonance problem has already been exhibited (Jupp, 1972) explicitly in terms of the 'mean' elements; to second order in the small parameter in the case of libration, and to first order in the case of deep circulation. Both representations possess a singularity when the 'mean' modulus of the Jacobi elliptic functions is unity; this corresponds to motions on or close to the separatrix of the phase plane of the dynamical system.

It is shown that, provided particular coefficients associated with the problem satisfy specific relations, the singularity is removed, and the final solution is applicable throughout the deep resonance region.

The solution is next expressed in terms of the initial conditions of the system. Again, in general, the solution has a singularity associated closely with the limiting motion, and the circulation form of the solution is restricted to deep circulation. It is shown that when the previously-mentioned coefficients satisfy particular constraints, the singularity is removed. Moreover, with the same constraints, the deep-circulation solution extends naturally to cover the entire circulation regime. It is of interest that these constraints are quite different from those associated with the 'mean' element formulation.

In the light of these results a global formal series solution in terms of initial conditions of the problem of the free rotation of a rigid body can simply be extracted from the general solution of the ideal resonance problem. The small parameter is provided by requiring that two of the principal moments of inertia differ by a small quantity. Such a solution can readily be checked with the well-known exact solution.

Tentative conclusions are drawn regarding the removal of the singularity in the two separate formulations, and some interesting aspects of the Lie series perturbation procedures are noted.

Reference

Jupp, A. H.: 1972, *Celes. Mech.* **5**, 8–26.

Y. Kozai (ed.), The Stability of the Solar System and of Small Stellar Systems, 209.

STATISTICS OF THREE-BODY EXPERIMENTS

M. J. VALTONEN
Institute of Astronomy, Cambridge, U.K.

Abstract. A large number of three-body interactions involving one initial binary have been studied by a numerical regularization technique. In each set of experiments some parameters have fixed values, whereas others are selected by uniform sampling of the corresponding distribution functions. Similar statistical results are obtained for different random number sequences at a lower integration accuracy.

This experimental approach permits an approximate determination of the final distributions of eccentricity, velocity and lifetime. These results show a strong dependence on the total angular momentum, total energy and the mass range, whereas other parameters are usually of secondary importance.

1. Introduction

Detailed numerical studies of the three-body problem reveal a considerable complexity of motions (e.g. Szebehely and Peters, 1967) which at first sight appear to limit the usefulness of a qualitative description of the dynamical behaviour. Thus each individual example looks different and the computed orbits are sensitive to numerical errors as well as to changes in the initial conditions. Nevertheless, characteristic trends begin to emerge when the results of many calculations are combined. Such investigations have mainly been concerned with planar motions in which the three particles are initially at rest (Agekyan and Anosova, 1967, 1968; Anosova, 1969a; Szebehely, 1972). The recent two-dimensional study by Standish (1972) consists of 800 examples, mostly with nonzero starting velocities which also enable the angular momentum dependence to be analysed.

In the present investigation we adopt the method of statistical sampling of initial conditions in three dimensions, the only restriction being that two of the particles form a binary initially. A total of 25000 examples have been calculated in an attempt to cover a wide parameter range systematically. This paper presents a preliminary analysis of the results. Although the accurary requirement has been relaxed somewhat with respect to 'exact' integrations in order to reach this goal, we first establish empirically that the adopted procedure does not invalidate the statistical results.

2. Initial Conditions and Numerical Method

The parameters of the initial system are specified as follows (Figure 1): A binary lies in the plane A with its major axis in the x-direction and its centre of mass at C. Only one of the binary orbits is shown. The mean anomaly of the binary is chosen randomly within the range $[0, 2\pi]$, while its eccentricity e is either zero or e^2 has a uniform random distribution in $[0, 1]$. A third particle approaches in the plane B with inclination i and longitude of the ascending node Ω. The latter is chosen randomly while the inclination is either zero (denoted a disk system) or $\cos i$ is randomized

Y. Kozai (ed.), The Stability of the Solar System and of Small Stellar Systems, 211–223.

over its entire range. The longitude of pericentre ω is also randomized, and in most cases the semilatus rectum s has a uniform random distribution between specified limits. The semimajor axis a_3 of the third particle is always such that the total energy $E<0$. The inverse of a_3 is randomized in many examples; in others a_3 is constant. There are also approximately 4000 examples where the third particle is initially close

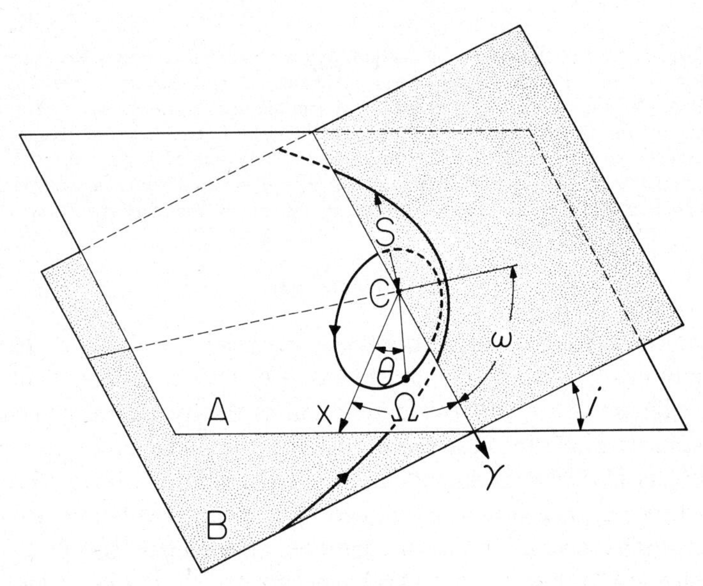

Fig. 1. Geometry of the problem. Symbols are described in the text.

to the binary and thus the orbital elements refer only to the instantaneous situation at the outset of integration. The masses denoted by m_1, m_2, m_3 have either fixed values or are chosen from a uniform random distribution within [0.1, 1] and normalized to make the total mass equal to unity. The semimajor axis of the initial binary is also taken to be unity. The system of units is then defined by taking the gravitational constant equal to unity.

The integration is usually started with the third particle at a distance of 20 units from the binary, unless its unperturbed orbit restricts it to shorter distances. In the latter case the integration starts with the third particle near the apocentre of its orbit. When a particle reaches 20 units from the centre of mass of the binary formed by the other two at any later time the integration is temporarily halted. The orbital elements are determined on the assumption of unperturbed two-body motion and the sign of the semimajor axis is used to decide whether escape has occurred. If the distant particle is still bound, it is moved analytically on its unperturbed orbit back to a distance of 20 units, and the binary is advanced to the corresponding phase after which the integration resumes. This procedure was also adopted by Standish (1972). Unless escape takes place before, each experiment continues until 1000 times the original binary period has elapsed. In some strongly bound triple systems an alternative time limit of 100 periods was adopted. A more complete description of the initial conditions is given elsewhere (Saslaw *et al.*, 1974).

The integration method uses a fourth-order polynomial (Aarseth, 1971) together with a two-body regularization technique (Heggie, 1973). In this way close approaches between two particles can be studied without numerical difficulties while critical triple encounters are relatively rare and hence of little significance.

For each type of system we perform a large number of experiments to obtain statistically significant results. Comparisons indicate that 200–300 combined examples give a satisfactory statistical description. Also we have repeated one set of experiments with two, three and four times the standard mean integration step, each time using a different set of random numbers, to investigate the effect of integration accuracy on the final distributions.

Table I shows the rms relative energy errors, $\Delta E/E$, excluding experiments with unperturbed two-body motion. The distributions in Tables II–IV show no clear trend

TABLE I

Summary of data for accuracy tests

Set	$\langle(\Delta E/E)^2\rangle^{1/2}$	Number of experiments	Completed (in per cent)
I	5×10^{-4}	173	95
II	6×10^{-4}	179	96
III	1×10^{-2}	162	97
IV	3×10^{-2}	208	96

TABLE II

Eccentricity distribution (in per cent)

Set	0–0.2	0.2–0.4	0.4–0.6	0.6–0.8	0.8–1.0
I	7	22	22	29	20
II	12	20	22	26	20
III	5	26	20	25	24
IV	11	15	25	24	26

when going from the highest to the lowest accuracy. We are therefore confident that the accuracy of set *I* is adequate for our investigation which also includes more difficult examples and this accuracy is used throughout. A more detailed discussion of the distributions shown in Tables II–IV is given below.

3. General Properties

When the third particle approaches on a nearly parabolic orbit one of three events may occur: (1) Capture, if no particle escapes within one original binary period after

TABLE III

Terminal velocity distribution (in per cent)

Set	0–0.2	0.2–0.4	0.4–0.6	0.6–0.8	0.8–1.0	> 1.0
I	12	27	22	18	12	9
II	14	29	26	16	5	10
III	13	27	29	17	9	4
IV	11	26	28	19	10	7

TABLE IV

Lifetime distribution (in per cent)

Set	0–2	2–4	4–6	6–10	10–20	20–30	> 30
I	14	18	15	17	20	10	6
II	13	23	16	17	18	10	4
III	12	25	15	19	17	6	6
IV	16	21	18	15	20	8	3

the first strong interaction. If a capture does not occur one has either (2) fly-by, if the third particle escapes or (3) exchange, if one of the binary particles escapes. We have not considered systems with weak interactions only, i.e. systems where the third particle having initially an elliptic orbit is still found in a similar orbit after 2000π time units. The relative frequency of the events shows a complicated parameter dependence, but the pericentre distance q of the third particle in its initial unperturbed orbit around the centre of mass of the binary is most important. In Figure 2 we show the probabili-

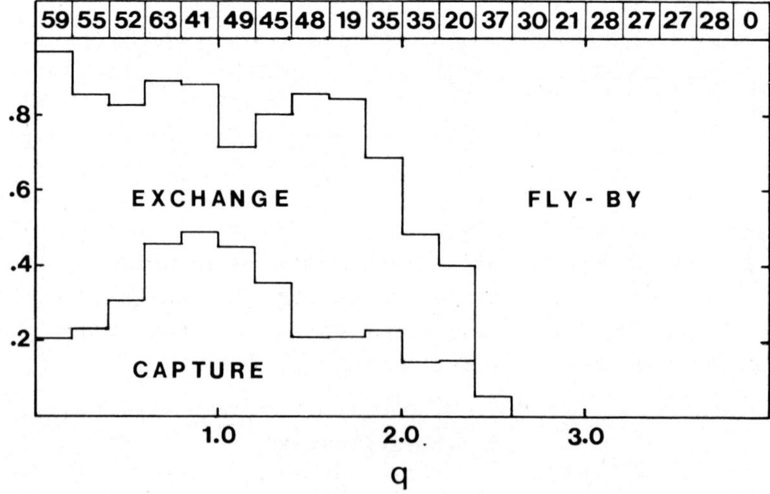

Fig. 2. Probability of the three types of motion for nearly parabolic encounters. Details are given in the text.

ties of each event by the area occupied, as a function of the pericentre distance. The numbers on top indicate the sample size. For this set $m_1 = m_2 = 0.25$, $m_3 = 0.5$, $a_3 = 300$, $e = 0$ and $\cos i$ is randomized. If the third particle mass is decreased, the probability of exchange diminishes and disappears almost completely when m_3 is the smallest of the three particles.

When a triple system breaks up the lightest of the three particles tends to be ejected, confirming previous investigations. Figure 3 shows the frequency of escape for a

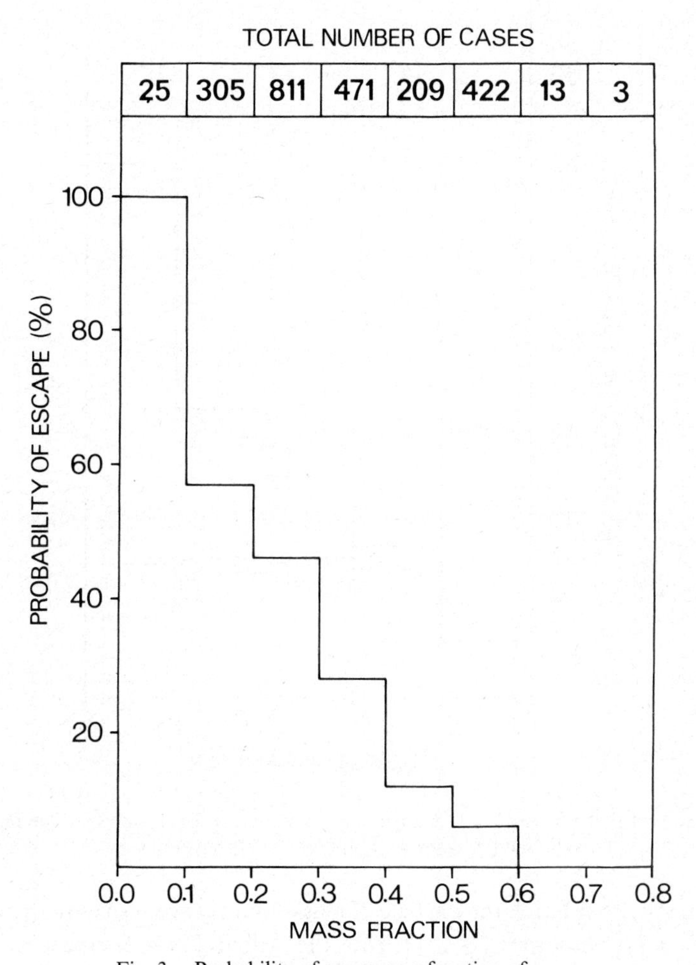

Fig. 3. Probability of escape as a function of mass.

particle of a given mass in both initially strongly bound systems and those created through a capture. The numbers on top again indicate the corresponding sample size. The escape probability depends on the adopted mass distribution but is not very sensitive to other parameters. It should be noted that there is a small but nonzero probability for a particle with more than half the total mass to escape. Figure 4 shows the probability of the incoming particle of mass 0.5 also being ejected after the interaction, as a function of impact parameter s (solid line). The binary masses are both 0.25 and e^2 is random within [0, 1] in $\frac{3}{4}$ of all experiments and $e=0$ in the remaining $\frac{1}{4}$. The two other lines show the probability of escape for a binary component of mass $m_2=\frac{2}{7}$, the other particles having $m_1=\frac{4}{7}$ and $m_3=\frac{1}{7}$. The dashed line refers to a set with $e=0$ (also used for Table I), while the dotted line corresponds to e^2 randomized

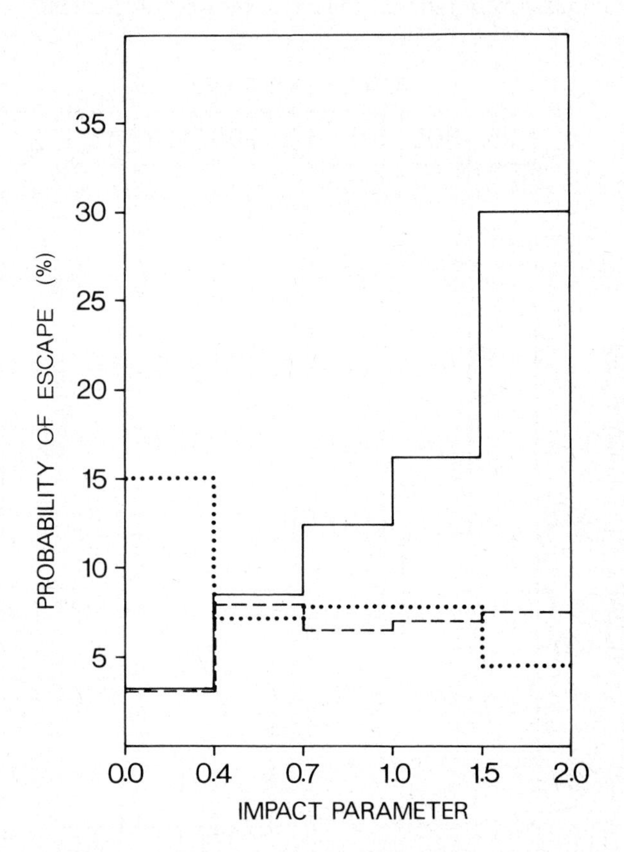

Fig. 4. Escape probability (in per cent) for a particle of given mass. The systems are specified in the text. Note that q of Figure 2 is approximately equal to $0.5s$.

in $[\frac{2}{3}, 1]$. On the other hand, the particle of mass $\frac{4}{7}$ hardly ever escapes. In all these sets $a_3=30$ and $\cos i$ is random in $[-1, 1]$. Thus the probability of escape for a given mass also depends on its initial role. The direction in which a particle is ejected during the breakup of a triple system is not always random. Figure 5 shows the distribution of the polar angle θ of the escaped particle, measured from the direction of the rotation axis of the initial binary, for a system with $a_3=30$, $m_1=\frac{4}{7}$, $m_2=\frac{2}{7}$, $m_3=\frac{1}{7}$, $e=0$ and $\cos i$ random in $[\frac{1}{3}, 1]$. Similar distributions for $\cos i$ randomized in $[-1, -\frac{1}{3}]$ show a random distribution of $\cos\theta$, and for $\cos i$ randomized in $[-\frac{1}{3}, \frac{1}{3}]$ one has a slight preference for the binary plane $\theta=90°$. This trend becomes more pronounced for low inclination systems. We have also measured the polar angle of escape from the direction of the total angular momentum of the triple system, as illustrated in Figures 6 and 7 on the right-hand side. On the left we show the distribution of polar angles of the incoming directions of the third particle. The set of experiments included in Figures 6 and 7 have $\cos i$ randomized within $[\frac{1}{3}, 1]$ and $[-1, -\frac{1}{3}]$, respectively. The final distributions are more concentrated towards the plane normal to the total angular

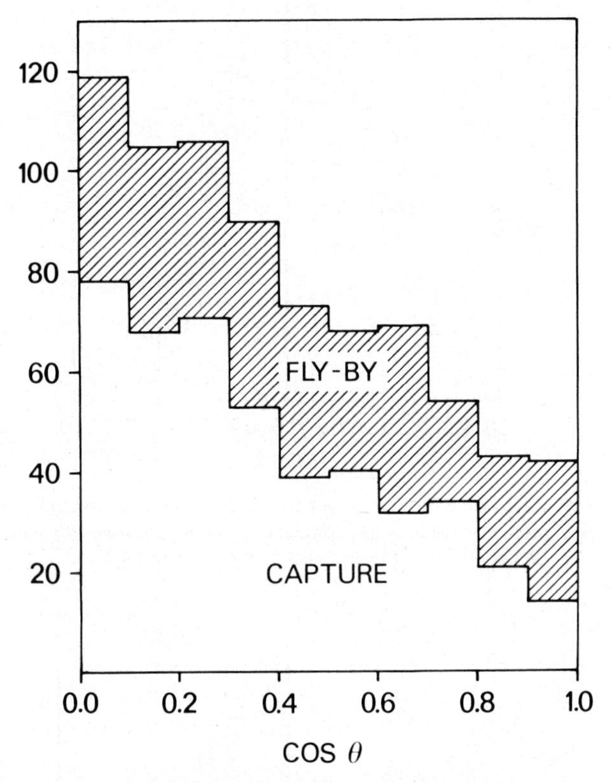

Fig. 5. Number distribution of the polar angle for the escaping particle. The angle θ is measured with respect to the initial rotation axis of the binary.

momentum than the initial ones. This effect disappears, however, when the third particle mass is increased to values comparable with the binary components. In all three previous figures the fly-by examples are shown as shaded areas to distinguish them from captures. No exchanges occurred in this set.

The escape energy is strongly peaked near zero as shown in Figure 8. The data are obtained from the same set as that of the three previous figures, with $\cos i$ random in $[-1, 1]$ and $e = 0$ (solid line). The dashed line refers to a similar set except e^2 is random in $[\frac{2}{3}, 1]$. The escape energy appears more likely to have a large value in the latter set. In general the escape energy is negatively correlated with the minimum of the moment of inertia relative to the centre of mass.

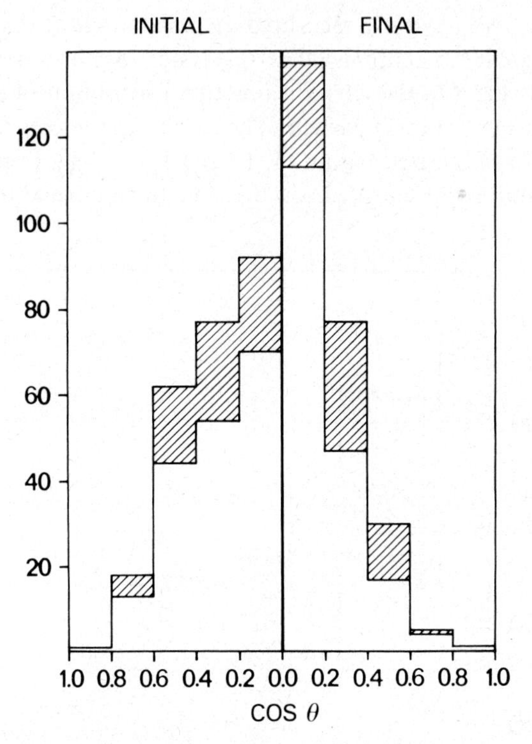

Fig. 6. Number distribution of the polar angle for the incoming (left) and escaping (right) particle. The angle θ is measured with respect to the direction of total angular momentum. Initial inclinations are in the range 0°–70°.5. Further details are given in the text.

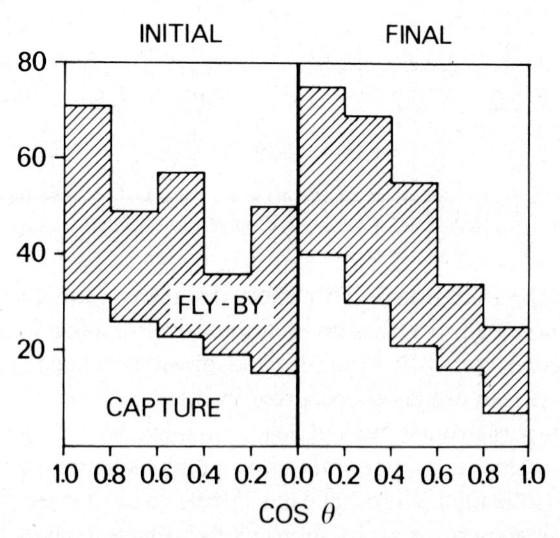

Fig. 7. Number distribution of the polar angle for the incoming (left) and escaping (right) particle. The angle θ is measured with respect to the direction of total angular momentum. Initial inclinations are in the range 109°.5–180°. Further details are given in the text.

4. Close Triples and Captures

Initially strongly bound systems and triples formed by captures are found to behave similarly if the main parameters are similar, and therefore they can be studied together. Here we concentrate on the final distributions of the eccentricity of the binary which is left behind after an escape, the distribution of the relative velocity (at infinity) of the escaping particle, and the distribution of lifetimes of triple systems measured in cros-

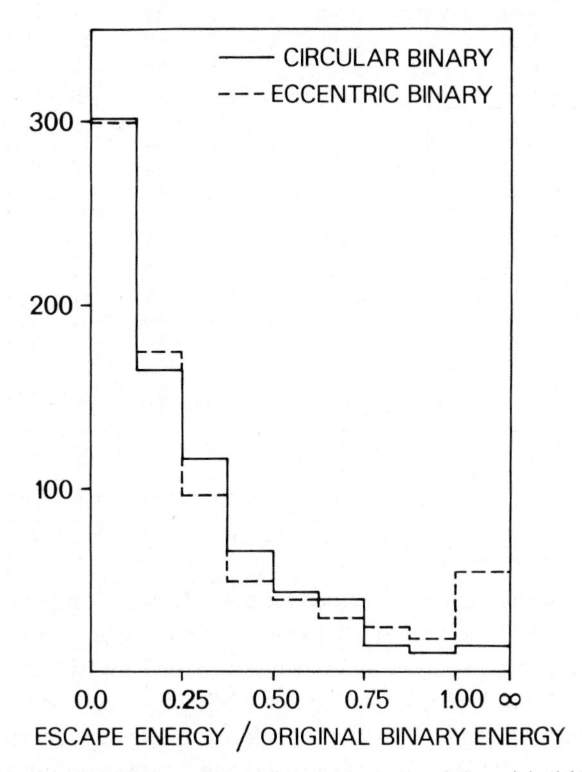

Fig. 8. Number distribution of escape energy in terms of the original binary energy. Details are given in the text.

sing times defined by $T_{cr} \equiv |2E|^{-3/2}$ in our units. Results are shown in Table V for a few sets chosen to demonstrate the effect of various parameters. The columns give the number of experiments included in each set, the mean energy $\langle E \rangle$ and mean total angular momentum $\langle h \rangle$ as well as the mean mass range $\langle m_{max}/m_{min} \rangle$ for each set; the times (in units of the crossing time) by which one quarter, one half or three quarters of the systems are terminated ($T_{1/4}$, $T_{1/2}$, $T_{3/4}$); and the mean of the terminal escape velocity $\langle v_\infty \rangle$.

The first six sets have $a_3 = 30$ and fixed masses in the order m_1, m_2, m_3 as indicated in the last column. In set 5, $e = 0$; in others e^2 is random within [0, 1] in $\frac{3}{4}$ of all examples and $e = 0$ in the remaining $\frac{1}{4}$. The impact distance s is always $\leqslant 2$, and $\cos i$ is random in $[-1, 1]$. Comparison of sets 1–4 shows that increasing the mass range makes the

TABLE V

Summary of results

Set	Number of experiments	$\langle E \rangle$	$\langle h \rangle$	$\langle m_{max}/m_{min} \rangle$	$T_{1/4}$	$T_{1/2}$	$T_{3/4}$	$\langle v_\infty \rangle$	Notes
1	327	−0.06	0.22	1.0	10	28	75	0.42	1, 1, 1
2	250	−0.05	0.25	1.5	8	22	54	0.38	1, 1, 1.5
3	263	−0.04	0.24	2.0	4	15	45	0.40	1, 2, 2
4	217	−0.04	0.25	2.0	4	13	42	0.37	1, 1, 2
5	483	−0.08	0.21	4.0	9	26	68	0.48	4, 2, 1
6	219	−0.12	0.19	20.0	4	13	40	0.63	20, 20, 1
7	216	−0.16	0.20	1.3	15	41	92	0.55	
8	377	−0.12	0.22	2.5	8	22	57	0.54	
9	340	−0.09	0.23	5.7	4	10	32	0.53	
10	240	−0.09	0.03	3.3	4	9	20	0.66	
11	469	−0.14	0.14	2.8	5	13	40	0.70	
12	503	−0.13	0.25	2.8	7	19	47	0.51	
13	240	−0.09	0.27	3.2	6	18	48	0.47	
14	441	−0.10	0.36	2.7	10	30	100	0.36	
15	374	−0.02	0.32	2.9	4	11	40	0.36	Random
16	118	−0.02	0.33	3.2	3	8	20	0.37	Disk
17	139	−0.02	0.42	2.7	4	8	30	0.32	Disk
18	120	−0.17	0.29	1.0	10	40	150	0.52	Disk

systems less stable while the escape velocity is not affected. Comparison of sets 3 and 4 indicates that the results are relatively insensitive to the intermediate mass. Sets 7–9 which contain a range of masses initially strongly bound show the same phenomenon. By comparing sets 6 and 8 we note that the instability increases with mass range even when the latter is large already. Furthermore, the escape velocity increases in this case.

Sets 10–14 demonstrate the effect of the total angular momentum h. These are also strongly bound systems with a range of masses. The stability increases with angular momentum while the escape velocity behaves in the opposite way. Comparison of sets 15–17 or 7 and 18 shows that the third dimension is not important from the point of view of stability or escape velocity. However, because of the small sample size it is not possible to rule out a small effect. In sets 15–17 the parameters are: $m_1 = m_2 = 1$, m_3 is random in $[1.2, 2.8]$, $-400 < a_3 < -6.7$, $s < 3$ and e^2 is random within $[0, 1]$ in $\frac{3}{4}$ of all examples and $e = 0$ in the remaining $\frac{1}{4}$. Set 15 has random orientations and sets 16 and 17 are disk systems. Finally set 18 consists of equal masses initially strongly bound in a disk.

Comparing sets 12–14 with 15–17 we see that the strongly bound systems are more stable and produce higher escape velocities than systems created by capture. The same effect can be seen by comparing sets 1 and 2 with 7 or sets 3 and 4 with 8. Figure 9 shows the actual distributions of the escape velocity and eccentricity for sets 10, 11, 13 and 14. These results from the low angular momentum set 10 ($h=0.027$) may be compared with earlier results of zero angular momentum systems (e.g. Standish, 1972; Szebehely, 1972). Although the initial conditions are slightly different, the final distributions of lifetime, eccentricity and velocity are in good agreement. Other similar calculations (Agekyan and Anosova, 1968; Anosova, 1969a, b) employ different escape criteria and cannot be compared directly with our results. However, the lifetime dependence on mass range and angular momentum is still in qualitative agree-

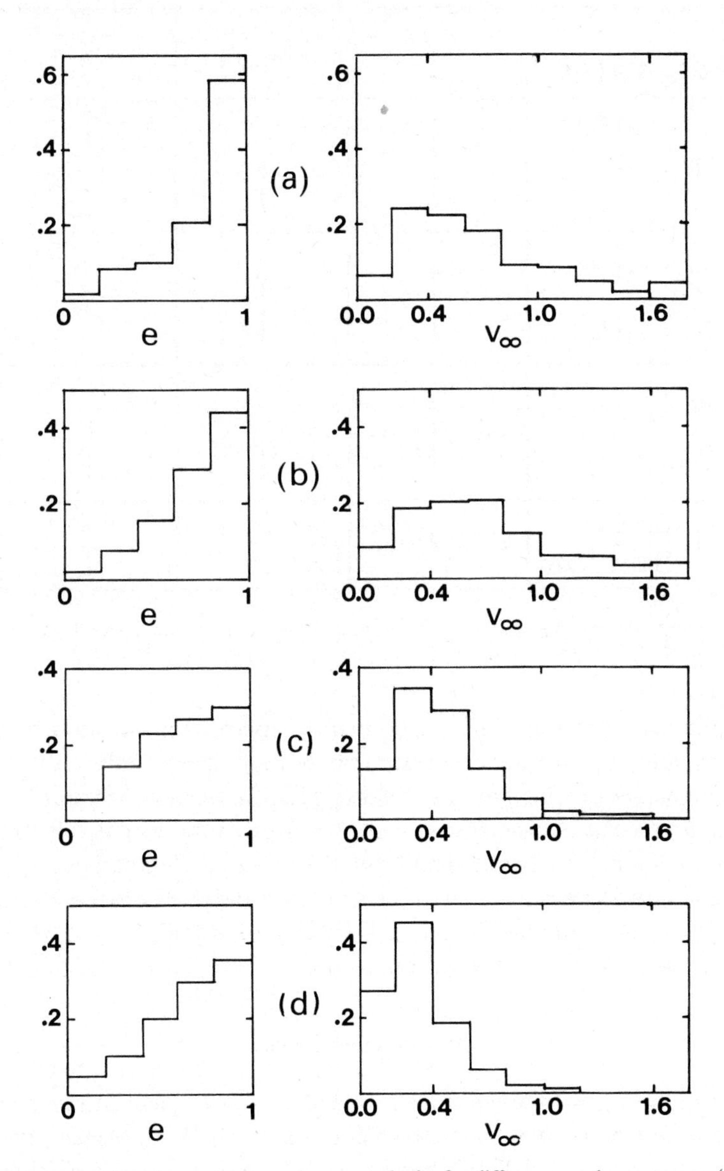

Fig. 9. Distributions of final eccentricity and escape velocity for different angular momenta. (a) $h=0.027$; (b) $\langle h\rangle=0.14$; (c) $h=0.27$; (d) $\langle h\rangle=0.36$. The results refer to sets 10, 11, 13, 14 of Table V.

ment. The effect of the main parameters is summarized in Figure 10. The arrows point in the direction of increase for each of the quantities, and the absence of an arrow implies approximate constancy over the corresponding range of parameter values. For example, the eccentricity does not vary appreciably when the angular momentum changes from 0.2 to 0.4 but decreases when h increases from 0.03 to 0.2. Question marks indicate a constancy which is not well established.

Parameter		T	v_∞	e
h	0.03 0.2 0.4	↓ (0.03–0.4)	↑ (0.2–0.4)	↑ (0.03–0.2)
E	-0.15 -0.02	↑	↑	?
$\frac{m_{max}}{m_{min}}$	1 5 20	↑ (1–20)	↓ (5–20)	↑ (5–20)
i	Disk Random	?		↑

Fig. 10. Qualitative dependence of lifetime, escape velocity and final eccentricity on the main parameters. Arrows point in the direction of increase as explained in the text.

After a preliminary analysis of the experiments it appears that strongly bound three-body systems and those created by capture can be described by only a few parameters, the most important of which are the total angular momentum (Standish, 1972), the total energy and the mass range (Heggie, 1972). Furthermore, a relatively small sample is sufficient to obtain statistically significant distributions. Comparisons of three- and two-dimensional systems have shown their similarity and therefore support the general validity of the results obtained before. A more extensive analysis is to be undertaken to study the significance of other parameters.

Acknowledgements

It is a pleasure to record the constant help and encouragement in this project from my supervisor, Dr S. J. Aarseth. I also wish to thank Dr W. C. Saslaw and Dr D. C. Heggie for assistance and helpful discussions. The author is supported by an Osk. Huttusen Säätiö Foundation Fellowship.

References

Aarseth, S. J.: 1971, *Astrophys. Space Sci.* **14**, 118.
Agekyan, T. A. and Anosova, Zh. P.: 1967, *Soviet Astron.* **11**, 1006.
Agekyan, T. A. and Anosova, Zh. P.: 1968, *Astrophys.* **4**, 11.
Anosova, Zh. P.: 1969a, *Astrophys.* **5**, 81.
Anosova, Zh. P.: 1969b, *Publ. Astron. Obs. Leningrad* **26**, 88.
Heggie, D. C.: 1972, Thesis, Cambridge University.
Heggie, D. C.: 1973, in B. D. Tapley and V. Szebehely (eds.), *Recent Advances in Dynamical Astronomy*, D. Reidel, Dordrecht, The Netherlands, p. 34.
Saslaw, W. C., Valtonen, M. J., and Aarseth, S. J.: 1974, *Astrophys. J.* (to be published).
Standish, E. M.: 1972, *Astron. Astrophys.* **21**, 185.
Szebehely, V.: 1972, *Celes. Mech.* **6**, 84.
Szebehely, V. and Peters, C. F.: 1967, *Astron. J.* **72**, 876.

DISCUSSION

R. H. Miller: How do your results compare with earlier works by Yabushita and others?

M. J. Valtonen: The work of Agekyan and Anosova was mainly concerned with planar systems and usually equal or fixed masses. The same applies to the work by Standish and Szebehely, while Yabushita was concerned with systems of positive total energy. Our results agree with the previous works as far as we have been able to check.

P. Stewart: The experiments were originally conceived for the purpose of explaining why radio sources have double structure. Do you have any thought on the more complex structure involving sometimes four radio components on each side of an optical galaxy?

M. J. Valtonen: In the case of radio-source models one is not restricted to three-body systems alone, and in fact models with four or more bodies have also been studied.

THE ROLE OF BINARIES IN CLUSTER DYNAMICS

D. C. HEGGIE

Trinity College, Cambridge and Institute of Astronomy, Cambridge, U.K.

Abstract. There is considerable empirical evidence on the behaviour of binary stars in computer simulations of isolated star clusters. Even when there are no binaries initially they form quickly if the mass spectrum of the stars is reasonably realistic. These phenomena are discussed by a combination of analytical and numerical techniques.

1. Introduction

There is now considerable empirical evidence on the behaviour of binary stars in computer simulations of isolated star clusters. In particular, Aarseth (1972) showed that, even when there are no binaries initially, they form quite quickly if the mass spectrum of the stars is reasonably realistic. Even for large clusters with up to 500 members, it is found that, within a few mean crossing times, one or two binaries have taken possession of most of the binding energy of the cluster. It takes longer if all masses are equal. Also, the evolution of these energetic binaries is accompanied by energetic escape of stars from the cluster.

In a thesis (Heggie, 1972) and also in a paper now under preparation by the author, these and other phenomena are discussed in detail by a combination of analytic and numerical techniques. Here some of the results will be summarized without detailed derivation, in order to emphasize qualitatively the implications of this work for our understanding of the dynamics of star clusters.

One question of importance concerns the relation between the evolution of the binaries and that of the cluster. Does the observed development of energetic binaries merely follow in response to the evolution of the cluster by other processes such as collisional relaxation, as held by Hénon (1972), or does the energetic behaviour of the binaries itself exert a substantial influence on changes in the structure of the cluster?

2. Binaries in Equilibrium

Arguing by analogy with the theory of chemical equilibrium, several authors (e.g. Jeans, 1929) have suggested that the distribution of binaries should relax to the Boltzmann form. The resulting distributions of ε, the pair binding energy, and of the eccentricity, e, take the forms

$$f(\varepsilon) \propto \varepsilon^{-5/2} \exp(\varepsilon/kT), \tag{1}$$

$$f(e) = 2e, \tag{2}$$

respectively, where $\frac{3}{2}kT$ is the mean random kinetic energy of the stars in the cluster. Equation (2) fits experimental data very well, one example being the result of Aarseth and Hills (1972), but Ambartsumian (1937) pointed out that it follows from many

Y. Kozai (ed.), The Stability of the Solar System and of Small Stellar Systems, 225–229. All Rights Reserved.

distributions besides the Boltzmann. The energy distribution (1) is in conflict with experiment, and the singularity for $\varepsilon \uparrow \infty$ is physically implausible.

Certainly the Boltzmann distribution is incorrect, but in addition the testimony of computer experiments is that, for energetic binaries at least, the phenomenon is not one of equilibrium. Interactions with other stars are crucial, and they lead to the evolution and, sometimes, the disruption of binaries. The analogy with atomic or nuclear collision processes is so close that the language of 'cross-sections', 'resonances', and so on, is appropriate.

3. Soft Binaries

The first case to be considered occurs when $\varepsilon \ll kT$. The internal energy of the binary is much less than the energy of the relative motion of the third body before collision, and such a binary is henceforth termed 'soft'. An equipartition argument suggests that, on average, kinetic energy is transferred to the internal energy of the binary, and so such binaries tend to disrupt (Gurevich and Levin, 1950).

For theoretical calculations the encounter may be treated as impulsive. Then the cumulative cross-section for energy changes greater than a certain amount may be computed for comparison with the results of specially-designed numerical experiments. The agreement is satisfactory, and also the theory confirms the conclusion of the above physical argument that soft binaries exhibit a tendency to disruption.

Soft pairs also tend to an approximate equilibrium distribution, the time scale for its attainment being about $0.06N(\varepsilon/kT)\, t_{cr}$, where N is the number of particles in the system, and t_{cr} is the mean crossing time. Since energy changes of order only ε are required for disruption of a soft pair with energy ε, this time is generally much less than a mean Chandrasekhar relaxation time, which is the time scale for changes of order kT.

4. Hard Binaries

The case when $\varepsilon \gg kT$, that of 'hard' binaries, is the other case for consideration. If the change in binding energy of the binary during the encounter is large and positive, the third body receeds with great energy afterwards. This is one source of the energetic escapers noted by Aarseth (1972) and others (e.g. Hayli, 1972). If the change in energy of the binary is negative and large enough, the third body will become bound to the binary. One component of the original binary may be ejected immediately in an 'exchange' event. Alternatively, from the work of Agekyan and Anosova (1967, 1968) it is known that most bound triple systems end with the release of an energetic escaper. The net effect of all these processes is that hard binaries tend to become harder.

Theory is more difficult than for soft pairs, but it can be developed roughly for ordinary close encounters and exchange events, and perturbation techniques are applicable for distant encounters. For example, it may be shown that binaries tend to form from the most massive stars.

The most difficult theory is that of bound triple systems. The rate at which they form can be calculated roughly, and then a detailed balance argument may be applied to find how fast these triple systems disrupt, if it is assumed that a bound triple system eventually 'forgets' most of its initial conditions except for the values of the classical integrals. This may be likened to the 'Bohr assumption' of nuclear physics (e.g. Blatt and Weisskopf, 1952). Here it is very successful in predicting the distribution of the energy of the escaping particle, as comparison with the results of Szebehely (1972) shows.

Wide encounters with a hard binary have one interesting property. Changes in the eccentricity are generally much bigger than changes in energy, and they are systematic if the third body is loosely bound to the binary. The latter phenomenon was noted in N-body experiments by Aarseth (1972).

It turns out that the average rate at which a hard binary gains energy is approximately independent of the binding energy of the binary. This result follows from two facts. First, the energy changes are typically of order ε itself, as a result of close encounters. Second, the total cross-section for close encounters is proportional to a, where a is the semimajor axis. It is not proportional to a^2, the area of the binary, because approaching particles converge towards a hard binary. Hence the rate of change of binding energy is proportional to $a\varepsilon$, which is independent of the energy. Note that this result applies to the *average* rate of change of energy; if a binary is too hard, a close encounter is so rare that none is likely to occur during the whole lifetime of the cluster.

5. Applications

It can be shown that the total time taken for the hard binaries to form and then absorb an amount of energy equal to the total energy of the cluster is about $(N^2/B)\, t_{cr}$, where B depends very sensitively on the structure of the cluster, but a value of about 100 is not untypical. Now the time taken for collisional relaxation to exchange the same amount of energy is about $(N/50 \log N)\, t_{cr}$ (Chandrasekhar, 1942). This is generally much less, and so we conclude that the evolution of binaries does not have a significant energetic effect on the overall evolution of the cluster provided that N is not very small. For example, during the final evolution of the core, which may be treated roughly as a small N-body system within the cluster, binary evolution is very important. Experimental results discussed by Hénon (1972) already pointed to the fact that binary evolution has negligible effect on the evolution of the cluster in its early stages, before core formation.

In each star cluster only one or two very hard binaries are produced. Yet, the proportion of visual binaries in the neighbourhood of the Sun is very high (van de Kamp, 1971). We conclude that only a few of these can have been formed by dynamical processes in star clusters, although such an explanation has been proposed (Gurevich and Levin, 1950; Kumar, 1972). The rest must have formed through processes occurring early in the lifetime of the stars, possibly at birth as discussed by Larson (1972) and others. Hence it may be expected that star clusters initially should contain large numbers of binaries.

If this is so, the previous estimate of the time taken for hard binaries to absorb the energy of the cluster is in need of revision. It becomes about $(N^2/70N_b(0))\, t_{cr}$, and if $N_b(0)$, the initial number of hard binaries, is comparable with N, then this time is of the same order as the relaxation time. It may be expected then that the evolution of hard binaries must play as important a part in the evolution of the cluster as collisional relaxation. The effects will be qualitatively different, since two-body encounters are elastic, while those involving binaries are typically superelastic.

It is urged that attention be paid in future to the dynamics of star clusters initially containing many binaries. They may be astrophysically more relevant than clusters without binaries, and their dynamics may be different.

Acknowledgements

I am grateful to several people, but especially to Dr S. J. Aarseth, for innumerable discussions on binaries, to Trinity College, Cambridge, for a Research Fellowship, and to the Institute of Astronomy, Cambridge, for its hospitality.

References

Aarseth, S. J.: 1972, in M. Lecar (ed.), *Gravitational N-Body Problem*, D. Reidel, Dordrecht, The Netherlands, p. 88.
Aarseth, S. J. and Hills, J. G.: 1972, *Astron. Astrophys.* **21**, 255–63.
Agekyan, T. A. and Anosova, Zh. P.: 1967, *Astron. Zh.* **44**, 1261–73.
Agekyan, T. A. and Anosova, Zh. P.: 1968, *Astrofiz.* **4**, 31–40.
Ambartsumian, V. A.: 1937, *Astron. Zh.* **14**, 207–19.
Blatt, J. M. and Weisskopf, V. F.: 1952, *Theoretical Nuclear Physics*, Wiley, London, p. 340.
Chandrasekhar, S.: 1942, *Principles of Stellar Dynamics*, University of Chicago Press, Chicago, p. 201.
Gurevich, L. E. and Levin, B. Yu.: 1950, *Astron. Zh.* **27**, 273–84.
Hayli, A.: 1972, in M. Lecar (ed.), *Gravitational N-Body Problem*, D. Reidel, Dordrecht, The Netherlands, p. 73.
Heggie, D. C.: 1972, Ph.D. Thesis, Cambridge University (unpublished).
Hénon, M.: 1972, in M. Lecar (ed.), *Gravitational N-Body Problem*, D. Reidel, Dordrecht, The Netherlands, p. 44.
Jeans, J. H.: 1929, *Astronomy and Cosmogony*, 2nd ed., Cambridge University Press, p. 302.
Kumar, S. S.: 1972, *Astrophys. Space Sci.* **17**, 453–58.
Larson, R. B.: 1972, *Monthly Notices Roy. Astron. Soc.* **156**, 437–58.
Szebehely, V.: 1972, *Celes. Mech.* **6**, 84–107.
van de Kamp, P.: 1971, *Ann. Rev. Astron. Astrophys.* **9**,

DISCUSSION

M. Lecar: Do you think that in actual globular clusters, initial binaries (i.e. binaries present at birth) significantly affect the time scale of the dynamical evolution of the cluster?

D. C. Heggie: Provided that the initial number of binaries that are hard (but not too hard) is significant in comparison with the number of stars, then their effect also will be significant. The initial number depends on details of the process by which binaries form, but the evidence suggests that the total number of all hard binaries will be large.

M. Hénon: You said that the time needed by a hard binary to swallow the whole energy of the system is of order $N^2 t_{cr}$, therefore, much larger than the relaxation time. But is not it true that in numerical experiments like those of Dr Aarseth a central hard binary usually absorbs a large fraction of the total energy?

D. C. Heggie: The statement on the time scale is true provided that the structure of the cluster is not too 'extreme'. For example, it is spatially extremely inhomogeneous in the late stages of core evolution when these hard binaries are observed to form. Alternatively, you can regard the core as a small n-body system within the cluster, and then the formula shows that binary evolution is quite rapid.

ON THE STABILITY OF SMALL CLUSTERS OR CLUSTER REMNANTS

L. O. LODÉN
Astronomical Observatory, Uppsala, Sweden

and

H. RICKMAN
Stockholm Observatory, Saltsjöbaden, Sweden

Abstract. There is strong evidence that the occurrence of mutually very nearby stars with identical spectral type and apparent magnitude is in many cases appreciably more frequent than one has reason to expect if there should only be a case of pure random combination. This phenomenon might be associated with the problem of stability of small stellar systems, if we can identify a certain number of the coincidences as some sort of cluster remnants. Besides the fact that there are several physical explanations of the coincidence phenomenon without consideration to dynamical stability conditions, there is also more than one stability aspect which may be regarded as reasonable. In the first instance we have to survey the possibilities that the stability of certain conglomerations of stars – particularly those with two components – may depend on the mass relation and will reach a maximum value for equal masses. As most straightforward we regard the conception that the most massive stars in a cluster (or even a multiple system) form the most stable and hence the most long-lived configuration, so that a cluster remnant most probably will consist of stars of rather equal mass.

1. Introduction

At visual inspection of objective-prism plates of Milky Way regions one frequently detects pairs – or occasionally higher multiples – of spectra of identical type and apparent magnitude and so close together on the plate that they form a conspicuous configuration (Lodén, 1969, 1973). In connection with a spectral survey of the southern Milky Way (Nordström and Sundman, 1973) about 2000 coincidence phenomena of this type with stars brighter than $m=14$ have been detected in the Carina–Crux–Centaurus–Norma region ($l=280°–330°$). The most common spectral type of the objects is A (36%), followed by B (31%) and F (13%). For stars of type B and A brighter than $m=13$ the number of detected coincidences overwhelmingly exceeds the one expected in a random distribution. For fainter stars the excess becomes less pronounced and finally it turns into a deficit. This, however, is to a large extent connected with detection problems, and the real distribution of the coincidences for faint stars is difficult to estimate from the present material. The angular separation between the components ranges between 0.′1 and 20′, and the correspondingly estimated linear separation in the transverse direction will range from 0.001 to 10 parsec.

2. Possible Explanations of the Phenomenon

There is no indication whatsoever that the coincidence phenomenon may be explained in terms of a unique astronomical feature. Probably there are a series of different circumstances, which in different cases cause two or more apparently equal stars to

Y. Kozai (ed.), The Stability of the Solar System and of Small Stellar Systems, 231–238.

appear close to the same line of vision from the observer. The following ones may be suggested:

(a) A projection effect of two stars with a certain difference in luminosity and a rather large mutual distance along the line of sight. In this case the stars would have no directly physical association with each other.

(b) A close but temporary encounter between two equal stars of different origin.

(c) A binary (or multiple) system with components either of accidentally coinciding types, or with a physically conditioned identity (Stock, 1972).

(d) Accidentally coinciding members of a cluster or association which is not dense enough to contrast against the general stellar background.

(e) A cluster remnant consisting of two or more stars of approximately equal mass.

It should be emphasized that this list is probably not complete, and there is reason to believe that all the above explanations, and possibly a few more, are relevant in at least some of the cases. However, the different cases do certainly not occur with the same frequency. In our opinion, the explanations (d) and (e) should predominate over the other ones (Lodén, 1973).

3. Explanations with Relevance to the Problems of Stability of Small Stellar Systems

It seems that the indicated explanation (e) above will be the most interesting one to investigate with consideration to stability problems. This is due to the circumstance that if the explanation in question is preponderant, a tendency towards favouring of equal components in cluster remnants can be considered as observationally supported.

Next we have to look for a reasonable physical explanation of such a phenomenon. It is obvious then that no cluster with a restricted number of members can be absolutely stable. That is, external forces caused by, for instance, differential galactic rotation, passing stars and gas clouds and perhaps nearby clusters, and in addition mutal interaction between the members themselves will cause stars to leave the cluster. It has also been shown (e.g. Michie, 1963) that this 'evaporation' to a maximum extent influences the low-mass stars.

We may therefore have the situation that originally rich clusters are now highly impoverished and contain, say five to ten stars, all of which are of great mass. This in turn implies that the masses (and hence the spectral types) of these stars are approximately equal. If, on the other hand, we had a cluster which originally contained only a few stars of various masses, it seems more uncertain whether the remnant of such a cluster should consist of the two most massive members. It would certainly be of great interest to investigate to what extent one could expect that situation.

It seems reasonable, however, to interpret the majority of the coincidences as some kind of cluster remnants, whether the case (d) or the case (e) is relevant. Obviously the explanation (d) does really exist. Coincidences have in fact been found in established clusters. A special observation programme intended to give a clue to the problem about the importance of explanation (d) is at present in progress.

Explanation (d) is also interesting with consideration to stability problems, because whether or not the objects are remnants of rich clusters they indicate the presence of a great number of small clusters in space. This number appears at maximum if the coincidences are only accidental, and becomes reduced if the coincidences are conditioned by some physical phenomenon, for example that two massive stars or two stars of equal mass may tend to occur close together in a cluster (or elsewhere). Such an effect would directly lead to the situation suggested above, i.e. a favouring of equal components in cluster remnants, because two massive stars at short distance compared to the size of the cluster may form a subsystem of relatively great stability and is likely to be found in the remnant of the cluster. It is therefore an urgent mission to find out theoretically whether this effect is to be expected or not. Before we proceed to a more detailed discussion about this point, we mention a few previous results of interest.

A depletion of low-luminosity – or low-mass stars in open clusters compared to the solar neighbourhood luminosity function has been found, and this depletion can be explained as a result of the 'differential evaporation' of low-mass stars according to Michie (1963). Oort (1957) drew attention to the fact that open clusters older than about 5×10^8 yr, i.e. those with earliest main-sequence members of type A1 or later, are almost missing. Spitzer (1958) found that this time is consistent with the time-scale for disruption of an open cluster by collisions with interstellar clouds. It is interesting to note that the A-type coincidences are the most common ones, and they might thus reveal the remnants of the disrupted clusters.

4. On the Existence and Persistence of Close Pairs of Massive Stars in Clusters

We now consider a question which might be important for the understanding of the coincidence phenomena: What are the theoretical conditions for the formation of close pairs of stars of equal mass in a cluster and how permanent would such a configuration be? It should be realized that this general problem is extremely difficult, and it would probably be necessary to study several cluster models with a computer and to analyze the results statistically in order to get a clear picture of the general situation. We will instead consider only some very simple concepts in a very simple model sketching the rough outlines.

To begin with, we consider two very massive stars instead of any two with equal mass. This is justified by the fact that the observed coincidences are formed by massive stars in a majority of the cases. Criticism may be raised against consideration of an isolated cluster since, in practice, there are no such ones, but on the other hand, if a certain result is found for an isolated cluster we may later on extend the discussion to concern a case of external influence.

Consider a cluster consisting of N stars of mass m and two stars of mass $m'>m$. The only quantities for which we have some relation are the total energy and angular momentum plus the virial of the cluster. This can be expressend through the quantity I given by:

$$I=\frac{1}{Nm}\sum_{i,j} m_i m_j r_{ij}^2. \tag{1}$$

In a steady-state cluster we should expect $I = \text{constant}$ (Chandrasekhar, 1942). If we separate I into three terms corresponding to (a) pairs of low-mass stars, (b) pairs of one high-mass and one low-mass star, and (c) the two high-mass stars, we find that the last term will be very small compared to the other two in realistic cases. Therefore a rearrangement of the stars so that the distance between the two high-mass members suddenly decreases would not have much influence on the virial of the cluster, and the steady-state situation will not be violated by such a rearrangement.

The total potential energy Ω, on the other hand, is more sensitive to changes in the contribution Ω' by the two massive members. For a cluster with an 'average radius' R, where the two high-mass members are situated at mutual distance r, the ratio of the two energies is:

$$\frac{\Omega'}{\Omega} \sim \frac{2}{N^2} \left(\frac{m'}{m}\right)^2 \cdot \left(\frac{R}{r}\right). \tag{2}$$

If we assume $N = 50$, $m' = 5m$, $r = 0.1R$, we get $\Omega' = 0.2\Omega$, i.e. a large fraction of the total potential energy of the cluster is stored in the mutual attraction of the two massive members. During the evolution of clusters it must frequently happen that two massive stars occur at such a short distance from each other, and the question is: What will happen then? Is their velocity of encounter so high that they will at once separate again? The answer is negative. If the encounter velocity is v the condition for immediate breakup of the system gives a maximum distance $r_{\max}$ depending upon v, and if we use the velocity dispersion of the low-mass cluster members, corrected for the fact that heavier stars move at correspondingly lower velocities in a steady state, we arrive at the condition for 'internal' stability of the pair:

$$r < r_{\max} \sim \frac{2R}{N} \left(\frac{m'}{m}\right)^2. \tag{3}$$

With the assumptions above we find that $r \ll r_{\max}$. Therefore, if the other members of the cluster shall split up this pair by their gravitational action, a large binding energy – of the order of Ω – has to be overcome.

We expect that the most active 'splitting-up process' will be close encounters with other members of the cluster. Some of these will increase the total energy of the pair and others will decrease it, so that the effect will be a dispersion of the energy increasing with time. The question is, if this dispersion grows rapidly enough to split up the pair before the whole cluster has been disrupted. It seems that in a dense cluster with many members the process proceeds so rapidly that one should, in general, not expect any stable pairs of massive stars. On the other hand, in more dispersed clusters or poor ones in the stage of breaking up such close pairs of massive members should be able to persist for much longer times.

In conclusion to this section we may state that if we consider dispersed clusters or eventually cluster remnants containing a small number of stars, it seems that theoretically we can expect coincidences of massive stars to appear more frequently than would be the case in a pure random configuration. We have no quantitative information about this, however, and furthermore there remains to study the effect of interaction between the star couple and other stars of approximately equal mass and the influence upon the close pairs of massive stars during the disruption of the cluster due to external forces.

5. Remarks About the Sketched Ideas

As critical remarks about the principal ideas in this contribution we mention the following: Only two of the explanations above of the coincidence phenomenon give significance to the discussion performed. Suppose, for instance, that most of the stars with coinciding spectral type constitute some sort of twins or triplets etc. created by cleavage into equal parts of rapidly spinning stars. In such a situation most of the ideas would have no foundation in reality. Besides this fact, it is not at all certain that open clusters constitute a unique type of stellar systems from physical point of view, particularly as far as the dynamical constitution and the stability are concerned. Different clusters may well show quite different dynamical properties, internally as well as in relation to the surrounding field and neighbouring clusters. An explanation in terms of clusters or cluster remnants should therefore not imply an attempt to force the groupings of stars into a certain theoretical pattern deduced from simplified models assuming the existence of a well-defined center of mass, extremely simplified mass distribution, a Maxwellian velocity distribution or other formalistic regularities.

It should also be remarked that the stability is never uniquely mass-conditioned, and that we shall expect a few cluster remnants that contain no particularly massive stars. Furthermore, we must again draw attention to problems connected with the persistence of close pairs of stars under external influences and in the internal gravitational potential of a small system of stars having approximatively equal mass. Besides, we have as yet no explanation to the extreme similarity between many coincidence components, considering the fact that in a real cluster there are stars of, roughly speaking, any mass.

We emphasize, however, that the discussion above is directly inspired by observational experience and thereby put in connection with open clusters. On the other hand, if some results can be reached about general stability aspects or if some 'stability rules' may be deduced from further studies of these coincidences, it is quite conceivable that these results might also have implications for other kinds of systems – such as associations or clusters of galaxies.

6. Problems Concerning Empirical Check

No theoretical treatment of the stability problem can be considered as meaningful until it gets into contact with observational evidence. In the case of open clusters, the empirical check most near at hand would be performed by means of a statistical study of the distribution of spectral types in established clusters – particularly as a function of the number of stars in the cluster. One would then get an idea about the change of the mass distribution with decreasing number of members. As far as possible, this should also be combined with an age classification of the clusters. A convincing result – positive or negative – of such a study would give a rather decisive information about the applicability of the theory.

Unfortunately, it is very difficult to estimate the number of stars in a cluster, as the apparent one is mainly conditioned by limiting magnitude and density of the general stellar background. It is particularly troublesome to obtain spectra of a reasonable number of cluster members. Consequently there is a considerable risk for selection effects.

Furthermore, the detection completeness is highly dependent upon the density and dimensions of the cluster. Small and poor clusters are much more difficult to discover

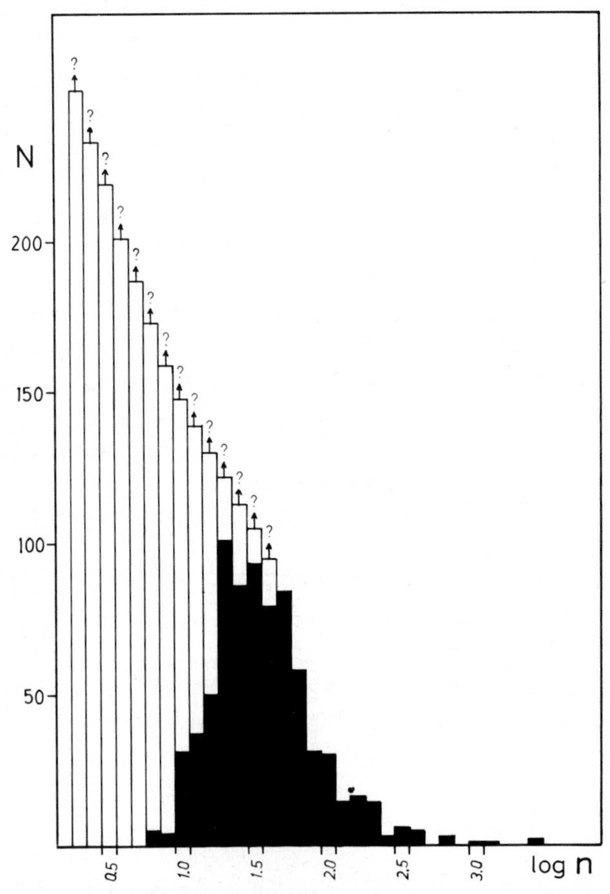

Fig. 1. Filled blocks show the apparent distribution of open clusters as a function of the logarithm of the indicated number of members. The open extensions indicate the suggested tentative extrapolation for the poor clusters which are difficult to detect.

than the large rich ones. If we make an attempt to give the apparent frequency of open clusters as a function of the quoted number of members according to the information found in the catalogue by Alter *et al.* (1970) we obtain the result shown in Figure 1. According to that figure there should be a maximum of clusters with about 25 members and a rapid decline towards the poorer ones. In our opinion, however, this picture is misleading because of the detection function and possibly there should in reality be a continuous increase of the frequency towards the two-star clusters ($\log n = 0.3$) as tentatively indicated in the figure.

It is therefore obvious that we cannot immediately apply the suggested checking procedure without a preliminary estimation of the true distribution of clusters with respect to their physical parameters.

7. Hints About Future Studies

For a fruitful continuation of the studies of cluster stability it is necessary to proceed with both observational and theoretical work. As far as the observations are concerned, we find it most urgent to continue the search for small star-poor clusters in order to get a better knowledge about the correlation between the number of members and the frequency of clusters than the one of today. Then there should also be performed a scrutiny of the stellar content with respect to spectral type for a representative selection of clusters so that we may obtain a better estimate of the correlation star number – spectral type.

On the theoretical side it is important to proceed with attempts to solve the stability problem and determine the evolution with time of a variety of the relevant parameters, in the first instance mass distributions of the cluster members and the topographic composition of the cluster (correlation between mass and coordinates). The cluster models subject to stability studies up to now have been very simplified and idealized and one has a feeling that they may well have been oversimplified and overidealized.

References

Alter, G., Balázs, B., and Ruprecht, J.: 1970, *Catalogue of Star Clusters and Associations*,
Chandrasekhar, S.: 1942, *Principles of Stellar Dynamics*, University of Chicago Press, Chicago, p. 200.
Lodén, L. O.: 1969, *Vistas in Astronomy* **11**, 161.
Lodén, L. O.: 1973, *Astrophys. Space Sci.* **24**, 511.
Michie, R. W.: 1963, *Monthly Notices Roy. Astron. Soc.* **126**, 331.
Nordström, B. and Sundman, A.: 1973, in C. Fehrenbach and B. E. Westerlund (eds.), 'Spectral Classification and Multicolour Photometry', *IAU Symp.* **50**, 85.
Oort, J. H.: 1958, *Pont. Acad. Sci.* **16**, 415.
Spitzer, L.: 1958, *Astrophys. J.* **127**, 17.
Spitzer, L. and Härm, R.: 1958, *Astrophys. J.* **127**, 544.
Stock, J. and Wroblewski, H.: 1972, *Astron. Astrophys.* **18**, 341.

DISCUSSION

R. H. Miller: Do you consider your results to indicate the presence of a hierarchic composition of the Milky Way?

L. O. Lodén: To some extent, yes.

S. R. Aarseth: According to my own calculations, the dynamical lifetime of a cluster will exceed that of stars earlier than A0. What is your comment to that?

L. O. Lodén: I admit that the situation is apparently critical for the B stars, but – on the other hand – we might have underestimated the influence of external forces.

THE DYNAMICAL EVOLUTION OF TRAPEZIUM SYSTEMS

C. ALLEN and A. POVEDA

Instituto de Astronomía, Universidad Nacional Autónoma de México, México, D.F.

Abstract. To study the dynamical evolution of trapezia, the equations of motion of the member stars in 30 different trapezia were integrated numerically. It was found that two-thirds of the trapezia still remain as such after 10^6 yr. Thus, the dynamical ages of trapezia containing O-stars are only slightly shorter than their nuclear ages. The properties of the binaries formed during the computations are studied. The numerical results are discussed from the observational point of view.

1. Introduction

Small stellar systems of the trapezium type pose interesting kinematic and dynamical problems, but little attention has, so far, been devoted to them. On the one hand, many workers have studied numerically the dynamical evolution of moderately rich and rich clusters (25 to several thousand stars), either by direct integration of the equations of motion, or by various Monte Carlo methods. On the other hand, the analytic solution of the two-body problem is well known and has been used to describe the motions of hierarchical triple or quadruple systems, since these systems are fairly stable over many periods of revolution. Between these extremes, only the unstable three-body problem has received significant attention. It seems therefore necessary, from the purely dynamical point of view, to fill the existing gap by studying the evolution of small stellar clusters (4 to 20 stars, say) of the trapezium type (see below for definitions of hierarchical and trapezium-type systems).

There are other reasons as well; observationally, trapezium systems seem to be not uncommon. The best known example is the Orion Trapezium, but already Ambartsumian (1954), in one of the first studies on trapezia, gave a catalogue of 108 such systems. More recently, a search of the *Index Catalogue of Double Stars* has revealed the existence of many more trapezia (Allen, 1974). Clearly, these samples contain valuable cosmogonical and dynamical information. In fact, as was first pointed out by Wallenquist (1944) and later by Markarian (1950, 1951) and Ambartsumian (1951a, b, 1954) multiple stars of the trapezium type are strongly related to young OB stars, and their space distribution is highly concentrated to the galactic plane.

Finally, several unproven statements about trapezium systems have been propagated in the literature. We cite two examples: (a) that trapezia are in expansion (Parenago, 1953; Ambartsumian, 1954; Sharpless, 1965); (b) that trapezium systems are extremely young (ages of about 10000 yr) (Parenago, 1953). Previous work (Poveda, 1973; Poveda *et al.*, 1974) dealt with point (a). One of the aims of the present paper is to clarify point (b).

The way we chose to study the dynamical evolution of trapezium systems is by direct numerical integration of the equations of motion of a sample of 30 different trapezia, over a time of 10^6 yr. In Section 2 the relevant definitions are given; in

Y. Kozai (ed.), The Stability of the Solar System and of Small Stellar Systems, 239–246.

Section 3 we discuss the numerical method and the initial conditions; in Section 4 our results are summarized and discussed; finally, some conclusions are given in Section 5.

2. Definitions

According to Ambartsumian (1954), a trapezium-type system, or trapezium, is defined as follows: let a multiple star system (of 3 or more stars) have components *a*, *b*, *c*,...; let *ab*, *ac*, *bc*,... denote the distances between the components; if three or more such distances are of the same order of magnitude, then the multiple system is of *trapezium type*; if no three distances are of the same order of magnitude, then the multiple system is of *ordinary type*. For the sake of clarity, we shall call these systems *hierarchical* systems. Two distances are 'of the same order of magnitude', in this context, if their ratio is greater than 1/3 but less than 3. Usually, only the distances projected on the plane of the sky can be measured. If one uses the projected distances in the definition of a trapezium, the sample of observed trapezia will contain, in addition to true physical trapezia, also *optical trapezia* and *pseudo-trapezia*. An optical trapezium is an apparent multiple system, whose components are not connected physically, and appear close together as a result of projection. A pseudo-trapezium is a multiple system of hierarchical type that appears to be of trapezium type due to projection. By means of an elaborate statistical analysis. Ambartsumian (1954) showed that the number of pseudo-trapezia is small among systems of spectral types O and B. We shall therefore concern ourselves solely with trapezia composed of massive stars.

For a clearer analysis of the dynamical evolution of trapezia, we find convenient to introduce three additional definitions:

If a system contains, among its components, three stars whose mutual distances are of the same order of magnitude, and such that the largest of these distances is less than 40000 AU, then the system is a *normal trapezium*.

If a system contains, among its components, three stars whose mutual distances are of the same order of magnitude, and such that the largest of these distances is greater than 40000 AU, then the system is a *wide trapezium*.

If a system satisfies the definitions of both normal and wide trapezium, then the system is a *mixed trapezium*.

3. Initial Conditions and Numerical Method

Since our aim is to simulate numerically the dynamical behaviour of real trapezia, we must choose initial conditions that correspond as closely as possible with the actual observational material.

The first thing that must be investigated is whether trapezia are systems with positive or negative total energy. In previous work (Poveda, 1973; Poveda *et al.*, 1974) we studied 33 trapezia for which there exist more than four observations (at different times) of position angles and separations; these systems were examined for changes in the separations between their components. The observations cover a long time span, generally over 100 yr. In *none* of the 33 trapezia was it possible to establish – not even marginally – the existence of an *overall expansion* or *contraction.* In 12 trapezia some change of separation was detected for one or two stars. In the case of the Orion Trapezium only Star E shows a detectable relative motion. If trapezia had positive energies, then their stars should have relative velocities of such magnitudes as to be easily observed over the time intervals considered above. To cite one specific example: for the Orion Trapezium the implied motion would be, in projection, 0.28″ in 130 yr, with respect to the center of mass. With exception of Star E, the motions of the other stars are a full order of magnitude smaller.

It was therefore concluded that observational evidence contradicts the idea of trapezia being expanding systems with positive energy. In view of the above results, we have to consider the dynamics of nonexpanding trapezia with negative total energy.

Our initial conditions and parameters are inspired by the observed properties of the Orion Trapezium. We take 6 components with a total mass of $170 M_\odot$ and choose our initial conditions as follows:

(i) Each trapezium has two stars of 50 $M_\odot$, two of $20 M_\odot$ and two of $15 M_\odot$, contained within a radius of 5000 AU. (This radius is our unit of length.)

(ii) Pairs of stars of equal masses were placed symmetrically on the coordinate axes x, y, z, at positions ± 1.0, respectively; to each of these symmetrical positions, a perturbation of random direction and random magnitude of up to 25% of the initial radius was superimposed; the perturbed positions are taken as initial positions. This procedure ensures that our systems will satisfy the definition of trapezia.

(iii) Each star was given a velocity of random direction and random magnitude with dispersions of less than 1 km s^{-1}; the velocities were then multiplied by an appropriate factor to force the system to satisfy the virial theorem; these virialized velocities are taken as initial velocities. This ensures that our systems will not be expanding or contracting initially.

A total of 30 different systems was integrated for a length of time of 10^6 yr, corresponding approximately to 30 mean crossing times.

The integrations were carried out with a computer programme developed by the Department of Aerospace Engineering and Engineering Mechanics of the University of Texas at Austin, which uses a 7th-order Runge–Kutta-Fehlberg technique, and regularizes the equations of motion of the closest pair by means of the Kustaanheimo–Stiefel transformation (Szebehely and Bettis, 1971).

Typically, we had errors in the total energy of the cluster of $\Delta E/E \simeq 3 \times 10^{-8}$ at the end of the run.

4. Results and Discussion

4.1. Dynamical Stability

The dynamical stability shown by the trapezia of our computer simulations is quite surprising. In the past, trapezia were thought to be very unstable dynamical systems, which after a few crossing times, either would evolve to a more stable hierarchical configuration, or else would eject stars until only a stable binary was left (Ambartsumian, 1954).

The present study shows, however, that even after 10^6 yr, or about 30 crossing times, out of an original sample of 30 trapezia, 19 systems are still of trapezium type. In other words, the probability that after 10^6 yr a trapezium remains as such is 63%. Closer examination of these 19 systems shows that they are not identical to the initial trapezia (see Table I). Specifically, only 6 systems have retained all their stars; the remaining

TABLE I

Trapezia that survive after 10^6 yr

Trapezium No.	$(r_{ij})_{max}$	Member stars	Classification
1	35.7	1, 2, 3, 4, 5, 6	M
2	39.3	1, 3, 4, 5, 6	W
3	21.3	1, 2, 3, 4	W
5	1.1	1, 2, 3, 4	N
7	56.9	1, 2, 4, 6	W
8	7.0	1, 2, 4, 5, 6	N
9	25.4	1, 2, 3, 4, 5	W
14	27.1	1, 2, 3, 4, 5, 6	M
15	69.4	1, 2, 3, 4	W
16	37.3	1, 2, 3, 4, 5, 6	M
17	29.5	1, 2, 3, 4, 5	W
18	18.2	1, 2, 3, 5, 6	W
19	21.6	1, 2, 3, 4, 5, 6	M
24	6.9	1, 2, 3, 4, 5, 6	N
25	13.1	1, 2, 4, 5, 6	M
26	14.4	1, 2, 3, 4	W
27	3.3	1, 2, 4, 5	N
28	39.1	1, 2, 4, 6	W
29	57.4	1, 2, 3, 4, 5, 6	W

Stars 1 and 2 are of $50 M_\odot$, stars 3 and 4 of $20 M_\odot$, stars 5 and 6 of $15 M_\odot$.
N: normal trapezium; W: wide trapezium; M: mixed trapezium.

13 ejected one or two stars each. Also, 15 of these 19 trapezia are what we have called wide trapezia. It is important to note that the evolution from initial trapezia to wide trapezia is *not* a process of expansion; rather, individual stars get thrown successively into weakly bound orbits. We suggest that the bright stars in galactic clusters are wide trapezia that evolved from normal trapezia originally immersed in the substratum of fainter stars that make up the bulk of the cluster. For example, the normal trapezium in Orion, which lies at the center of a rich cluster of infrared stars, may evolve, with the passage of time, into a system resembling the Pleiades, where the brightest stars form a wide trapezium of the type described above.

It is surprising that a total of 9 systems continue to be trapezia of dimensions comparable to their initial dimensions even after 10^6 yr of dynamical evolution. Put in another way, the probability that a normal trapezium continues to be a normal trapezium after 10^6 yr is 30%.

Let us turn our attention now to the remaining 11 systems, that is, the systems that were no longer trapezia at the end of the computations. Out of these 11 systems 3 dissolved completely, that is, they ejected stars with positive energies, until only a binary was left. In all three cases the binary was composed of the two most massive stars. The remaining 8 cases had evolved, at the end of the computations, into hierarchical systems. Stated differently, the probability that a trapezium dissolves in 10^6 yr is 10%; the probability that a trapezium becomes a hierarchical system in 10^6 yr is 27%. The previous results are schematically represented below:

30 trapezia + 10^6 yr =
- 19 trapezia
  - 15 wide trapezia (Nos. 1, 2, 3, 7, 9, 14, 15, 16, 17, 18, 19, 25, 26, 28, 29)
  - 9 normal trapezia (Nos. 1, 5, 8, 14, 16, 19, 24, 25, 27)
  - 5 mixed trapezia (Nos. 1, 14, 16, 19, 25)
- 3 dissolved systems (Nos. 4, 23, 30)
- 8 hierarchical systems (Nos. 6, 10, 11, 12, 13, 20, 21, 22)

The above results lead us to expect that in an unbiased sample of normal trapezia, the great majority of the systems should be younger than 10^6 yr; in this interval of time no significant evolution of the member stars is to be expected. Therefore, the ratio of O to B stars in such a sample of trapezia is just the intrinsic ratio in which these stars are born; in contrast, the ratio of O to B stars in the general field or in hierarchical systems should be different, because it is a function also of the ratios of their nuclear ages.

4.2. Comparison with Observations

When attempting to compare our theoretical results with observations, it is important to bear in mind that observational material on trapezia comes from catalogues of double and multiple stars. The inclusion of trapezia in these catalogues will depend on rather artificial selection criteria, like the angular separation of the closest pair (Heintz, 1973). Thus, we would not expect many of the observational counterparts of our wide trapezia to be included in the catalogues. Rather, from the observational point of view, many of these systems should be considered as dissolved, even though, theoretically, the distant stars are still bound to the system. The consideration that trapezia are not isolated systems, and therefore their distant stars are subject to perturbations by galactic tidal forces, passing field stars, etc. lends support to this conclusion. Therefore, our results should be examined anew, from the observational point of view. For this purpose, we introduce an arbitrary, but reasonable, cutoff distance of 40000 AU. When a star (or, exceptionally, a close pair) reaches a distance from all the other stars of 40000 AU we stop considering it as a member of the trapezium, even though it may still be weakly bound. Elimination of these stars yields results as summarized below:

30 trapezia + 10^6 yr =
- 9 normal trapezia (Nos. 1, 5, 8, 14, 16, 19, 24, 25, 27)
- 13 dissolved systems (Nos. 2, 3, 4, 7, 9, 15, 20, 21, 22, 23, 26, 28, 30)
- 8 hierarchical systems (Nos. 6, 10, 11, 12, 13, 17, 18, 29)

4.3. Binaries

It was found that, at the end of the computations, 33 binaries had been formed. We want to draw attention to the fact that stars 1 and 2, which are the most massive ones, show a strong tendency to form a binary (see Table II). In fact, in 21 out of 30 trapezia, stars 1 and 2 formed the tightest pair. In all of the trapezia that ejected four stars with positive energies, the left-over binary is composed of stars 1 and 2. In 77% of the trapezia that can be considered as dissolved from the observational point of view the same thing is true.

The orbital elements of the 33 binaries are given in Table II. As a group, these binaries are characterized by the following average values of their elements:

$$\langle a \rangle = 0.7000; \qquad \langle \varepsilon \rangle = 0.6581.$$

TABLE II

Orbital elements of binaries

Trapezium No.	Members of binary	a	ε
1	1–2	0.6078	0.2357
2	1–3	0.1354	0.9928
3	1–2	0.2468	0.2356
4	1–2	0.2293	0.6121
5	1–4	0.1969	0.6182
6	1–2	0.1301	0.5107
7	1–2	0.1758	0.9972
8	1–2	0.1450	0.8233
	4–5	3.8050	0.6568
9	2–3	2.0318	0.8863
10	1–2	0.3253	0.9094
11	1–2	1.1059	0.3953
	4–6	0.5834	0.1958
12	1–2	0.2683	0.5211
13	1–2	0.1976	0.2228
14	1–2	0.4860	0.8403
15	1–2	0.1966	0.9551
16	2–6	0.7765	0.7620
17	1–2	0.3313	0.7710
18	1–2	0.3041	0.7131
19	1–4	3.6753	0.9523
20	1–2	0.1894	0.6273
	4–6	3.8602	0.5767
21	1–2	0.2660	0.2140
22	1–2	0.2586	0.3024
23	1–2	0.2644	0.9213
24	1–2	0.4929	0.9463
25	1–2	0.3271	0.7992
26	1–2	0.2877	0.9616
27	2–5	0.0620	0.3457
28	2–4	0.1380	0.9413
29	1–2	0.8114	0.6371
30	1–2	0.1893	0.6377

It is to be noticed that the value of the mean eccentricity of the systems agrees very closely with 2/3, which is the value of the mean eccentricity expected for an ensemble of binaries whose distribution function in phase space is a function only of the energy of the pair (Ambartsumian, 1937). Furthermore, the distribution of eccentricities of our pairs follows approximately the Jeans–Ambartsumian law $N(\varepsilon)\,d\varepsilon = 2\varepsilon\,d\varepsilon$. No correlation was found between the periods and eccentricities of the 33 binaries. This result is in agreement with the observed properties of visual binaries (Heintz, 1969).

5. Conclusions

Trapezium-type systems are much more stable than previously thought; it was found that after 10^6 yr, or about 30 crossing times, 63% of the trapezia still remain as such. Therefore, the dynamical lifetimes of trapezia containing O stars are not much smaller than their nuclear ages.

In the same interval of time, only 10% of the trapezia dissolve, leaving behind a binary star; about 27% of the trapezia evolve into ordinary hierarchical systems.

A very strong tendency was shown by the two massive stars to join into a bound pair.

The distribution of the eccentricities of the binaries resembles the Jeans–Ambartsumian law; the mean eccentricity agrees with the value predicted by this law.

Acknowledgements

We gratefully acknowledge that the computations were done at the Centro de Servicios de Cómputo, Universidad Nacional Autónoma de México. It is a pleasure to thank Dr Sebastian von Hoerner for suggesting an algorithm to determine the state of hierarchization of a multiple system. We also thank Mrs L. Parrau and Mr M. Tapia for computational assistance.

References

Allen, C.: 1974, in preparation.
Ambartsumian, V. A.: 1937, *Astron. J. U.S.S.R.* **14**, 207.
Ambartsumian, V. A.: 1951a, *Publ. Acad. Sci. Armenia* **13**, 97.
Ambartsumian, V. A.: 1951b, *Publ. Acad. Sci. Armenia* **13**, 129.
Ambartsumian, V. A.: 1954, *Contr. Obs. Byurakan* **15**, 3.
Heintz, W. D.: 1969, *J. Roy. Astron. Soc. Canada* **63**, 275.
Heintz, W. D.: 1973, cited in *J. Roy. Astron. Soc. Canada* **67**, 67.
Markarian, B. E.: 1950, *Bull. Obs. Byurakan* **5**, 3.
Markarian, B. E.: 1951, *Bull. Obs. Byurakan* **9**, 3.
Parenago, P. P.: 1953, *Astron. J. U.S.S.R.* **30**, 249.
Poveda, A.: 1973, reported by W. D. Heintz in *J. Roy. Astron. Soc. Canada* **67**, 65.
Poveda, A., Allen, C., and Worley, C.: 1974, in preparation.
Sharpless, S.: 1965, *Vistas in Astronomy* **8**, 127.
Szebehely, V. and Bettis, D.: 1971, *Astrophys. Space Sci.* **13**, 365.
Wallenquist, A.: 1944, *Uppsala Ann.* **1**, No. 5.

DISCUSSION

R. H. Miller: If an observer looked at one of your trapezium systems at an age of 10^6 yr, could he conclude that the age was only 10^4 yr?

A. Poveda: On projection our clusters will not show a general expansion, but the random motions of the stars. Therefore, an expansion age of 10^4 yr could not be inferred from our calculations, as it cannot be inferred from observations either.

S. J. Aarseth: I would like to mention a new regularization method developed by Zare and myself for the general three-body problem. All motions can be studied, except the case of triple collisions, and perturbations from other bodies can be included. Furthermore, Heggie has developed a global regularization method for the general *N*-body problem in which all binary collisions can be regularized.

THE STABILITY OF TRIPLE STELLAR SYSTEMS

T. A. AGEKJAN and J. P. ANOSOVA

Astronomical Observatory, University of Leningrad, Leningrad, U.S.S.R.

The motion of triple stellar systems whose components are at rest at the initial epoch is studied. In this study all the components have equal masses, and at the initial epoch the component A is at $(-\frac{1}{2}, 0)$ on the (x, y)-plane, B is at $(\frac{1}{2}, 0)$, and C is anywhere in the region (s) bounded by the positive x-axis, y-axis and a line expressed by $(x+\frac{1}{2})^2 + y^2 = 1$. By this way all the possible initial configurations are given.

If the distance between any two components exceeds D, the largest possible value of the smallest distance, the system is called in the state of ejection at this epoch. There are two kinds of ejections, one with escape (disintegration) and one with return. It can be shown that the ejection starts when the distance of at least one component to the center of masses exceeds $3^{1/2}G/(-E)$, where E and G are, respectively, the energy of the system and the gravitation constant.

To each initial configuration corresponds the time of disintegration when the ejection with escape starts and the number of returns for the ejection with return.

The region (s) is divided into smaller regions (s') by lines of instability. When the initial position of C crosses one of this lines the number n changes. The value of the change is not restricted. On the other hand the time of disintegration is a continuous function of the initial coordinates of C within the region (s') and is infinity on the line of instability.

The regions (s') are divided into more smaller regions (s'') by lines of the change of the number n to one unity. When the initial position of C crosses one of this lines the number n becomes greater or smaller to one unity, the time of disintegration remains continuous. Within the region (s'') the number n is constant.

The area of regions (s') and (s'') are approximately proportional to n^{-2}.

Y. Kozai (ed.), The Stability of the Solar System and of Small Stellar Systems, 247. *All Rights Reserved.*

SUR UN INVARIANT INTÉGRAL DU PROBLÈME DES *n* CORPS: CONSÉQUENCE DE L'HOMOGÉNÉITÉ DU POTENTIEL

L. LOSCO

Laboratoire de Mécanique Théorique, Université de Besançon, 25030 Besançon Cédex, France

Abstract. An integral invariant is a generalization of first integrals to differential forms. Although this mathematical technique is more difficult, the integral invariants allow to obtain new properties for systems which have already well-known first integrals. Integral invariant of first order correspond to a 'local first integral' near any solution of motion. In this work I obtain an '11th local first integral' for the gravitational n-body problem, or any homogeneous n-body problem as planetary systems. As this local first integral contains a secular term, a discussion of the stability is obtained. The integral invariant is used for the construction of very particular solutions (Levi Civita's or Poincaré's singular solutions). These solutions realize conditional maximum or minimum of the contraction of the system.

1. Introduction

Un invariant intégral est une forme différentielle qui se conserve au cours du mouvement: c'est une généralisation naturelle de l'intégrale première. Pour les problèmes dont les intégrales premières connues sont limitées, les invariants intégraux peuvent amener à des conséquences très utiles. Dans une récente publication (Losco, 1973) je mets en évidence pour tout problème de n corps de positions $(q_1 \ldots q_n)$ de variables conjuguées $(p_1 \ldots p_n)$ dans un champ de forces homogène de $d^0 k$, l'existence de l'invariant intégral

$$\omega = (1 + \tfrac{1}{2}k)\, p\, \mathrm{d}q - \mathrm{d}(pq) + (1 - \tfrac{1}{2}k)\, t\, \mathrm{d}h,$$

où h est l'énergie.

Cet invariant intégral est déjà signalé dans Poincaré (*Méthodes nouvelles de la mécanique céleste*, tome 3), qui ne l'a exploité que pour le problème des deux corps.

Dans ω apparaît

$$pq = \sum_{i=1}^{n} m_i q_i \dot{q}_i = \frac{\mathrm{d}}{\mathrm{d}t}\,(\tfrac{1}{2} m_i q_i^2).$$

Posons $I = m_i q_i^2$ (moment d'inertie en O fixe ou G, viriel), $pq = \frac{1}{2}(\mathrm{d}I/\mathrm{d}t) = \frac{1}{2}\dot{I}$.

Si $k = -1$:

$$\omega = \tfrac{1}{2} p\, \mathrm{d}q - \tfrac{1}{2}\mathrm{d}\dot{I} + \tfrac{3}{2} t\, \mathrm{d}h,$$

ω ne correspond pas en général à une intégrale première sauf si $k = 2$ $(I = 4ht + I_0)$.

L'invariant complet (Cartan, 1971) est aussi très utile:

$$\omega^* = \omega - kh\, \mathrm{d}t.$$

Extensions. Par la même méthode (Losco, 1973):

Y. Kozai (ed.), The Stability of the Solar System and of Small Stellar Systems, 249–255.

(a) Le lagrangien est homogène de degré 2 par rapport à q et $\dot{q}$, alors on a

$$\omega = p\,\mathrm{d}q - q\,\mathrm{d}p.$$

Ceci s'applique aux mouvements relatifs à un système d'axes tournants dans un potentiel du second degré.

(b) Le potentiel dans le plan tournant est du type spirale:

$$L = \tfrac{1}{2}(\dot{r}^2 + r^2\dot{\theta}^2) + \omega r^2\dot{\theta} + \tfrac{1}{2}\omega^2 r^2 + U_0(r) + A(r)\cos[\phi(r) - m\theta].$$

Supposons que U_0 et A sont homogènes du second degré et que $\phi(r) = \alpha \log r$:

$$\omega = 2p_\theta\,\mathrm{d}\theta - r\,\mathrm{d}p_r - (\alpha/m)\,\mathrm{d}p_\theta + p_r\,\mathrm{d}r.$$

2. Application à la stabilité

Poincaré montre que si $\omega = a_i\,\mathrm{d}x^i$ est un invariant intégral d'ordre 1 de $\mathrm{d}x/\mathrm{d}t = X$, si $x(t)$ est une solution particulière et $x(t)+\xi(t)$ une solution voisine, $\xi(t)$ étant alors solution des équations variationnelles:

$$\frac{\mathrm{d}\xi^i}{\mathrm{d}t} = \sum_j \frac{\partial X^i}{\partial x^j}(x(t))\,\xi^j$$

(équations fondamentales pour l'étude de la stabilité), alors $\phi(\xi, t) = a_i(x(t))\,\xi^i$ est une intégrale première des équations variationnelles: $\phi(\xi, t)$ est une intégrale première 'locale' au voisinage de toute solution particulière.

$$\varphi = \tfrac{1}{2}kp\xi - qn + (1 - \tfrac{1}{2}k)\,t\,[H(q+\xi, p+\eta) - H(q, p)]$$

est intégrale première des équations variationnelles du problème posé au 1.

Si on réduit par $H = h$, on a l'intégrale première locale

$$\bar{\phi} = \tfrac{1}{2}kp\xi - q\eta.$$

La forme de ϕ est particulièrement intéressante car elle contient un terme séculaire $(1 - \frac{1}{2}k)\,t\,\delta h$. Pour le problème des n corps δh n'est jamais nul car il n'y a pas de configuration d'équilibre. Supposons $k \neq 2$, et supposons que l'on étudie le voisinage d'une trajectoire $(q(t), p(t))$ bornée, comme ϕ se conserve au cours du mouvement et que $(1 - \frac{1}{2}k)\,t\,\delta h \to \infty$, on en déduit l'instabilité dès que $\delta h \neq 0$ et $k \neq 2$.

Ainsi toutes les solutions bornées, paramétrées par le temps, du problème des n corps sont instables: toute solution d'énergie différente s'éloigne: ce résultat est valable, par exemple, pour la configuration équilatérale de Lagrange. Le raisonnement est en défaut si $\delta h = 0$ ou $k = 2$.

Cependant si on considère le simple problème keplérien, on sait que les ellipses sont stables du point de vue géométrique: si M décrit une ellipse E, M' décrit une ellipse E' voisine et le raisonnement précédent indique que si $\delta h \neq 0$ il y a décalage horaire sur E' par rapport à E: il est donc intéressant de poser le problème de la stabilité 'orbitale', stabilité de l'arc géométrique (orbite) indépendamment du paramétrage temporel.

Supposons que M décrive T, M' décrit une orbite adjacente T', M' s'écarte de M du fait du décalage horaire si $\delta h \neq 0$. Si il y a stabilité on peut trouver sur T' M_1 voisin de M et atteint à $t' = t + \Delta t$. Nous sommes ainsi amenés à comparer deux points M et M_1 atteints à des instants différents: il faut donc considérer l'invariant complet ω^*. Faisons donc un changement temporel $\mathrm{d}t = \mu\,\mathrm{d}\tau$, ceci afin de freiner M' ou de l'accélérer afin de le comparer à M.

Par une extension du lemme on trouve que

$$\phi = (1 + \tfrac{1}{2}k)\, p\xi - (p\xi + q\eta) + (1 - \tfrac{1}{2}k)\, t\, \delta h - kh\, \Delta t$$

est intégrale première des équations variationnelles paramétrées par τ, t étant la loi horaire sur T, $t + \Delta t$ celle sur T'.

Ainsi, si il y a stabilité orbitale on sait comment se comportent les lois horaires de deux orbites voisines:

$$\Delta t/t \sim (1/k - \tfrac{1}{2})\, \mathrm{d}h/h.$$

Remarquons que l'on peut vérifier aisément cette relation pour le problème keplérien à l'aide des formules

$$n^2 a^3 = \text{cte}, \qquad a = -\mu/2h, \qquad n(t - t_0) = l,$$
$$\Delta t/t \sim -\delta n/n = \tfrac{3}{2}\delta a/a = -\tfrac{3}{2}\delta h/h.$$

Par conséquent le comportement est analogue à celui du problème keplérien dans un espace de phases de dimension $6n$.

Le problème de la stabilité orbitale peut se poser de façon 'conditionnelle' (Losco, 1968). Si $q = q(t)$ est solution, $Q_\varepsilon = (1 + \varepsilon)\, q[(1 - (1 - \tfrac{1}{2}k)\, \varepsilon)\, t]$ est aussi solution. Par conséquent, toute solution particulière q peut être plongée dans un ensemble de solutions Q_ε d'énergie $(1 + k\varepsilon)\, h$: nous sommes dans les conditions d'application de la stabilité conditionnelle: si chaque solution Q_ε est stable pour les mouvements de même énergie, la solution $q = q(t)$ est orbitalement stable. L'intérêt est de se ramener à $\delta h = 0$, sans décalage horaire.

L'étude précédente étant due à l'homogénéité du potentiel, elle reste par conséquent valable pour le problème restreint de n corps soumis à leur attraction et à celle d'une masse fixe placée en O, mais par contre n'est plus valable pour le problème restreint classique des trois corps car il n'y a plus homogénéité.

3. Caractérisation de solutions particulières grâce à Γ

Levi Civita caractérise des solutions particulières d'un système canonique de hamiltonien H de la façon suivénte: si $F_1, \ldots, F_p$ sont p intégrales premières d'un système différentiel canonique de hamiltonien H, l'ensemble des points où, sur la surface d'équations $F_1 = C_1, \ldots, F_p = C_p$, l'énergie H est extrémum, est un ensemble invariant formé de solutions remarquables appelées solutions de Levi Civita. Ainsi, par exemple, pour le point matériel dans un plan soumis au champ de forces $U(r)$, les solutions circulaires sont des solutions de Levi Civita associées à l'intégrale première

$r^2\dot{\theta}=C$. On exprime analytiquement la condition d'extrémum lié de H par la condition des multiplicateurs de Lagrange:

$$\mathrm{d}H=\lambda_1\,\mathrm{d}F_1+\cdots+\lambda_p\,\mathrm{d}F_p,$$

ou mieux, grâce au produit extérieur:

$$\mathrm{d}H\wedge\mathrm{d}F_1\wedge\ldots\wedge\mathrm{d}F_p=0.$$

Poincaré a introduit lui aussi des solutions remarquables: les solutions singulières, que l'on retrouve dans le livre de Whittaker (1917). Ces solutions singulières sont associées, plus généralement que les solutions de Levi Civita, à des invariants intégraux $\omega_1\ldots\omega_p$ d'ordre 1, formes différentielles invariantes de degré 1: elles sont caractérisées par le fait que le long de ces solutions singulières $\omega_1\ldots\omega_p$ sont lineéairement dépendantes:

$$\omega_1\wedge\ldots\wedge\omega_p=0.$$

J'ai étudié en détail ces solutions dans ma thèse (Losco, 1972) en mettant en évidence de telles solutions pour des problèmes concrets de la mécanique céleste, mais pour la première fois je vais mettre en évidence de telles solutions pour un invariant intégral, avec une interprétation concrète d'extrémum conditionnel sur $\dot{I}$: vitesse de contraction ou de dilatation.

3.1. Solutions singulières $\omega\wedge\mathrm{d}H=0$

Ces solutions vérifient $(\frac{1}{2}p\,\mathrm{d}q-\mathrm{d}(\frac{1}{2}\dot{I}))\wedge\mathrm{d}H=0$, donc:

$$\mathrm{d}(\tfrac{1}{2}\dot{I})=\tfrac{1}{2}p\,\mathrm{d}q+\alpha\,\mathrm{d}H.$$

Elles réalisent parmi les solutions de même énergie, de même configuration, à chaque instant, l'extrémum de $\dot{I}$ (que j'appelle vitesse de contraction),

$$\mathrm{d}H=-\lambda(\tfrac{1}{2}p\,\mathrm{d}q-\mathrm{d}(\tfrac{1}{2}\dot{I}))=\lambda(\tfrac{1}{2}p\,\mathrm{d}q+q\,\mathrm{d}q).$$

Or, $\mathrm{d}H=\dot{q}\,\mathrm{d}p-\dot{p}\,\mathrm{d}q$. Donc:

$$\mathbf{V}_i=\lambda\mathbf{M}_i, \tag{1}$$

$$\mathbf{\Gamma}_i=\tfrac{1}{2}\lambda\mathbf{V}_i. \tag{2}$$

(1) montre qu'il s'agit de solutions par homothétie; (2) et (1) impliquent $M_i(t)=$ $=A_i(3t+2\alpha)^{2/3}$.

Un calcul direct permet de caractériser ainsi toutes les solutions par homothétie, d'énergie nulle: ce sont les solutions de Levi Civita $\omega\wedge\mathrm{d}H=0$.

Pour le problème plan des 3 corps les seules solutions par homothétie sont les solutions équilatérales (Annexe) ou alignées. Il existe de nombreuses solutions par homothétie: Walvogel (1972) a mis en évidence récemment des solutions par homothétie pour un nombre de corps arbitraire et dans l'espace.

3.2. SOLUTIONS DE LEVI CIVITA $\omega \wedge \mathrm{d}H \mathrm{d}C=0$

(a) *Problème plan*

Soit $C=m_i \mathbf{GM}_i \wedge \mathbf{V}_i$ le moment cinétique par rapport à GZ.

$$\mathrm{d}C=(\mathbf{z} \wedge \mathbf{GM}_i)\, \mathrm{d}(m_i \mathbf{V}_i)+(m\mathbf{V}_i \wedge \mathbf{z})\, \mathrm{d}(\mathbf{GM}_i).$$

Donc les solutions de Levi Civita sont telles que

$$\mathrm{d}H=\lambda(\tfrac{1}{2}P\, \mathrm{d}q+q\, \mathrm{d}p)+\mu\, \mathrm{d}C,$$

$$\dot{M}_i=\lambda M_i+\mu z \wedge M_i, \tag{3}$$

$$\dot{V}_i=\tfrac{1}{2}\lambda V_i-\mu V_i \wedge z. \tag{4}$$

(3) montre qu'il s'agit de solutions par similitude; (3) et (4) aboutissent à

$$\dot{q}=\frac{2}{3t+2\alpha}\, q+\frac{\beta}{3t+2\alpha}\, z \wedge q. \tag{5}$$

Il s'agit de solutions spirales: l'orbite est une spirale logarithmique. Ces solutions réalisent à chaque instant l'extrémum de la contraction parmi les solutions de même configuration, de même énergie, de même moment cinétique. Or de telles solutions ne peuvent exister (sauf si $\beta=0$: (3.1)), car les mouvements par similitude plans sont nécessairement kepleriens.

Ceci montre que ω est indépendant de H et C.

(b) *Problème de l'espace*

Soit $Gx_1x_2x_3$ un repère galiléen, $C_1C_2C_3$ les moments cinétiques sur ces axes, les solutions singulières sont caractérisées par:

$$0=\omega \wedge \mathrm{d}H \wedge \mathrm{d}C_1 \wedge \mathrm{d}C_2 \wedge \mathrm{d}C_3,$$

$$\Omega$$

$$\dot{M}_i=\lambda M_i+(\mu_1 x_1+\mu_2 x_2+\mu_3 x_3)\, M_i, \tag{6}$$

$$\dot{V}_i=\tfrac{1}{2}\lambda V_i+\Omega \wedge V_i.$$

D'après (6) ce sont encore des mouvements par similitude,

$$\ddot{M}_i=-\tfrac{1}{2}\lambda V_i+\Omega \wedge V_i=\dot{\lambda} M_i+\lambda \dot{M}_i+\dot{l} \wedge M_i+\Omega \wedge \dot{M}_i=$$
$$=\dot{\lambda} M_i+\lambda^2 M_i+\lambda\Omega \wedge M_i+\dot{\Omega} \wedge M_i+\Omega \wedge \dot{M}_i.$$

Donc:

$$\dot{\lambda}=-\tfrac{3}{2}\lambda^2, \qquad \lambda=2/(3t+2\alpha), \qquad (\dot{\Omega}+{}^{\mathbf{3}}\lambda\Omega) \wedge M_i=0.$$

Si la configuration n'est pas alignée: $\dot{\Omega}+\frac{3}{2}\lambda\Omega=0$, Ω a par conséquent une direction fixe $\mathbf{z}$: $\Omega=[\beta/(3t+2\alpha)]\, z$. Or, un calcul immédiat montre que:

$$\sum m_i \mathbf{M}_i \wedge \mathbf{\Gamma}_i=\beta(3t+2\alpha)^{-2/3}\, (m_i x_i^2 \mathbf{z}-\sum m_i(x_i z)\, \mathbf{x}_i+m_i(x_i z)\, \mathbf{x}_i \wedge z=0).$$

Projetons sur z:

$$0=(3t+2\alpha)^{-3/2}\,\beta\left(\sum m_i x_i^2 - m_i(x_i z)^2\right).$$

Donc soit $\beta=0$ (3.1), soit la configuration est alignée.

Montrons qu'il n'est pas possible d'avoir de tels mouvements, même en configuration alignée: Si la configuration est alignée sur Ω, comme $\dot{M}=\lambda M+\Omega\wedge M$, le mouvement est par homothétie (3.1); si la configuration n'est pas alignée sur $\mathbf{\Omega}$, prenons des axes i, j, k tels que $\mathbf{i}$ porte les points M_i, (i, j) est le plan $(\mathbf{q}, \mathbf{\Omega})$.

$$\ddot{M}_i\wedge M_i=\tfrac{1}{2}\lambda(\Omega\wedge M_i)\wedge M_i+(\Omega\wedge(\Omega\wedge M_i))\wedge M_i.$$

Un calcul rapide de $\ddot{M}_i\wedge M_i$ montre que nécessairement $\Omega=0$ ou $\theta=0$. Cette réponse négative est regrettable mais intéressante en elle-même, car elle aurait pu donner une interprétation variationnelle des solutions spirales comme réalisant une contraction maximum pour un moment cinétique et une énergie donnée à un instant t à n corps en attraction newtonienne.

3.3. Les solutions par homothétie, par similitude

Cherchons plus généralement quels sont les mouvements qui réalisent l'extrémum de $\dot{I}$ pour une énergie donnée h, une configuration q donnée à l'instant t. $\mathrm{d}\dot{I}=\mathrm{d}H+ +\beta_i\,\mathrm{d}q^i$ traduit la condition liée à satisfaire. Il suffit que $\dot{q}^i=q^i/\alpha$.

Par conséquent, si pour une configuration et une énergie donnée à t on cherche la distribution de vitesses réalisant l'extrémum de $\dot{I}$, on trouve la distribution de vitesses par homothétie.

Tous les mouvements par homothétie réalisent donc l'extrémum de $\dot{I}$ à chaque instant parmi les mouvements qui auraient la même énergie et la même configuration:

$$\mathrm{d}\dot{I}\wedge\mathrm{d}H\wedge\mathrm{d}q^1\wedge\cdots\wedge\mathrm{d}q^n=0.$$

Par un calcul tout à fait analogue, on trouve que c'est une distribution de vitesses par similitude qui réalise à l'instant t l'extrémum de I pour h, C, q données. Tous les mouvements par similitude réalisent cet extrémum à tout instant:

$$\mathrm{d}\dot{I}\wedge\mathrm{d}H\wedge\mathrm{d}C\wedge\mathrm{d}q^1\wedge\cdots\wedge\mathrm{d}q^n=0.$$

Donc:

(a) Les solutions par rotation sont des solutions de Levi Civita réalisant l'extrémum de h pour C donné.

(b) Les solutions par homothétie réalisent l'extrémum de $\dot{I}$ pour h, q données. Seules sont de Levi Civita les solutions d'énergie nulle ($\omega\wedge\mathrm{d}H=0$).

(c) Les solutions par similitude réalisent l'extrémum de $\dot{I}$ pour h, C, q données. Seules seraient de Levi Civita des solutions spirales.

Annexe. Recherche des solutions par similitude plane pour trois corps

Carathéodory a montré en 1933 que les seules solutions par similitude planes sont les solutions de Lagrange. Nous retrouverons plus simplement ce résultat:

Posons: $Z=z_2-z_1$, $z_3-z_1=aZ$. $a=\alpha+i\beta$ vérifie alors

$$\text{(E)}\qquad m_1(1+a)\left(1-\frac{1\ldots}{|1+a|^3}\right)+m_2a\left(1-\frac{1}{|a|^3}\right)+m_3a(1+a)\left(\frac{1}{|1+a|^3}-\frac{1}{|a|^3}\right)=0.$$

Cherchons m_1, m_2, m_3 tous trois positifs satisfaisant à (E). Posons

$$A=1-1/|1+a|^3,\qquad B=1-1/|a|^3,\qquad C=1/|1+a|^3-1/|a|^3=B-A,$$
$$m_1(1+\alpha)A+m_2\alpha B+m_3(\alpha+\alpha^2-\beta^2)(B-A)=0,$$
$$m_1\beta A+m_2\beta B+m_3(\beta+2\alpha\beta)(B-A)=0.$$

Sauf lorsque $\beta=0$ (configurations alingnées de Lagrange) et $A=B=0$ (configuration équilatérale de Lagrange):

$$m_1=\lambda B(B-a)(\alpha^2+\beta^2),$$
$$m_2=-\lambda A(B-A)[(1+\alpha)^2+\beta^2],$$
$$m_3=\lambda AB.$$

Donc $B(B-A)$, $-A(B-A)$, AB sont tous du même signe. —A et B sont demême signe, contraire à celui de $B-A$, ce qui est impossible.

Références

Cartan, E.: 1971, *Leçons sur les invariants intégraux*, Hermann, Paris.

Losco, L.: 1968, *Bull. Astron.* Sér. 3, **III**, Fasc. 4, 433–42.

Losco, L.: 1972, *Solutions particulières et invariants intégraux*, Thèse Besançon.

Losco, L.: 1973, *C. R. Acad. Sci. Paris* **277**, Sér. A, 323–25.

Poincaré, H.: 1957, *Les méthodes nouvelles de la mécanique céleste*, Dover publications, New York.

Waldvogel, J.: 1972, *Celes. Mech.* **5**, 37.

Whittaker, E. T.: 1917, *A Treatise on the Analytical Dynamics*, University Press, Cambridge, U.K.

Wintner, A.: 1941, *The Analytical Foundations of Celestial Mechanics*, Princeton University Press, Princeton, N. J., U.S.A.

DISCUSSION

T. Inoue: Vous avez bien trouvé la onzième intégrale première. Est-ce que c'est possible de me montrer une forme concrète de cette intégrale (dans le cas du problème des trois corps)?

L. Losco: $F=p\xi/2-(p\xi+q\eta)+2t\,\delta h/2$, si j'ai un mouvement particulier que je connais $q=q(t)$, $p=p(t)$, et sont petits pour que, $q+\xi$ et $p+\eta$ explorent le voisinage du mouvement particulier.

S. F. Mello: Est-ce que vous croyez qu'en changeant légèrement la formulation du problème restreint, il serait possible de servir de vos résultats pour étudier les mouvements autours de L_4 et L_5?

L. Losco: Le potentiel du problème restreint n'étant pas homogène, les résultats ne sont pas applicables.

F. Nahon: La onzième intégrable première locale, déjà signalée par Cartain, peut servir à l'étude de l'inégalité de Sundman et à l'étude de la diffusion dans les problèmes voisins des problèmes intégrables.

NUMERICAL EXPERIMENTS ON EXPANDING GRAVITATIONAL SYSTEMS

G. JANIN and M. J. HAGGERTY

European Space Operations Centre, Darmstadt, Germany, and The University of Texas at Austin, U.S.A.

It has been suggested that particle positions in expanding N-body gravitational systems may become increasingly correlated. Such incoherent density fluctuations could develop into bound clusters. Similarly superclusters of such clusters may be dynamically created. We present results of explorations of this idea, using computer simulation. Accurate integrations are made of gravitationally interacting systems of a few hundred particles.

The effect of the rest of the universe on the system is simulated by adding a smooth time-dependent gravitational field. The results are compared with the evolution of similar systems with more discrete particles placed outside. During the computer experiments a reduction of the number of bodies is sometimes made by reducing small subclusters to single particles at their centres of mass.

With near balancing kinetics and potential energies for the system, the rate of formation of well-isolated small clusters is lower than postulated for the simplest dynamical models of creation of hierarchical structures.

If the initial total potential energy slightly exceeds the initial expansion kinetic energy in magnitude, then our results suggest that a hierarchical density distribution is in the process of being created.

If the total energy is positive, then our results suggest a two-level hierarchy for the density distribution.

The correlations between binding energies, angular momenta and masses of small isolated clusters agree well with observations and with some previous theoretical and computational results. Indirect evidence is presented for the rapid production of massive but less conspicuous clusters.

A description of this study will be published in *Astronomy and Astrophysics.*

DISCUSSION

R. H. Miller: What do you mean by several orders of subclustering for initial energy <0?

G. Janin: Instead of speaking of several orders of clustering, one can introduce the less ambiguous notion of hierarchical clustering, meaning that the interaction between clusters or superclusters is similar to the interaction between particles in a cluster.

A. Poveda: If you have cases with negative initial energy, do you mean that after some time the initial expansion is reversed to a contraction of the whole system?

G. Janin: We have cases with slightly negative initial total energy. In these cases a part of the system keeps expanding.

D. C. Heggie: In the case when only one order of subclustering occurs, what is the distribution of the sizes of the subclusters? Do you get only pairs, or are there also triples, and so on?

G. Janin: In a typical case, if one defines the five following categories of clusters containing 1 star, 2–3, 4–7, 8–15, 17–31 stars, after two crossing times the proportion between the population of two successive categories is nearly 2:1.

Y. Kozai (ed.), The Stability of the Solar System and of Small Stellar Systems, 257. *All Rights Reserved.*

NUMERICAL EXPERIMENTS ON THE STABILITY OF SPHERICAL STELLAR SYSTEMS

M. HÉNON
Observatoire de Nice, 06300 Nice, France

The concentric shell model is used to investigate numerically the stability of spherical steady-state stellar systems. Polytropic models with an isotropic velocity distribution are found to be stable almost down to the limiting index $n = \frac{1}{2}$. 'Generalized polytropes', with a distribution function depending on energy and angular momentum, show instability when n is low and the velocity distribution is radially elongated.

The full text of the paper has been published in *Astronomy and Astrophysics* **24**, 229 (1973).

DISCUSSION

P. Bouvier: When you speak of a stable system, do you mean stable with respect to any kind of perturbations?

M. Hénon: No. The stability considered here is concerned only with perturbations preserving spherical symmetry.

Y. Kozai (ed.), The Stability of the Solar System and of Small Stellar Systems, 259. *All Rights Reserved.*

ON THE 'THERMODYNAMICS' OF SELF-GRAVITATING N-BODY SYSTEMS

R. H. MILLER

Dept. of Astronomy and Astrophysics, University of Chicago, Chicago, Ill. 60637, U.S.A.

Abstract. Results of some simple 'thermodynamic' experiments on self-gravitating n-body systems are reported for a variety of boundary conditions. Systems placed in specularly reflecting enclosures did not show any unusual behavior, even though a variety of conditions was tried in an attempt to start a 'gravothermal castastrophe'. Similarly, there was no tendency to transfer energy between 'hot' and 'cool' subclosures within a given cluster. However, systems in 'isothermal' enclosures gave up energy to the enclosure at a surprisingly high rate, and sustained the energy-transfer rate as long as the experiment was continued. An explanation of these different behaviors was sought and found in an examination of the premises that underlie certain attempts to construct a thermodynamics for self-gravitating systems. Conventional application of the H-theorem implies violations of the n-body equations of motion and predictions not consistent with observation. Both the 'gravothermal catastrophe' and the experiments in an 'isothermal' enclosure share this violation of the equations of motion. A new formulation that allows for all the interactions in an n-body system shows that isolated n-body systems need not form binaries or condense into other subaggregates. The virial theorem follows as an ensemble average over the microcanonical ensemble.

1. Introduction

The work reported here represents a different approach to the study of systems of self-gravitating mass points from the mathematical methods you have heard so nicely described at this symposium by Moser and Pollard, among others. The mathematical approach is an attempt to describe *all* solutions of the n-body equations of motion; it must be capable of treating anything an n-body system can do, and it founders on the technicalities presented by singular cases. The methods of statistical mechanics provide an alternative approach by means of which it is possible to sidestep measure-zero effects like the singularities that plague the mathematical approach. This is accomplished by averaging over the parameter space (or phase space) in such a way that measure-zero effects are unimportant. The methods promise an overview without requiring detailed solutions. But the usual price is paid – equilibrium situations are stressed, or for nonequilibrium systems, the best that can be done is to describe the approach to equilibrium. Many of the obvious questions cannot be answered within this framework.

Not only does statistical mechanics seem to be a tool that should be useful in stellar dynamics, but stellar dynamics can provide a useful testing-ground for some unsolved problems in statistical mechanics. Galaxies, clusters of galaxies, and star clusters, are not uniform; they are groups of stars or galaxies that stand well separated from their neighbors. Under the presumption that what we see is some kind of equilibrium state, it is clear that equilibrium of a stellar dynamical system is *not* the uniform state of maximum entropy that characterizes equilibrium in most cases that we know how to treat by statistical mechanics. But many naturally occurring systems are nonuniform; when we look about us, very few of the systems that we see in nature

Y. Kozai (ed.), The Stability of the Solar System and of Small Stellar Systems, 261–272.

are uniform. Biological systems, especially, are nonuniform. (For an interesting discussion of the statistical mechanics of biological systems, see Prigogine *et al.*, 1972). Thus stellar systems provide a challenge: they represent what is quite likely the simplest kind of naturally occurring system for which the equilibrium state is not uniform; they are certainly the simplest one-component system that is known to display this property. The study of stellar dynamical systems promises to provide clues, not only to the understanding of the beautiful objects we see in the sky, but also of things closer to home in which nonuniformity seems to be one of the principal attributes.

Unfortunately, the statistical approach turns out (for systems of negative total energy) to emphasize an uninteresting state in which the energy is concentrated in a binary with all the remaining $(n-2)$ particles at infinite separation, a situation that is not of much interest for stellar dynamics. The statistical method founders on a different technicality from the mathematical approach. The arguments leading to this conclusion will be reviewed in this paper.

2. *N*-Body Calculations

This program started from an attempt to study the 'thermodynamics' of a self-gravitating n-body system by using that system as the 'thermodynamic fluid' in a Carnot engine. The technique of numerical experiments was used, in which n-body systems were simulated in a computer. There are many respects in which the concepts of thermodynamic systems and of stellar dynamical systems are incompatible; these gave rise to technical difficulties that will not be gone into here. The essence of a Carnot engine is that the system must be placed in an enclosure whose walls can be made adiabatic or isothermal.

The system was placed in a box, which worked as follows. After each integration step, each particle was checked to see if it were outside the box. Any particle outside the box was reprojected into the box with a new kinetic energy randomly chosen to have an exponential distribution with mean value T. Changes in kinetic energy (and the change in angular momentum) were tallied. The particle was reprojected from the location it had when it was discovered to be outside the box. A cubic box, with boundaries at ± 1 in all three directions, was used. The box was endowed with a 'temperature', T, that governed the mean kinetic energy of the reprojected particles. It was also endowed with a 'heat capacity', according to which its temperature can change in proportion to the net energy interchange between the stellar system and the box. Most experiments were run with a rather large heat capacity, and the resultant temperature changes (for the box) were negligible. This approximated the isothermal enclosure.

It would be possible to design an adiabatic enclosure for these systems; however, adiabatic enclosures were not used for the experiments reported here.

A particularly simple condition results if the box temperature is zero. Any particle that ventures outside the boundary is stopped, its kinetic energy removed, and the particle is then released to fall from the point at which it was caught. This practically assures that the particle will fall back through the most dense part of the cluster, where it can interact very strongly with the remaining cluster members. This 'cold box' condition produced the interesting results shown in Figure 1, where the energy delivered to the box by the stellar system is shown as a function of time. The kinetic

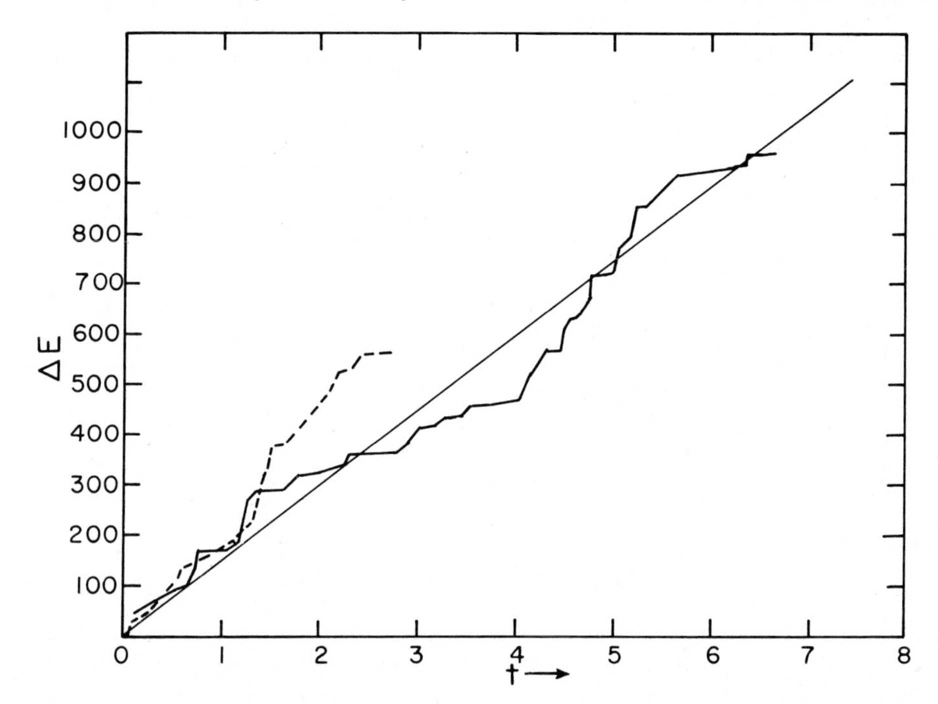

Fig. 1. Energy transfer from a 32-body system to a cold enclosure. The total initial kinetic energy was about 250. The time units on the abscissa were nearly a crossing time of the initial state. The two tracks represent two distinct calculations.

energy of the 32-body initial state was about 250, the potential energy about -500, and the total energy about -250. The time units of the abscissa were about one crossing time of the initial cluster.

The remarkable features of Figure 1 are the rate at which energy is given up to the box and the fact that the rate does not diminish appreciably as the process continues, even though the total amount of energy transferred is quite large (four times the total kinetic energy in the initial state). The solid curve represents one calculation – the dashed curve was another calculation from similar starting conditions for comparison purposes. The two curves give some idea of the reproducibility of the results from experiment to experiment.

The situation is qualitatively the same with the box at other temperatures: in Figure 2, the box temperature was about twice the mean kinetic energy of particles in the initial cluster. After a short dip, in which the cluster received energy from the box, the energy transfer began, and continued much as it did with the cold box, but at about half the rate. Apparently, the box prevents the potential energy of the cluster from becoming much larger (it is always negative, so a larger potential energy means that the particles are farther apart), while the hotter box increases the mean kinetic energy of the particles. After some time, the box and the cluster have about the same temperature, and then the transfer begins. It is not clear what determines the rate of energy transfer, once it has begun.

A different set of experiments was carried out in which the cluster was confined

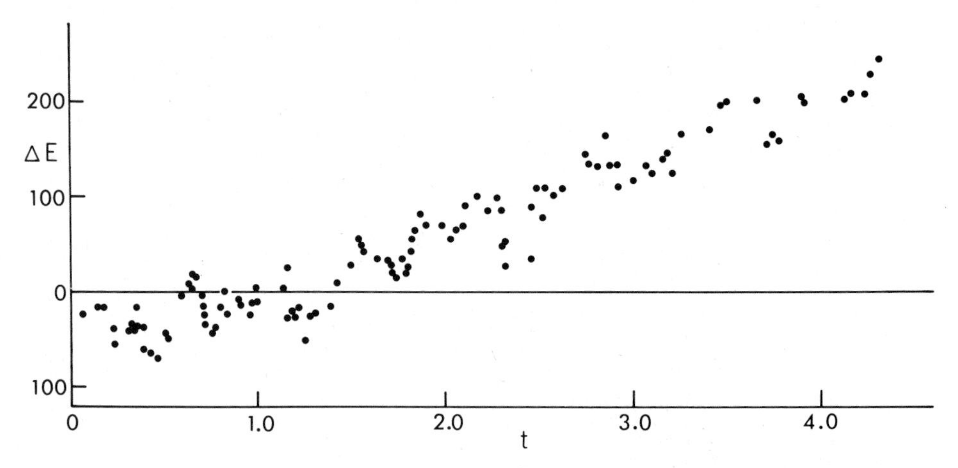

Fig. 2. Energy transfer from a 32-body system to an enclosure with nonzero temperature. The time units on the abscissa are nearly a crossing time of the initial state.

to a specularly reflecting spherical enclosure. No example tried has shown a behavior like that associated with the energy transfers of the cases run in the isothermal enclosure. The specular boundary condition prohibits energy (and angular momentum) transfer to the enclosure, but the transfers in the isothermal enclosure were associated with a shrinking of the cluster and an increase in kinetic energy that would be easy to identify if they did occur in the specular enclosure. Nothing surprising happened in the runs with the specular enclosure, except for the sharp contrast shown with the earlier runs with the isothermal box.

A third set of experiments was carried out in which the members of a given cluster were arbitrarily assigned to one of two subclusters for certain summaries such as total kinetic energy, total potential energy, and so on. The mechanism by which 'negative specific heat' is argued to lead to the formation of core-halo structures in the 'gravothermal catastrophe' (Lynden-Bell and Wood, 1968) is based on energy transfers between 'hotter' and 'cooler' subsets of particles within a given cluster. The terms 'hotter' and 'cooler' refer to particles with high and low kinetic energy respectively. One way to mimic these notions in a machine calculation is to separate the particles of an ordinary calculation into two classes according to their kinetic energy. This set of experiments makes use of a normal calculation for a star cluster, and is free of the unnatural boundary conditions of the other experiments. Whenever a fairly complete summary of cluster properties is run, the particles are sorted into two subsets according to kinetic energy. Along with the usual cluster properties, the kinetic and potential energy of each subcluster is tallied as if the rest of the particles did not exist; the potential energy of interaction of the two subclusters is also tallied. A given particle may move freely from one to the other of the two subsets. The separation applies only to the tallies; no distinction is made for the integration. This model appears to be very much in the spirit of discussions on 'negative specific heat' and of the 'gravothermal catastrophe'.

No strong tendency to transfer energy from the 'hot' to the 'cool' subcluster is evident in these experiments. The low-kinetic-energy subcluster nearly obeys the virial theorem as if the other subcluster were not present.

An alternative way of looking at these experiments that is free of some of the arbitrariness of division into two subclusters is to study the distribution of particle kinetic energy at various times. This distribution definitely did not show a tendency to segregate into extreme examples of high and low kinetic energy; high kinetic energies increased sharply during close encounters, but apart from this effect, no trend toward relatively distinct subsets of high- and low-kinetic-energy particles appeared. The experiments are consistent with the maintenance of an isothermal structure. More complete descriptions of the experimental results have been published elsewhere (Miller, 1973).

Among the various computer experiments, that with the isothermal enclosure is clearly the anomalous case. Both the examples with the specular reflector and those in which the system was regarded as if made up of two different subsystems showed no tendency to shrink into a small, 'hot' system that coexists with another extended, 'cool' system (core-halo structure). We anticipate the logical chain of arguments of this paper to comment on the reasons for this different behavior. The essential difference between the examples run with the isothermal enclosure and the other cases seems to be that the boundary conditions for the isothermal case are not describable by a Hamiltonian. Thus, there is no Liouville theorem for this case. By contrast, both the specular reflector and the subdivision of the particles into classes yield problems that are describable by Hamiltonians, and for which there is a Liouville theorem. This is manifestly the case for the subdivision; the specular reflector can be described as a Hamiltonian system by introducing an infinite (positive) potential at the position of the wall. The condensation into a compact system does not occur for the systems describable by a Hamiltonian; there is no such constraint on the problem with the isothermal enclosure. Similarly, the 'gravothermal catastrophe' proceeds in violation of the Liouville theorem (in $6n$ dimensions).

A collapse, if it is to occur, must do so on a secular (relaxation) time scale, and not on a dynamical time scale. Unfortunately, n-body calculations are not well suited to an examination of this question since the secular and dynamical time scales are not well separated for systems that contain fewer than about 1000 particles. The time-scale of energy transfer to the enclosure in the experiments with the isothermal enclosure may have been dominated either by dynamical or by relaxation time-scales, or by still another (undefined) time-scale. However, it is suggestive that the rate of energy transfer did not change with the dynamical time-scale of the cluster as the cluster shrunk. As the cluster got smaller in the configuration space, and the dynamical time-scale got shorter in the units of the calculation, the rate of energy transfer stayed about the same. It would have decreased if measured in terms of the dynamical time-scale, but it would not decrease nearly as strongly if measured in terms of a relaxation time-scale. The energy transfer, and the shrinkage of the residual cluster, seems to have been dominated by the relaxation time scale, as expected.

It was not possible, in these experiments, to define a thermodynamic temperature of a stellar system by the operational method of placing the system in contact with a foreign body, and allowing the two to reach equilibrium. No equilibrium was reached. Again, this can be attributed to violations of the Liouville theorem. It might be possible to construct the 'heat bath' in such a way that the total system is describable by a Hamiltonian; there must then be a Liouville theorem for the total system. But the phase volume accessible for the stellar component need not be conserved. It thus appears to be impossible to design a system that would permit an operational definition of 'temperature' for a stellar system.

3. *H*-Theorem

The connection between statistical and thermodynamic descriptions is usually made through identification of the Boltzmann H with entropy, with the subsequent association of the H-theorem with the increase of thermodynamic entropy in irreversible processes; of through some other equivalent assumptions. This same connection, by means of the H-theorem, has been made for stellar dynamical systems (Antonov, 1962; Lynden-Bell and Wood, 1968), although it is unlikely that there is an H-theorem for stellar dynamical systems. Prigogine and Severne (1966) have shown that there is no H-theorem for infinite, spatially uniform, self-gravitating systems, but because such systems are Jeans-unstable, the question is still open for finite systems (Prigogine and Severne made this point in their paper).

With the usual projected distribution functions normalized to unity, the functions

$$
\begin{aligned}
H_B &= -\int \mathrm{d}^3x \, \mathrm{d}^3v \, f_1 \log f_1 , \\
H_G &= -\int \mathrm{d}^{3n}x \, \mathrm{d}^{3n}v \, f_n \log f_n ,
\end{aligned}
\tag{1}
$$

are the same as the definitions of 'entropy' used in discussions of information theory (Shannon, 1948; Khinchin, 1957). The subscripts B and G signify that these are essentially the Boltzmann and Gibbs H-functions of statistical mechanics. The special form of Equation (1) seems to have been invented to yield systems for which a variational calculation would produce Maxwellian velocity distributions. The function, H_G, is a constant of the motion as a consequence of the Liouville theorem, and so is usually regarded as not very interesting. An essential inequality is proved in books on information theory:

$$nH_B \geqslant H_G, \tag{2}$$

with equality if and only if the n-particle distribution can be written as a product of single-particle distributions:

$$f_n(1, 2, \ldots, n) = f_1(1) \cdot f_1(2) \cdot \ldots \cdot f_1(n). \tag{3}$$

It is straightforward to extend this inequality to show that increases in H_B require increased deviation from the form of Equation (3) in the sense of requiring more correlation, in order to be consistent with the Liouville theorem (and thus with the constant value for H_G). We will not go into the details here, as the argument is presented in a forthcoming publication (Miller, 1974).

There is, in general, no limit to the allowed increase in H_B through increased correlation. In the usual problems of statistical mechanics, this increase in correlation appears in momentum space along with an increase in the mean square particle momentum (or kinetic energy). Equilibrium, as defined by the H-theorem, occurs when the kinetic energy of the particles has reached the maximum value attainable, a maximum being assured by the assumption of a finite bound to the amount of energy available from the potential energy (internal energy) sources available to the system. There is no such bound for stellar dynamical systems, so other means must be used to halt the process if a finite limit is to be reached. The particle correlation does not mean that particles are necessarily close together; rather it implies that knowledge of the phase of one particle tells something about where other particles are to be found. Even if there is no upper bound to values of H_B, as is the case in stellar dynamics, evolution under the H-theorem should proceed toward states of stronger correlation, and away from uncorrelated states such as those described by Equation (3). If a maximum for H_B can be attained, it is reached by maximizing correlations.

Particle correlations make a contribution to the potential energy of a stellar system. With pair correlations written as an additive part to the two-particle distribution:

$$f_2(1, 2) = f_1(1) \cdot f_1(2) + g(1, 2), \tag{4}$$

the total potential energy breaks up into the sum of two pieces. There is the usual part obtained from the f_1's

$$V_1 = -Gm^2 \frac{n(n-1)}{n} \int \mathrm{d}^3x^{(1)}\, \mathrm{d}^3v^{(1)} f_1(1) \int \mathrm{d}^3x^{(2)}\, \mathrm{d}^3v^{(2)} \frac{f_1(2)}{|x^{(2)} - x^{(1)}|}, \tag{5}$$

and a second part due to the pair correlation:

$$V_c = -Gm^2 \frac{n(n-1)}{2} \int \mathrm{d}^3x^{(1)}\, \mathrm{d}^3v^{(1)} \int \mathrm{d}^3x^{(2)}\, \mathrm{d}^3v^{(2)} \frac{g(1, 2)}{|x^{(2)} - x^{(1)}|}. \tag{6}$$

This is not an idle exercise: real stellar systems show appreciable pair correlation, just as do n-body systems in a computer (Miller, 1971, 1972). The contribution of V_c to the total energy can become quite large (half of the total potential energy or more).

Any argument in which V_c is ignored in the bookkeeping for the energy is tantamount to the assumption of an n-particle distribution of the form of Equation (3). While it is possible to construct n-particle distributions that yield zero results for V_c, these distributions are sufficiently pathological that a rather strong argument is required to justify ignoring the contribution of V_c to the total energy of a stellar system. It is not a completely trivial exercise to construct f_n's that yield negligible values for V_c. Arguments to the effect that the H-theorem correctly indicates evolutionary trends in stellar systems, but in which V_c is ignored in calculating the total system energy, are internally inconsistent because those arguments, on the one hand, call for increased particle correlation while, on the other hand, one of the principal results of particle correlation is ignored.

Indications that correlations are important in self-gravitating systems are not new. Prigogine and Severne (1966) constructed a weak coupling kinetic theory of binary interactions which predicted an irreversible growth of correlational energy. Their model was spatially uniform and of infinite extent, and the arguments did not hinge on H_B. Similarly, Chandrasekhar has repeatedly stressed the dependence of the potential energy on pair configuration distributions (Chandrasekhar and Lee, 1968; Chandrasekhar and Elbert, 1971).

If actual stellar systems evolved toward increased H_B, they would undergo a secular trend toward states characterized by a preponderance of binaries, since this is the easiest way to increase correlation. There is no observational evidence that cluster evolution proceeds in this way. Evolution towards greater values of H_B also runs against the virial theorem. As correlation increases, the magnitude of the potential energy increases; the total kinetic energy, K, must also increase in order to conserve total energy. The ratio, $-V/K$, must approach 1, not the value of 2 appropriate to the virial theorem. Both observational data and n-body calculations seem to confirm the 2 of the virial theorem over the 1 of a variational calculation based on the H-theorem.

Evolution toward greater values of H_B does not seem to provide a useful technique for studying stellar system's on two grounds: because H_B can become infinite without local maxima, it does not predict stable equilibria; and the evolution predicted is not in agreement with observation. It also does not predict metastable equilibria, but this requires arguments not presented in this paper (Miller, 1973). At any rate, such evolution, far from being a 'catastrophe', must proceed on a secular (relaxation) time-scale. Note that nothing has been said about those versions of the H-theorem that make use of coarse-graining, and that we have *not* shown that there is no H-theorem for self-gratitating systems.

4. The Microcanonical Ensemble

An alternative approach to the statistical mechanics of a stellar dynamical system is through the construction of one of the ensembles commonly used in statistical mechanics. The microcanonical ensemble is the only ensemble appropriate to stellar systems, which are presumed to obey the n-body equations of motion and to conserve total energy. The canonical and grand canonical ensembles are not suitable because the notion of a heat bath is alien to stellar systems and because the long range forces preclude treatment of subdivisions of the total system. The difficulties with numerical experiments, described in Section 2, underline these features.

The microcanonical ensemble is a useful illustration, and brings out some unexpected features of stellar dynamical systems. The volume of (the $6n$-dimensional phase space included in) the region between two neighboring hypersurfaces of total energy E and $E+\mathrm{d}E$ is $\sigma(E, R)\,\mathrm{d}E$, with

$$\begin{aligned}\sigma(E, R) &= C_N(R)\int_{-E}^{\infty} \mathrm{d}(-V)(-V)^{2-3N}(E-V)^{(3N-5)/2} = \\ &= \text{const.}(R^3)^{N-2}(-E)^{(1-3N)/2}.\end{aligned} \tag{7}$$

The coefficient $C_N(R)$ arises because this phase volume becomes infinite if infinite configuration volume is available; the R appearing in Equation (7) is a cutoff radius, inside which the entire cluster is presumed to lie. The integrand of Equation (7) may be regarded as a probability distribution function for $(-V)$. More negative values of V lead to rapidly growing phase volumes in the momentum space through the $(E-V)^{(3N-5)/2}$ term. Since $(-V)$ may become infinite by letting two or more particles come close together in the configuration space, an infinite volume (at quite a high order infinity!) opens up in the momentum space. The discussion is usually terminated at this point with the observation that there is infinite phase volume because of close binaries. However, the configuration volume available for such particle aggregates diminishes faster than the momentum space opens up, leaving a net decrease in the available phase volume. The most probable values for $(-V)$ are those in which the two terms just play off against each other. This yields a virial theorem. The details of this calculation appear elsewhere (Miller, 1973, 1974).

The integral of Equation (7) is the leading term of a sequence of similar terms that arise from the calculation of the configuration volume inside a surface $(-V)=$constant. But the volume inside such a surface can be infinite because an arbitrarily large amount of potential energy is available by letting two particles come close together; all the remaining particles can then be removed to infinity. All but three particles can be removed to large distance, but the volume inside $(-V)=$constant is infinite for three particles by leaving two close together and removing the third, and so on. The contributions of all these combinations may readily be summed. The growing infinity must soon overtake any finite contribution from interesting parts of the $(-V)$

surfaces near the origin; thus for large enough distances (R large), the volume associated with $\mathrm{d}(-V)$ tends asymptotically to the form given in Equation (7). The remaining terms of the sequence, which have been ignored in Equation (7) are those with three particles near each other, with two sets of binaries, with four particles, and so on. Thus the phase volume in the microcanonical ensemble is dominated by a state with a single binary having all the energy, with the other particles at rest at infinite separation, the uninteresting state referred to earlier.

The expression for phase volume in Equation (7) can be taken into one of the usual definitions of 'entropy' through the relation $\exp(S)=\sigma(E)$ (see, for example, Landau and Lifshitz, 1969; but also note the admonitions in their Section 8 on the inapplicability of statistical methods to problems involving gravitation). From this, a 'thermodynamics' can be constructed, yielding a (positive) 'temperature', $T=2(-E)/(3N-1)$, related to the total system energy and not to the kinetic energy alone. The specific heat associated with this temperature is negative. Other thermodynamic functions may be worked out as well, but that does not seem to be a fruitful undertaking.

Dynamical formation of binaries is not a preferred process in this formulation: systems do not tend to form many binaries because there is not a preponderance of phase volume accessible. The same argument applies to higher particle aggregates as well. The principle underlying these assertions is that situations with more phase volume accessible are more probable; this seems to be one principle of statistical mechanics that it should be possible to carry over into stellar dynamics with some confidence. The results quoted here apply to cases where all particles have equal mass. With unequal masses, there is considerably more phase volume available to a configuration in which the two most massive particles form a binary; the tendencies in this direction, as described at this meeting by Heggie, might profitably be considered from this viewpoint.

In the study of star clusters, we do not want a true ensemble average or (assuming ergodicity) a true time-average; either of these would only tell about the state with a single binary and infinite dispersion of the other particles. The desired averages are over systems with most of the particles near the origin. For example, the 'virial theorem' that can be obtained from this formulation actually results from the lack of interaction of the particles removed to infinite dispersion. All the energy (both kinetic and potential) resides in the binary; the other particles make no contribution. But binary systems are known to obey the virial theorem, so the stated result is not surprising. However, the formulation also yields expressions for the probability distribution of the virial ratio as a function of particle number; this is an interesting result.

The development of the microcanonical ensemble does not proceed in the direction that would cause the most rapid increase of H_B, as might be expected if the H-theorem were valid. The argument is based on the nature of the infinity in the phase volume. Larger values of H_B imply that there are conditional probabilities such that some knowledge of the state of the system allows more precise statements to be made about the total state than could be made in the absence of that knowledge. But the final state of the microcanonical ensemble is the antithesis of that condition: the preponderance of phase volume is dominated by states such that knowledge of the phase of one particle only carries information of order $(1/n)$ about which particle should be near some other particle. While H_B might become arbitrarily large with correlation order of $(1/n)$, it could become much larger still with correlations of order unity (in the particle number).

The microcanonical ensemble does not provide a valid counterexample to the H-theorem. However, the H-theorem is not useful if there is no maximum to H_B (so that no terminal equilibrium state can be predicted) and the evolution does not even proceed in such a way that the state of the system can be correctly predicted at later times by solving for a maximum of H_B while H_B is still finite.

There is a vaguely unreal feeling to the arguments based on ensemble averages over the microcanonical ensemble, and the time averages that are equivalent if the system is ergodic. The arguments leading to the assertion that larger phase volumes are more probable imply that there is some mechanism available to permit the system to evolve in that direction even if it has started out in some other direction. For example, suppose the system formed *two* binaries at some early stage. Then the system begins to dissolve. The dynamics says that if all the other particles are essentially infinitely far apart, and if the two binaries are similarly at infinite separation, there is no way that the two binaries can exchange energy to reach the more probable terminal state with only one binary. But it is equally difficult to imagine that the system, while it is still relatively compact with interactions available to redistribute energy among the constituent parts, can know that it should arrange itself to have only one binary, in order to be properly prepared for future developments. Further, the arguments, like those of the mathematical treatment of n-body systems, presuppose infinite time, and so allow states of little physical interest to dominate the ensemble or time-averages. The time-scales for development of such systems is unrealistically long from a physical point of view.

No equilibrium solution has been found by any of the methods used here. Some equilibrium solutions are known, but they represent a 'set of measure zero' relative to all possible configurations, so they might understandably escape detection by these methods. The criteria used to define equilibrium may be too strict; certainly it is not realistic to demand stability over times greatly in excess of the age of the universe. It seems likely that we are faced with a situation in which there may be no equilibrium in the mathematical sense, but in which the natural processes leading to the dissolution of clusters are so slow that clusters represent an equilibrium in a practical sense. If so, we need mathematical tools that allow us to calculate the properties of interesting subsets of all possible systems – those with all the particles near each other, but without need to be preoccupied with pathological collisions. An alternative is that other processes (dissipation due to interstellar gas, etc.) might be more important than we have believed them to be, and that some other mechanism may be responsible for the apparent equilibria observed in natura. The challenging puzzle still stands.

References

Antonov, V. A.: 1962, *Vest. Leningrad Gos. Univ.* **7**, 135.
Chandrasekhar, S. and Elbert, D. D.: 1971, *Monthly Notices Roy. Astron. Soc.* **155**, 435.
Chandrasekhar, S. and Lee, E. P.: 1968, *Monthly Notices Roy. Astron. Soc.* **139**, 135.
Khinchin, A. I.: 1957, *Mathematical Foundations of Information Theory*, Dover Publications, New York.
Landau, L. D. and Lifshitz, E. M.: 1969, *Statistical Physics*, Addison-Wesley, Reading, Mass.
Lynden-Bell, D. and Wood, R.: 1968, *Monthly Notices Roy. Astron. Soc.* **138**, 495.
Miller, R. H.: 1971, *Astrophys. J.* **165**, 391.
Miller, R. H.: 1972, *Astrophys. J.* **172**, 685.
Miller, R. H.: 1973, *Astrophys. J.* **180**, 759.
Miller, R. H.: 1974, *Advances in Chemical Physics* (to be published).
Prigogine, I. and Severne, G.: 1966, *Physica* **32**, 1376; *Bull. Astron.* (3) **3**, 273.
Prigogine, I., Nicolis, G., and Babloyantz, A.: 1972, *Physics Today*, November, 1972, pp. 23–28; December 1972, pp. 38–44.
Shannon, C. E.: 1948, *Bell Syst. Tech. J.* **27**, 379; **27**, 623; reprinted in C. E. Shannon and W. Weaver, *The Mathematical Theory of Communication*, University of Illinois Press, Urbana, Ill.

DYNAMICS AND CLUSTERS OF GALAXIES

D. G. SAARI

Dept. of Mathematics, Northwestern University, Evanston, Ill. 60201, U.S.A.

Abstract. Under the assumption that the inverse square central force law is a good approximation to the gravitational force, at least for large distances, the different possibilities for the evolution of the Universe are sketched. Several of the possibilities lead naturally to a dynamical classification of clusters of galaxies in an expanding universe. In one of the classifications the galaxies must define configurations which are functions of the masses. The virial theorem approach of determining masses of galaxies in a cluster is briefly examined. Some tentative statements concerning a dynamical explanation of the local hypothesis for quasars are advanced. Finally, the role of mathematical probability in predicting the behavior of the Universe is discussed.

1. Introduction

In earlier papers (Saari, 1971a, b to be referred to as Papers 1 and 2) the general qualitative and asymptotic behavior as time approaches infinity of all solutions of the n-body problem was derived and outlined. Here n is an arbitrary but fixed positive integer. That is, in these papers we gave a mathematical description of Newton's universe for large values of time. In this talk I would like to review some of these results, but attempt to do so in a fashion where the mathematical solutions are related to possible astronomical interpretations. That is, we assume that Newton's law is a good approximation for the gravitational force, and then we describe the evolution of the Universe.

In the first sections we shall concentrate on giving different possible dynamical interpretations for clusters of galaxies. This discussion includes a disintegrating system as one type of 'cluster of galaxies'. As a byproduct of this discussion, we shall make some brief comments about the related problem of determining masses of the clusters via the virial theorem.

It will turn out that in one classification of clusters of galaxies the galaxies must tend toward the vertices of well-defined configurations. These configurations are determined by the masses of the galaxies. In some settings the fact that these configurations are functions of the masses may be exploited to yield a method which would either determine whether a cluster is complete (and if not, it would give a prediction scheme which would indicate possible locations where one would expect to find a member galaxy), or determine the masses of the member galaxies up to certain proportionality parameters.

With a partial solution of the equations of motion at hand, a natural question would be to ask if there is anything in the dynamics of the n-body problem comparable to the behavior of quasars, namely, can quasars be explained in terms of Newtonian mechanics? In Section 4, we shall offer some extremely tentative statements.

This classification of motion is the result of a mathematical study of the equations of motion of the inverse square central force law for n point masses. However, any realistic model for the Universe would include forces other than the gravitational attraction between masses. In fact, some of the forces probably would dominate the

Y. Kozai (ed.), The Stability of the Solar System and of Small Stellar Systems, 273–284. *All Rights Reserved.*

gravitational force for 'local' distances. Fortunately, the results given here are stable respect to perturbations of this type. The conclusions stated here hold equally well for any force law which is dominated by the inverse square term for large distances. That is, it holds for force laws of the form

$$\mathbf{f}(\mathbf{r}, \mathbf{v}, t) = \mu r^{-3}\mathbf{r} + \boldsymbol{\varepsilon}(\mathbf{r}, \mathbf{v}, t),$$

where $r^2|\boldsymbol{\varepsilon}(\mathbf{r}, \mathbf{v}, t)| \to 0$ as $r \to \infty$, and $\boldsymbol{\varepsilon}(-\mathbf{r}, \mathbf{v}, t) = -\boldsymbol{\varepsilon}(\mathbf{r}, \mathbf{v}, t)$.

Notice that for local distances the dominating force could be most anything, even a repulsive force law. Indeed, it need not define a conservative system. Also, note that we require the forces to start acting like a central force law only for 'large distances'. In fact, the results hold with only minor modifications even when the inverse square term is changed to an inverse q force law where $1 < q < 3$. Consequently, the conclusions of this study apply to models which allow for oblateness effects, approximations to relativity, and some nongravitational forces. The major requirement is that the resulting differential equations have unique solutions which exist for $t \geqslant 0$.

2. Clusters of Galaxies

It follows from Kepler's equations that one of three things can occur in the two-body problem as time approaches infinity: (a) the motion is bounded, (b) the motion is parabolic, where the distance between particles separates like $t^{2/3}$, and (c) the motion is hyperbolic, where the distance between particles separates like t.

To see this, recall that the conservation of energy integral for the two-body problem is $\mathbf{v}^2 = 2(\mu r^{-1} + h)$, where $\mathbf{r}$ and $\mathbf{v}$ are respectively the position vector and velocity vector of the second particle relative to the first. If constant h is negative (elliptic motion), then $\mu r^{-1} + h \geqslant 0$, or $r \leqslant \mu h^{-1}$. This is conclusion (a).

Define $I = r^2$. Then $\ddot{I} = 2(\mathbf{v}^2 + \mathbf{r}\cdot\ddot{\mathbf{r}})$. From the conservation of energy integral and the equations of motion, this equation can be expressed as

$$\ddot{I} = 2(\mu r^{-1} + 2h) = 2\mu I^{-1/2} + 4h.$$

If constant h is positive (hyperbolic motion), then $\ddot{I} \geqslant 4h$. By integrating both sides of this inequality twice we have that $I \geqslant 2ht^2 + 0(t^2)$. This means that $\ddot{I} = 4h + 0(t^{-1})$. By integrating this last expression twice we obtain $r^2 = I = 2ht^2 + 0(t \ln t)$. This is conclusion (c).

Finally, if $h = 0$ (parabolic motion), then $\ddot{I} = 2\mu I^{1/2} > 0$. If I were bounded above for all positive time, then we see from this last inequality that $\ddot{I}$ would be bounded below by a positive constant. By integrating this new inequality twice, we obtain the contradiction that if I is bounded, then it goes to infinity faster than some constant multiple of t^2. Therefore, we conclude that I is unbounded. However, from the facts that $\ddot{I}$ is positive and I is unbounded we can show that $I \to \infty$ and that after some time $\dot{I}$ is positive. Therefore, by integrating $\dot{I}\ddot{I} = 2\mu I^{-1/2}\dot{I}$, we see that $\dot{I}^2 = 8\mu I^{1/2} + c$, where c is a constant of integration. Since $I \to \infty$ and $\dot{I}$ is positive, this can be expressed as $\dot{I}I^{-1/4} = (8\mu)^{1/2} + 0(1)$ as $t \to \infty$. Integrating this last expression leads to conclusion (b).

In the n-body problem the motion is, as one would expect, more complicated. If any two particles are chosen from the n particles, then their mutual distance either behaves as described in (a), (b), or (c), or it belongs to two other possible types of motion as given in Paper 1. These other types of motion will be discussed in Section 4. However, as shown in Papers 1 and 2, the important fact is that *all solutions of the n-body problem* which exist for all positive time (Saari, 1971c) *consist of various combinations of these five types of motion.* This is independent of the value of the total energy of the system!

For the remainder of this section we will concentrate on those important solutions where the motion is a combination of cases (a), (b), and (c). That is, the distance between any two particles is bounded, expanding like $t^{2/3}$, or expanding like t. (In the general n-body problem the terms 'parabolic' and 'hyperbolic' seem to be out of place, so we drop them and identify the motion via its major characteristic – the distances separate respectively like $t^{2/3}$ or like t.)

While this description is in terms of the behavior of any two particles, it can be translated immediately into a discussion of the total system. What happens is that several masses may have their mutual distances bounded after some time. These masses define, in a natural fashion, a group. Any two particles chosen from different groups must separate either like $t^{2/3}$, or like t. Since the groups remain (by construction) relatively bounded entities, it follows that the separation between the two groups in question is respectively like $t^{2/3}$ or like t. The separation can be measured from the centers of mass of the groups. Also, since $t^{2/3}/t \to 0$ as $t \to \infty$, after some time a clear distinction between the two rates of expansion would appear. A restricted case of this general picture can be found in Figure 1.

From these three types of motion there are seven qualitatively different possible classes of solutions. They are found from the various possible combinations, and they range from the case where the distances between all particles remain bounded for all time to the case where all three types of motion occur. The discussion of these cases is somewhat similar, and to eliminate repetition only the inclusive solution, which exhibits all three types of motion, will be discussed in detail. A similar discussion would follow for the remaining six types of solutions.

The first problem is to give an interpretation for relatively bounded motion, i.e. a group. There are two possibilities, and we accept both of them. The first is that these groups correspond to galaxies. Of course, since there are no upper or lower limits (other than 1 or n) for the number of particles belonging to a group, some of the groups simply may be escaping particles. We shall treat them as 'one-particle galaxies'.

It is possible that some of the observed clusters of galaxies are bounded. In this case some of the groups would correspond to 'bounded clusters of galaxies'. Therefore, a circle in Figure 1 could be interpreted as being either galaxies or a bounded cluster of galaxies.

The next motion is where the distance between objects expands like $t^{2/3}$. That is, particles not belonging to the same group will separate like $t^{2/3}$ (or faster). It is easy to show that this separation can be translated to the centers of mass of the groups; namely, the centers of mass of the groups separate like $t^{2/3}$. For purposes of identifica-

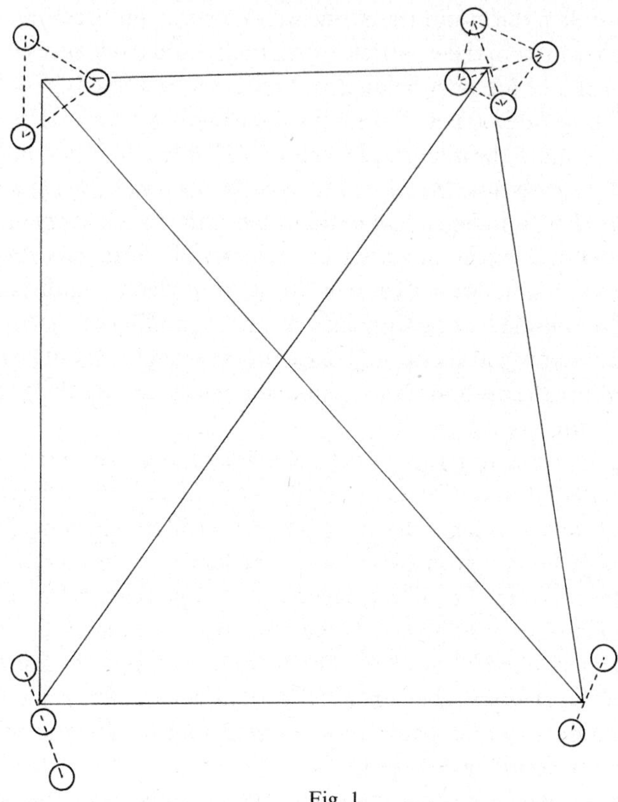

Fig. 1.

tion we denote motion in this setting as *subsystems*. It was shown in Paper 1 that in subsystems the velocities between the centers of mass of the groups go to zero like $t^{-1/3}$.

If the groups are interpreted as being galaxies, then a second type of cluster of galaxies appears. In this case they are galaxies where the distances between them are separating like $t^{2/3}$. That is, in this case the system is disintegrating, and the cluster of galaxies is a subsystem of individual galaxies. Since the magnitudes of the velocities become very small, it would seem to be observationally difficult, if not impossible, to distinguish this case from a 'bounded' cluster of galaxies.

A more interesting and intriguing case is when one or more of the groups in a subsystem is a bounded cluster of galaxies. Indeed, we would obtain a cluster of clusters of galaxies, or an *hierarchical cluster of galaxies.* Again, since the velocities between the centers of mass of the groups are very small, it would seem to be observationally difficult to distinguish this case from a bounded cluster of galaxies.

Of course, there remains the final possibility that a subsystem has only one bounded object. Under the first interpretation of bounded motion, this would yeild a field galaxy. Under the second interpretation, this would give an isolated cluster of galaxies.

The remaining type of motion between particles has the distance between objects expanding like t. In the same fashion as above, this can be described in terms of the centers of mass of the subsystems separating like $\mathbf{c}t$ where $\mathbf{c}$ is a constant vector whose value depends on the choice of the subsystems. Thus, this gives us a description of the expansion between clusters of galaxies. Since $t^{2/3}/t \to 0$ as $t \to \infty$, even a disintegrating cluster of galaxies given by the above subsystem interpretation will, for all time and from large distances, remain as an observational entity. As will be seen in the next section, even the configuration will remain the same!

This completes the picture for large values of t. However, for an evolving system there is a second possible interpretation for this expansion which is asymptotic to t. Quite simply, the system may not be old enough to show the sharp distinctions between the different possible types of motion given in Figure 1. That is, since t is 'small', $t^{2/3}/t$ is still 'large' for this particular cluster. Consequently, there remains the possibility that a cluster of galaxies is a new system where the rate of expansion of some of the galaxies is like t, that is, the bounded motion is separating like t, but the act of disintegration occurred recently. Note that with this interpretation, over long periods of time the cluster of galaxies would *not* remain as an 'observational entity', but it would evolve either into new clusters or field galaxies as described above. The time span necessary for this to occur is, of course, much too large to be of practical 'observational' value and hence the use of quotation marks. But it is important in the study of the evolution of the system and the 'stability' of clusters of galaxies to distinguish between these two types of disintegrating clusters.

3. Central Configurations and Subsystems

A surprising fact about subsystems is that the centers of mass of the groups must tend toward the vertices of well defined (expanding) configurations which *depend on the masses of the groups.* Thus, clusters of galaxies forming a disintegrating cluster of galaxies, or an hierarchical cluster of galaxies, would tend to form configurations which depend on the masses of the galaxies.

Assume that we have a subsystem which has p groups. In other words, assume that we have p galaxies, p bounded clusters of galaxies or p combinations of these two classes. Let M_s be the total mass of the sth group, $s = 1, 2, \ldots, p$, and let $\mathbf{r}_s$ be the vector position of the center of mass of the sth group relative to the center of mass of the subsystem.

It turns out that the expansion $t^{2/3}$ is slow enough to allow the attracting force due to the other groups to strongly influence the eventual direction of $\mathbf{r}_s$. This forces the position vector and the gravitational force vector to tend to line up along the same line. In fact, as shown in Paper 2, they tend to do so in such a fashion that if G is the gravitational constant, then

$$\lambda M_s \mathbf{r}_s + \sum GM_s M_k (\mathbf{r}_k - \mathbf{r}_s)/r_{ks}^2 = \mathbf{e}_s, \quad s = 1, 2, \ldots, p, \tag{1}$$

where λ is proportionality constant independent of subscript s, and $\mathbf{e}_s$ is an error term which goes to zero faster than $\lambda |\mathbf{r}_s - \mathbf{r}_j|$ as $t \to \infty$. It follows from Paper 2 that $\lambda = \frac{2}{9} t^{-2}$ and $\mathbf{e}_s$ is such that $t^{4/3} \mathbf{e}_s \to \mathbf{0}$ as $t \to \infty$.

Now it turns out that the centers of mass of the groups tend to the vertices of configurations defined by Equation (1) when the error term is set equal to zero. These configurations, known as *central configurations*, are determined by the masses of the groups (see Wintner, 1941). That is, this equation shows that the centers of the mass of the groups must form an expanding central configuration. For example, if $p=3$, then the configuration must be either an (expanding) equilateral triangle or one of the Euler straight line solutions where the relative distances are determined by the value of the masses. The main point is that for *clusters of galaxies defined by subsystems of individual galaxies, the galaxies must tend to form some central configuration.* The same statement holds for an hierarchical cluster of galaxies.

The case $p=3$ is misleading by the regularity of the configurations. For $p>3$ the configurations are by no means so regular and obviously recognizable. In fact, only some of them are known.

The fact that the subsystems tend to define configurations could possibly be exploited in several ways. It could be used (at least theoretically) to distinguish the various types of clusters described in the previous section. The problem is to determine some way to differentiate between a bounded and disintegrating cluster of galaxies. If the masses and the configurations of the groups come 'close' to satisfying Equation (1), then I suggest that there would be additional reason to suspect that the cluster is disintegrating and that the expansion is like $t^{2/3}$.

Secondly, if it is known (or suspected) that a cluster of galaxies is defined in terms of a subsystem, and if the masses of the galaxies and the configuration they form (in three-dimensional space) is known, then by substituting these values into Equation (1) it can be determined whether the cluster is complete. If the equation is not satisfied, then the cluster is not complete. A computer search scheme could be devised to determine the necessary location of additional masses needed to satisfy the central configuration equations.

Finally, if the system is known to be complete, and the configuration formed by the groups is known, then in some cases the central configuration equations can be solved for the individual masses of the groups up to certain proportionality parameters. (Unfortunately, this last approach will not work in all cases. For example, it fails in the equilateral triangle case previously cited.)

The observational difficulty with the above program is that the configuration must be known in three dimensional space; whereas, observationally only the projection of the configuration can usually be determined. (Because of these and other problems I hesitate to even attempt to relate the above to such observed objects as the 'quadrangles'.) Also, the number of galaxies in a cluster may be so large as to make this theoretical scheme impractical. However, these questions are outside the realm of my experience, and I leave them to other investigators.

4. Hubble's Constant and Quasars

As in the case with most models of the Universe, the Hubble constant for the n-body problem assumes a functional form. The function is particularly simple, and it is exactly what one would expect from the solution of the two-body problem. In the case of subsystems it is, $v/r=\frac{2}{3}t^{-1}+0(t^{-1})$ where t is time. The term $0(t^{-1})$ denotes error terms of magnitude such that $t0(t^{-1})\to 0$ as $t\to\infty$. Here v and r are respectively the speed and distance between the centers of mass of two groups in a sub-system. Hubble's constant for clusters of galaxies, or any other objects where the expansion between them is like t, would be $v/r=t^{-1}+0(t^{-2})$. The term $0(t^{-2})$ denotes error terms such that $t^2 0(t^{-2})$ is bounded as $t\to\infty$. This functional form for the Hubble constant is in terms of the instantaneous values of velocity, distance, and time. The constant is obtained by substituting the current 'age of the Universe' into the equation. (The reason for the quotation marks about the phrase 'age of the Universe' is that the inverse square force law model does not naturally admit, or define, a starting value for the birth of the Universe. Unless we restrict attention to a set of initial conditions of measure zero (Saari, 1971c), where as we go backward in time we find a point from which all particles were expelled – A newtonian 'big bang' – there is no natural point on the time scale to assign the origin.)

Can the Hubble relationship be violated? Expressed in other words, is there anything in the dynamics of the n-body problem which would correspond to quasars, in particular to the 'local' hypothesis? That is, is there motion where the velocities are so large that the value of v/r is much larger than accepted values for Hubble's constant? Unfortunately the answer must be in the form – maybe. There is a possible motion in the n-body problem, called oscillatory motion, which does satisfy the condition that at different times the velocities of some of the bodies may be too large to satisfy a Hubble's constant relationship. This may or may not be the dynamical equivalent of quasars.

There are several problems involved. The first is the very difficult mathematical problem of existence. Such motion has been shown to exist (Sitnikov, 1960) in only a very specialized setting where the velocities of the particles are nowhere near the required magnitudes needed to allow a dynamical explanation for the red shift of quasars and a local distance of the object. However, the question of existence is so difficult that the Sitnikov example, by its physically timid (but mathematically clever and technically difficult) construction, would be expected to yield same results, namely low velocities. (It also may be a characteristic of the three-body problem, rather than the n-body problem.) Theoretically, other types of oscillatory motion may exist where the velocities must be very large. While I have been unable to prove the existence of such motion where the velocities are of the required magnitude, I have likewise been unable to show that it does not occur!

Although the existence question still remains open, several properties of this motion are known, and they will be presented here. This motion is characterized by an oscillating behavior between particles. Namely, three masses m_j, m_i, m_k can be found such that (i) $\limsup\,(r_{ik}+r_{kj})=\infty$, (ii) $\limsup\,(r_{ik}/r_{kj})>0$, and (iii) $\liminf\,(r_{ik}/r_{kj})=0$.

Any particle m_s with the property that index s can be substituted for one of the above indices is said to participate in the same oscillatory motion.

We distinguish two cases. The first is where there is some distance between particles such that $\lim \sup(r_{kj}t^{-1})=\infty$. This motion is the cause of the difficult question of escape. This motion is also related to the question of non-collision singularities. For additional discussion we refer the reader to Saari (1973b).

The second type of oscillatory motion is where $\lim \sup(r_{kj}/t)=0$ for all particles m_k, m_j participating in the same oscillatory motion. In this case the particles defining the same oscillatory motion would appear in the sketch of the evolving Newtonian universe in the same way that a subsystem does. The centers of mass of subsystems and oscillatory motion would separate like a constant vector multiple of t. Therefore, if oscillatory motion did correspond to quasars, then theoretically we would expect to find quasars separated from the clusters of galaxies as described in Section 2, that is, at least for large values of time.

Additional research has indicated, but not proved conclusively, that the particles with large velocities are associated with larger masses (or clusters of mass). In fact this seems to be the mechanism which allows such large velocities to occur at local distances. The particles commute between the larger masses, which may be quite far apart, picking up velocity with each close passage. Observationally this means that if quasars correspond to oscillatory motion, then some of the quasars should be observed near larger objects.

While the above two paragraphs seem to agree with some of the observed characteristics of quasars (Arp, 1971), there is nothing in this dynamical description which would explain the lack of observed blue shifts for quasars. (Personally, I am skeptical that this is an explanation for quasars. However, it does yield a model employing known physical principles which may [or may not] admit distances much closer than those indicated by their red shift and Hubble's constant.)

A discussion of oscillatory motion in the three- and n-body problems can be found in Saari (1973b). It is shown here that in the three-body problem oscillatory motion has its expansion bounded above by a constant multiple of $t^{2/3}$.

The remaining class of motion is called pulsating motions. Its definition and some of its properties can be found in Paper 1. It can be shown that the distances between particles separate no faster than $0(t^{2/3})$. (This is true also for the second type of oscillatory motion.)

Therefore, if pulsating motion exists, it would play the same role as a group or 'galaxy' in the above sketch of the Universe.

5. The Virial Theorem

For at least the last decade the virial theorem approach to determine the masses of galaxies has been under attack. For example, it is known that unstable systems will give exaggerated values for the masses. With a partial solution of the n-body problem available this statement can be derived in a rigorous fashion, and we do so here.

Of course, there are other serious difficulties involved in using the virial approach, such as the need to use the instantaneous values for U and T rather than their time averages. Here T is the kinetic energy of a system and

$$2U = \sum_i \sum_j GM_iM_j/r_{ij}.$$

(In Paper 2 the assumption that the instantaneous values and the time averages agree over a period of time was shown to imply that $T = -h$ and $U = -2h$ for all time. Here, h is the total energy of the system. Furthermore, it is now known (Saari, 1973b) that if T and U are equal to constants over a period of time, then the constants must be respectively $-h$ and $-2h$ for all time. Finally, it is conjectured that this implies that the motion behaves like a rigid body. It is true for the three-body problem.)

In studying the effects the dynamics have on the virial theorem approach to mass determination, we shall ignore all other approximations. That is, we shall assume that all additional approximations leading to the simplification of the problem can be made without error. With this assumption the error in the value of the masses introduced by the hypothesis $2T = U$ will be examined under the different dynamical interpretations of clusters of galaxies.

Assume first that we have a bounded cluster of galaxies and that the masses in the definition of T and U are the masses of the galaxies. Then by Paper 2 and Section 2 of this note, $T - U$ is eventually negative. Now, the way the masses are determined goes as follows. Depending on the technique employed, certain assumptions are introduced to obtain the ratios of masses of the galaxies. This reduces the problem to a quasi-binary problem. That is, the 'virial' equation is changed from $2T = U$ to $2V^2 = 2M/R$, where V and R are defined as

$$V^2 = 2T\left(\sum M_s\right), \qquad R = \left(\sum M_s\right)^2/2U.$$

M is the total mass of the system, and it is the unknown. Let M^* be the value of the total mass obtained from the assumption $2T = U$, and let M denote the correct value of the total mass. That is $M^* = 2RV^2$.

The same assumptions used in the above reduction apply to the energy relationship converting $T - U < 0$ to $V^2 - M/R < 0$, or $M > RV^2$. That is, $M^* < 2M$. Consequently, if the system is bounded, then the error introduced by the assumption $2T = U$ is bounded above by a factor of 2.

The situation changes if the cluster of galaxies is defined by the subsystem description. It was shown in Paper 1 that in this case the energy relationship assumes the form $T - U \to 0$. That is, $V^2 - M/R \to 0$ or $M^* = 2RV^2 \to 2M$. Hence the error introduced by $2T = U$ in this setting is approaching a factor of 2.

Finally, if a cluster of galaxies is defined by an expansion like t, or a mixture of expansions of the type $t^{2/3}$ and t, then $T - U$ approaches a positive constant. Hence, $V^2 - M/R \to 2H/M > 0$. In this setting, $M^* = 2RV^2 \to 2M + 2HR/M$. Since R expands like t, it follows that the error in the computed value M^* will become at least twice as large as the correct value M. Actually the error will become *arbitrarily large* with the actual value of the error depending on the value of H and the age of the system.

6. Improbability of Certain Motions

In dynamical astronomy and celestial mechanics there are several statements to the effect that the set of initial conditions leading to a certain type of behavior has measure zero. That is, in a measure-theoretic sense it is mathematically improbable that such motion will occur. Statements of this type include collisions in the n-body problem (Saari, 1971c, 1973a), capture and escape (Chazy, 1922; Hopf, 1930), and motion which includes $t^{2/3}$ expansions in the two- and three-body problems. The next step is to assert that such motion probably does not occur in the Universe; consequently, it can be ignored.

While this last step is well understood to be incorrect, the argument periodically reappears in the literature. Therefore, we outline here some of the objections to this statement. The ideas advanced here are strongly motivated by a paper by Schwarz (1962) aptly titled 'The Pernicious Influence of Mathematics on Science'.

For the above assertion to be correct there remains the important step and basic problem of showing how the mathematical probability of an event corresponds to the observational (or experimental) probability of the same event. They need not be the same! The major problem is, of course, that while for technical reasons the Lebesgue measure is a natural measure to use in the study of the n-body problem, there is no reason to believe that nature is laboring under the same technical constraints. She may be fooling us with a different distribution function.

Even if the Lebesgue measure is the proper measure, the Universe may exhibit prejudices for certain initial conditions which are imposed upon it by different physical constraints due to conditions of its creation, etc. To illustrate this further, recall that in the n-body problem the set of initial conditions corresponding to zero angular momentum has measure zero. However, if the Universe does date its birth from some sort of 'big bang', then as the position vectors are traced backwards in time ($t \to 0$) to their common origin we have a mathematical collapse of the system, and all the position vectors tend to zero. This means that the total angular momentum is zero. Thus the motion is contained in a set of measure zero. (This could be viewed as a 'collision' at some past time. Again this means that we are restricted to a set of measure zero.) Now, phenomena peculiar to zero angular momentum would be mathematically improbable but they may be observationally and physically abundant. (Indeed, we should be using conditional probability.)

In other words, the constraints given by the creation of the Universe, if simply its survival to the present date, may actually restrict the initial conditions of the mathematical model to sets of measure zero or lower dimensional manifolds. Consequently, certain motions cannot be ruled out of the real world simply by mathematical probabilistic statements, they can only be ruled out by observational evidence proving that such predicted behavior does not, has not, and will not occur. Hence in dynamical astronomy the problems of capture, collision, etc. are solved in a mathematical probabilistic sense, but they must be considered open questions in a physical probabilistic sense! (The collision problem uses point masses, hence it would not apply for this reason alone.)

7. Comments

The above description given in Sections 2–5 depends on the behavior of solutions of the n-body problem for large values of time. Clearly our Universe has not reached that advanced age where all particles (stars or galaxies) have committed themselves or displayed their future course of expansion. Some clusters of galaxies may have only recently begun the process of disintegration, and they would not show the clear distinction between the different rates of expansion as indicated in Figure 1. Some groups of particles may at some future time evolve into several galaxies, and each of these might separate from each other. All of this leads to a far more complicated picture than the one indicated in Sections 2 and 3.

Of course, certain characteristics of the expansion would begin to manifest themselves at an earlier stage of the evolution, however, not with the same distinction as will eventually be the case. For example, the configurations discussed in Section 3 depend on the rate of expansion and the masses of the groups. If certain groups are separating from each other like $t^{2/3}$, then we would expect these groups to start to form central configurations. (The error of deviation from the correct configuration would be much larger than will be the case in the future.) This is independent of the fact that at some time in the future some gathering of particles may separate into two or more smaller groups with separations like $t^{2/3}$ or t. This is because the force law which gives these configurations, depends on the present location of accumulations of mass – not some future location or future behavior of the masses nor current velocities within the groups.

So although the Universe may be quite young, certain characteristics of the above description may already be displayed, and the above discussion may be of value in interpreting present observations.

Of course, the above is predicated on the unproved assumption that the inverse square force law is a valid approximation to the gravitational forces which governs the motion of the Universe. As was stated in the introduction, the force law used in this note can be quite general, and it is permitted to run the spectrum between the inverse force law and the inverse cube force law with fairly large perturbations permitted. All of this causes only minor modifications in the above discussion. The question still remains, can the gravitational force be approximated by these force laws? While we cannot answer this question, we can at least offer the following corollary. *If it can be shown that the dynamics of the Universe cannot now or ever in the future be described in terms of the above qualitative discussion, then it follows that any inverse force law between the inverse and the inverse cube force laws is not dominant for large distances.*

Finally, using the fact that the model allows for fairly large perturbations of the gravitational force, a prejudice among some investigators of dynamical astronomy against parabolic orbits, or in our setting, against motion separating like $t^{2/3}$, will be examined. The argument for their position usually centers around two points. The first is that the motion may be improbable in a measure-theoretic sense. The second is that 'small' perturbations disrupt this motion into expansions such as t, or into bounded motion. Hence, due to their sensitive nature, they would not, in general, survive. That is, one would not expect such motion to exist in the real universe.

The first statement was examined in a more general setting in the previous section. If the second holds, it can have some interesting consequences. As was previously noted, the actual central force law employed can deviate from the theoretical inverse square law by 'almost' as much as the law itself. Viewing these deviations as perturbations of the system, we have that perturbations which do not allow for this expansion are in reality quite significant forces, as significant as the inverse square term itself. If these forces do occur in nature, then they must be studied and understood for they significantly alter our understanding of gravitational forces! (In the same sense, if approximation schemes such as numerical integration cannot obtain these orbits, then the errors introduced must be considered as being quite large in a qualitative sense.)

It may be that oscillatory motion, $t^{2/3}$ expansion, or other motions discussed in this note do not occur in the universe; however, this is outside the realm of theory, and it must be observationally determined.

Acknowledgement

This research was partially supported by NSF contract GP-32116.

References

Ambartsumian, V. A.: 1961, *Astron. J.* **66**, 536–41.
Arp, H.: 1971, *Science* **174**, 1189–2000.
Chazy, J.: 1922, *Ann. Sci. Ecole Norm.* **39**, 29–130.
Hopf, E.: 1930, *Math. Ann.* **103**, 710–19.
Saari, D. G.: 1971a, *Trans. Am. Math. Soc.* **156**, 219–40.
Saari, D. G.: 1971b, *Astrophys. J.* **165**, 399–407.
Saari, D. G.: 1971c, *Trans. Am. Math. Soc.* **162**, 267–71.
Saari, D. G.: 1973a, *Trans. Am. Math. Soc.* **181**, 351–368.
Saari, D. G.: 1973b, *J. Diff. Eq.* **14**, 275–292.
Saari, D. G.: 1974, *SIAM J. Appl. Math.* **5**.
Schwarz, J.: 1962, in E. Nagel, P. Suppes, and A. Tarski (eds.), *Logic, Methodology and Philosophy of Science*, Stanford University Press, Stanford, Calif., pp. 356–60.
Sitnikow, K.: 1960, *Dokl. Akad. Nauk SSR* **133**, 303–06.
Wintner, A.: 1941, *The Analytical Foundations of Celestial Mechanics*, Princeton University Press, Princeton, N.J., U.S.A., pp. 274–75.

DISCUSSION

J. Moser: Can one give a detailed analytic description of the escape, in particular for $n=3$, of a bounded pair is present?

D. Saari: Yes. In this case the escaping particle would correspond to a 'one-particle' galaxy. For $n=3$ it would be possible to obtain a very detailed description.

J. Moser: Do you have a guess which of the asymptotic behaviors described by you is predominant?

D. Saari: I would guess that pulsating motion does not exist. If oscillatory motion with the property $\lim\sup r_{kj}/t=\infty$ does exist, then I would guess that it would be rare, i.e. of measure zero. This last problem is related to the problem of noncollision singularities.

DYNAMICAL FRICTION EFFECTS ON THE MOTION OF STARS IN ROTATING SPHERICAL CLUSTERS

A. S. BARANOV and Yu. V. BATRAKOV
Institute of Theoretical Astronomy, Leningrad, U.S.S.R.

Abstract. Effects of dynamical friction on star orbits in a spherical cluster uniformly rotating with small angular velocity about a fixed axis are considered, deformations of the cluster due to the rotation being neglected. The test star is supposed to move in a noncircular restricted orbit under the influence of both the attraction of the cluster with the smoothed-out distribution of stellar matter and dynamical friction due to random encounters of the test star with other stars of the cluster.

The approximate formula for dynamical friction has been deduced, the encounters being supposed to be the binary ones. The differential equations for the osculating elements of the star orbit have been obtained for the two cases of the density distribution – the uniform and the exponential ones. The numerical results demonstrate the complicated character of dynamical friction effects on the evolution of the orbit. The orbit tends to become circular, and its inclination decreases. These effects are proportional to the mass of the test star. This leads to the conclusion that dynamical friction contributes noticeably to the concentration of massive stars near the center of the cluster.

1. The Formulation of the Problem

The field of force in the star cluster may be described as a superposition of the regular field defined by the continuous distribution of stellar matter in the cluster and of the irregular field defined by chance stellar encounters. When considering the star orbits the greatest attention is usually paid to the regular component of the force field, and the effect of stellar encounters as a rule, is entirely neglected. On this scheme each star follows the determinate trajectory under the sole influence of the smoothed-out gravitational field of the potential of the system as a whole, which is a function of the space coordinates only, the total energy of the star being unchanged, and the motion of the star being uniquely defined by its orbital parameters.

It is clear, however, that this determinate process of the motion of the star along a specified trajectory in the regular field will be disturbed when the test star passes close to other stars. Due to the accidental encounters with stars of the cluster an additional force arises, the magnitude, the direction and the duration of which depend on the initial conditions characterizing each single encounter. The knowledge of the position of the star and of the disturbing force acting on the latter at some instant of time gives only a certain probability of each possible position at the following instant of time. Therefore, the changes of the orbit under the influence of the encounters are random.

Relatively distant encounters of the test star with other stars of the cluster occur much more often than the close ones, and although the effect of a single distant encounter is quite small, the accumulated effect of distant encounters may result in an appreciable change of the star orbit. The negligibly small probability of the close encounters also enables us to disregard discontinuities in the behaviour of the stellar

Y. Kozai (ed.), The Stability of the Solar System and of Small Stellar Systems, 285–296.

velocity as a random function of time and to consider the changes of the stellar velocity in the irregular field as a continuous random process.

As a result of the action of the irregular forces in gravitating systems, in particular, dynamical friction appears in the direction against the relative velocity of the test star. Dynamical friction in gravitating systems was for the first time investigated in detail by Chandrasekhar (1943b), but in his research the test star was assumed to move in a straight line in the structureless infinite medium. In real stellar systems the star orbits cannot, however, be considered rectilinear and must be restricted in space.

In this paper we deal with the influence of dynamical friction on the motion of the test star in a spherical cluster uniformly rotating with a small angular velocity around a fixed axis passing through the center of masses of the cluster. The deformation of the cluster caused by its rotation is considered to be neglected, when its influence on the test star motion is concerned. Furthermore, we shall assume that there are no external forces acting on the motion of the test star.

2. Dynamical Friction in Stellar Clusters

If one neglects the probability of the close encounters it is necessary to consider the action of the irregular forces in the stellar cluster as a random continuous Markoff process (for the fixed state at present, the state of the system in future does not depend on its state in the past). In this supposition the force of dynamical friction can be expressed as

$$\mathbf{F}=m\langle\Delta\mathbf{v}\rangle/\Delta t, \tag{1}$$

where m is the mass of the test star, $\langle\Delta\mathbf{v}\rangle/\Delta t$ is the average increment of the velocity of this star during the time interval Δt, which is sufficiently long for many encounters to occur during this time, but sufficiently short for the velocity of the star not to change appreciably (Chandrasekhar, 1943a, b). We shall suppose, as is usually the case (Chandrasekhar, 1942), that each encounter may be idealized as an independent two-body problem. For the sake of simplicity we shall suppose that all the stars in the cluster have the same mass, $\bar{m}$, and the mass m of the test star, which differs, in general, from the mass $\bar{m}$, is small as compared with the total mass of the cluster $N\bar{m}$ (N is the number of stars in the cluster). According to these assumptions the increment of the velocity per unit time may be represented in the form (Rosenbluth *et al.*, 1957)

$$\langle\Delta\mathbf{v}\rangle/\Delta t=\Gamma \operatorname{grad} h, \tag{2}$$

$$\Gamma=4\pi G^2\bar{m}^2 \ln\left[\frac{L\sigma^2}{G(m+\bar{m})}\right], \tag{3}$$

$$h(\mathbf{v})=\frac{m+\bar{m}}{\bar{m}}\int\frac{\psi(\mathbf{v}')}{|\mathbf{v}-\mathbf{v}'|}\,d\mathbf{v}' \tag{4}$$

where G is the constant of gravitation, L is the maximal encounter parameter in the two-body problem, which may be considered as the average distance between the stars (Chandrasekhar, 1942), σ^2 is the mean square velocity of the stars in the cluster, $\psi(\mathbf{v}')$ is the distribution function of the velocities, $\mathrm{d}\mathbf{v}'$ is a volume element in the velocity space. The dependence of the logarithm on the velocity in (3) is weak, therefore, the quantity Γ may be assumed to be constant. The denominator in (4) represents the velocity of the test star relative to an arbitrary cluster star. Equations (2)–(4) are valid in any coordinate system, but for the sake of convenience when calculating the function $h(\mathbf{v})$ the system of coordinates referred to the centroid of the given point is used.

We shall now suppose that the distribution function of the velocities is Maxwellian:

$$\psi(v') = D\left(\frac{3}{2\pi\sigma^2}\right)^{3/2} \exp\left(-\frac{3v'^2}{2\sigma^2}\right), \tag{5}$$

where D denotes the number of stars per unit volume. Substituting (5) in (4) and using integral representation,

$$\frac{1}{|\mathbf{v}-\mathbf{v}'|} = \frac{1}{\sqrt{\pi}} \int_{-\infty}^{+\infty} [\exp(-|\mathbf{v}-\mathbf{v}'|^2)\, \xi^2]\, \mathrm{d}\xi$$

(ξ is an auxiliary variable of integration) after some minor rearranging, we find

$$\operatorname{grad} h = \frac{m+\bar{m}}{\bar{m}} \frac{D}{v^3} [\phi(s) - \phi'(s)]\, \mathbf{v}. \tag{6}$$

Here s is a variable defined by

$$s^2 = 3v^2/2\sigma^2,$$

and $\phi(s)$ and $\phi'(s)$ are respectively the error function and its derivative;

$$\phi(s) = \frac{2}{\sqrt{\pi}} \int_0^s \exp(-\xi^2)\, \mathrm{d}\xi, \qquad \phi'(s) = \frac{2}{\sqrt{\pi}} \exp(-s^2).$$

Using (2), (3) and (6), we obtain from (1)

$$\mathbf{F} = -4\pi G^2 m\bar{m}(m+\bar{m}) \left\{ \ln\left[\frac{L\sigma^2}{G(m+\bar{m})}\right]\right\} \frac{D}{v^3} [\phi(s) - s\phi'(s)]\, \mathbf{v}.$$

The function $v(s) = \phi(s) - s\phi'(s)$ increases monotonically from zero to unit as s changes from zero to infinity. For the sake of simplicity in what follows we shall assume that the velocity of the test star is not too different from σ. We are interested in the approximate estimate of dynamical friction effects for comparatively small time intervals at which changes of function $v(s)$ are quite small. Expanding $v(s)$ in powers of the deviations of the velocity of the test star from the root mean square velocity of the cluster stars and taking into account only the first term of the expansion, we obtain

$$\mathbf{F} = -F_0 \frac{D}{v^3} \mathbf{v}, \tag{7}$$

where F_0 is a constant, which has the form

$$F_0 = 4\pi G^2 m\bar{m}(m+\bar{m}) \left\{ \ln \left[\frac{L\sigma^2}{G(m+\bar{m})} \right] \right\} [\phi(\sqrt{\tfrac{3}{2}}) - \sqrt{\tfrac{3}{2}}\, \phi'(\sqrt{\tfrac{3}{2}})].$$

It must be emphasized here that the right-hand side of (7) does not take into consideration unavoidable fluctuations of the force. The greater the mass of the test star and its velocity, the more correct it is to restrict ourselves to the determinate part of **F** given by (7). The estimates below do not claim to be quite accurate but they do reveal the main tendency in the evolution of the star orbits, the kinetic energy of which exceeds the average kinetic energy of the stars in the cluster.

3. Dynamical Friction Effects on the Motion of the Star Inside a Rotating Homogeneous Spherical Cluster

Let us consider a spherical cluster with constant density $D = D_0$ and suppose that in a certain appropriately chosen fixed frame of reference (x, y, z) with the origin in the center of the sphere, the cluster rotates with a uniform angular velocity ω about the z-axis. The potential U for the case under consideration takes the form

$$U = -\tfrac{1}{2}\alpha r^2, \tag{8}$$

where r is the distance of the test star from the center of the cluster, and

$$\alpha = \tfrac{4}{3}\pi G \bar{m} D_0. \tag{9}$$

It is well-known that the trajectory of the star in the force field (8) is an ellipse, the center of which coincides with the center of the sphere, and the period of the revolution of the star τ around the center of the sphere in the motion along this ellipse depends neither on the dimensions nor on the compression of the latter and is equal to $2\pi/\alpha^{1/2}$. The equation of the trajectory takes the form

$$r^2 = \frac{q^2}{1 + \kappa \cos 2\varphi}, \tag{10}$$

where

$$q = r_p(1+\kappa)^{1/2}, \qquad \kappa = \frac{r_a^2 - r_p^2}{r_a^2 + r_p^2}, \tag{11}$$

where r_p and r_a are respectively the pericentric and apocentric distances of the test star, φ is the angle between the radius vector and the direction to the pericenter. The quantity κ characterizes the compression of the ellipse and in many respects is analogous to the eccentricity.

The undisturbed star orbit in the gravitational field (8) is completely defined by the two isolating integrals of the motion: the energy integral (scalar) and the momentum integral (vector), which respectively take the form, in terms of peri- and apocentric distances,

$$\begin{aligned} \dot{r}^2 + \alpha r^2 &= \alpha\left(r_p^2 + r_a^2\right), \\ \mathbf{r} \times \dot{\mathbf{r}} &= \mathbf{c}, \end{aligned} \tag{12}$$

where dots denote the derivatives with respect to time. The following designations are used in (12):

$$\mathbf{c} = c \begin{Bmatrix} \sin i \sin \Omega \\ -\sin i \cos \Omega \\ \cos i \end{Bmatrix}, \qquad c = \alpha^{1/2} r_p r_a \tag{13}$$

$$\mathbf{r} = r \begin{Bmatrix} \cos u \cos \Omega - \sin u \sin \Omega \cos i \\ \cos u \sin \Omega + \sin u \cos \Omega \cos i \\ \sin u \sin i \end{Bmatrix}. \tag{14}$$

Here, as usual, i is the inclination of the orbit, Ω is the longitude of the ascending node, $u = \varphi + \varpi$, and ϖ is the distance of the pericenter from the node.

For investigating the disturbed motion of the star in a rotating cluster it is necessary to keep in mind that $\mathbf{v}$ in (7) is the relative velocity, and it is necessary to calculate it according to the formula

$$\mathbf{v} = \dot{\mathbf{r}} - \mathbf{w}, \tag{15}$$

where $\mathbf{w}$ is the linear velocity of the rotation of the cluster at the test star position.

To take into account the effects of the disturbing force (7) on the motion defined by (8), (10) and (12) we shall now consider r_p, r_a, i, Ω, ϖ to be the osculating variables. It means that the space coordinates and the components of the velocity in the disturbed motion at each instant of time are calculated according to the formulae of the undisturbed motion. We shall compose the differential equations for these variables.

Using the momentum and the energy integrals of the undisturbed motion (12) and the equation of the disturbed motion

$$\ddot{\mathbf{r}} = \operatorname{grad} U + \mathbf{F}/m,$$

we find the following equations for r_p and r_a:

$$\begin{aligned} \dot{r}_p &= \frac{1}{r_a^2 - r_p^2}\left[\frac{r_a}{\sqrt{\alpha}\, m}(\mathbf{r} \times \mathbf{F}) \frac{\mathbf{c}}{c} - \frac{r_p}{\alpha m}(\dot{\mathbf{r}} \cdot \mathbf{F})\right], \\ \dot{r}_a &= \frac{1}{r_a^2 - r_p^2}\left[\frac{r_a}{\alpha m}(\dot{\mathbf{r}} \cdot \mathbf{F}) - \frac{r_p}{\sqrt{\alpha}\, m}(\mathbf{r} \times \mathbf{F}) \frac{\mathbf{c}}{c}\right], \end{aligned} \tag{16}$$

where

$$\alpha = \tfrac{4}{3}\pi G\bar{m}D_0.$$

$$(\mathbf{r}\times\mathbf{F}) = -\frac{F_0 D_0}{v^3}\{\mathbf{c} - \omega[-\mathbf{i}xz - \mathbf{j}yz + \mathbf{k}(x^2+y^2)]\},$$
$$(\dot{\mathbf{r}}\cdot\mathbf{F}) = -\frac{F_0 D_0}{v^3}[\alpha(r_p^2 + r_a^2 - r^2) - \omega\sqrt{\alpha}\, r_p r_a \cos i]. \tag{17}$$

Here **i**, **j**, **k** are the unit vectors along the axes, x, y, z.

Along with the equations for r_p and r_a characterizing the dimensions and the compression of the orbit we find the equations for variables i, Ω and ϖ defining the orientation of the orbit:

$$\dot{i} = \frac{(\mathbf{r}\times\mathbf{F})\,\partial\mathbf{c}/\partial i}{mc^2} = -\frac{F_0 D_0 \omega r^2 \cos^2 u \sin i}{mcv^3}, \tag{18a}$$

$$\dot{\Omega} = \frac{(\mathbf{r}\times\mathbf{F})\,\partial\mathbf{c}/\partial\Omega}{mc^2} = -\frac{F_0 D_0 \omega r^2 \sin u \cos u \sin^2 i}{mcv^3}. \tag{18b}$$

For the change of ϖ we have (Subbotin, 1968):

$$\dot{\varpi} = -(\dot{\varphi}) - \cos i\dot{\Omega}, \tag{19}$$

where $\dot{\Omega}$ has been defined by relation (18b), and for the first term in the right-hand side (19) from the equation of the trajectory and the momentum integral we find:

$$(\dot{\varphi}) = \frac{q^2}{2\alpha^{1/2}\kappa r_p r_a r^2}\left[\frac{r^2 \sin 2\varphi}{q^4}(q^2\dot{c} - 2cq\dot{q}) - \sin 2\varphi\dot{c} + (\mathbf{r}\cdot\mathbf{F})\frac{\cos 2\varphi}{m}\right]. \tag{20}$$

The notation $\dot{\varphi}$ in (19) and (20) indicates that the derivative $\dot{\varphi}$ should be calculated with respect to time entering in the osculating orbital parameters only. For i and $\dot{q}$ according to (13) and (11) we have

$$\dot{c} = \alpha^{1/2}(\dot{r}_p r_a + r_p \dot{r}_a),$$
$$\dot{q} = \left(\frac{2}{r_p^2 + r_a^2}\right)^{1/2} \frac{(\dot{r}_p r_a + r_p \dot{r}_a) - r_p r_a(r_p \dot{r}_p + r_a \dot{r}_a)}{r_p^2 + r_a^2}.$$

The relations (16), (18) and (19) obtained above are the differential equations of interest for the osculating orbital parameters.

To estimate the evolution of the orbit under the influence of dynamical friction it is necessary to find the changes of the osculating parameters of the orbit for one revolution of the star around the center of the cluster. Approximate values of these changes can be found by integrating the right-hand sides of the above differential equations with respect to time. So, for instance, the change of r_p for one revolution is found with the formula

$$\Delta r_p = r_p(\tau) - r_p(0) = \int_0^{\tau} \mathfrak{f}\, dt = c^{-1}\int_0^{2\pi} \mathfrak{f} r^2\, d\varphi, \tag{21}$$

where $\mathfrak{f}$ is the right-hand side of Equation (16) for r_p. The changes of the remaining elements can be obtained in exactly the same way. While calculating the integrals of the type (21) it is necessary to take into account that the integrands are π-periodic with respect to φ and, moreover, possess the properties of evenness or oddness. The use of these properties enables us to reduce the interval of the integration by a factor 4 and avoid the calculation of the integrals which are equal to zero. In particular, it turns out that

$$\Delta\varpi = -\cos i \Delta\Omega . \tag{22}$$

It should be mentioned that the right-hand sides of the equations for i and Ω keep the angular velocity of the rotation of the cluster ω as a factor. If ω is zero, that is, the cluster does not rotate, the parameters i and Ω are constants, and the motion takes place in an invariable plane. The changes r_p and r_a in this case do not depend on i, Ω and ϖ. In the general case of the rotating cluster the quantities i, Ω and ϖ vary with time, and r_p, r_a depend on i and ϖ, but all these quantities do not depend on Ω (the problem is an axially symmetrical one).

The changes of the orbital parameters for one revolution of the star around the center of the cluster were obtained by numerical estimation of the integrals of type (21) for a number of the values of the parameters i, ϖ, $\omega/\alpha^{1/2}$ and for different values of the ratio $m/\bar{m}$ in two cases, namely, that of the globular cluster and that of the open cluster. To be more definite, we assume that the test star moves at the periphery of the cluster. According to the virial theorem we find $\sigma^2 = 0.5GN\bar{m}\varrho^{-1}$, where ϱ denotes the radius of the cluster. Besides (Chandrasekhar, 1943a), $L = 0.559\,36 D_0^{-1/3}$. As a typical example let us take the following values of the main parameters for the globular cluster: $N = 3\times10^5$, $\bar{m} = m_{\odot}$, $\varrho = 35$ pc, $r_p = 27$ pc, $r_a = 30$ pc. For the open cluster we accepted the values $N = 3\times10^2$, $\bar{m} = m_{\odot}$, $\varrho = 3.5$ pc, $r_p = 2.7$ pc and $r_a = 3$ pc.

The corresponding numerical values of Δr_p, Δr_a and Δi are given in Table I for a number of variants. Due to lack of space the full volume of data obtained is not presented here and we shall restrict ourselves to some qualitative conclusions. The computations show some peculiarities in the evolution of the star orbits. First of all one notices that r_p and r_a decrease in all cases under consideration and that r_a decreases faster than r_p. This means that the dimensions of the orbit decrease, and the orbit tends to become circular. The changes of r_p and r_a in the rotating cluster almost always surpass the corresponding values for the nonrotating cluster at least in the case of the direct motion of the test star ($0 \leqslant i \leqslant \frac{1}{2}\pi$). In the rotating cluster the greatest are the changes of r_p and r_a in the case $i = 0$, that is, for the motion in the equatorial plane of the cluster. The changes of r_p and r_a depend essentially on the magnitude of the ratio $m/\bar{m}$ of the mass of the test star to the average mass of the cluster stars. When the value of this ratio increases the changes of r_p and r_a increase too.

Another important peculiarity of the evolution of the orbit in the rotating cluster is the decrease of inclination with the passage of time. The velocity of decrease of inclination is greatest for the polar orbits ($i = \frac{1}{2}\pi$), while for the equatorial orbits it becomes zero. The magnitude of the change of the inclination is quite significant. As

TABLE I

Changes of orbital parameters r_p, r_a, i per one orbital revolution of the test star with $m=4\bar{m}$ due to dynamical friction in a cluster (abbreviations: H homogeneous, E exponential, O open, G globular)

Type of cluster	ω[a]	i[b]	$-(\Delta r_p/r_p)\times 10^4$	$-(\Delta r_a/r_a)\times 10^4$	$-\Delta i\times 10^5$ (rad.)
HO	0	0	534	660	0
HO	0.1	0	633	810	0
HO	0.1	$\frac{1}{2}\pi$	531	653	50.2
HG	0	0	2.31	2.85	0
HG	0.1	0	2.87	3.50	0
HG	0.1	$\frac{1}{2}\pi$	2.30	2.82	2.17
EO	0	0	353	575	0
EO	0.1	0	425	677	0
EO	0.1	$\frac{1}{2}\pi$	351	573	133
EG	0	0	1.52	2.48	0
EG	0.1	0	1.84	2.93	0
EG	0.1	$\frac{1}{2}\pi$	1.51	2.47	0.58

[a] The angular velocity of the cluster ω is taken in terms of the mean orbital motion of the test star.
[b] For the nonequatorial orbits the argument of pericenter ϖ is taken to be zero.

a result of such evolution the plane of the orbit tends step by step to coincide with the equatorial plane of the cluster. The speed of the change of the inclination depends essentially both on the value of the ratio $m/\bar{m}$ and on the initial value of ϖ. As to the changes of Ω and ϖ they depend on the quantity ϖ too and in the order of magnitude they are comparable with the changes of the inclination.

The numerical data of Table I show also that the changes of the orbital parameters in an open cluster are noticeably greater than in a globular one.

All this entitles us to draw the following conclusions: Dynamical friction effects on the star orbits in a rotating homogeneous spherical cluster contribute to concentrating the massive stars both in the vicinity of the center of the cluster and in the equatorial plane of the cluster. The spherical spatial distribution of the stars must transfer step by step to the ellipsoidal one.

4. The Motion of the Star in a Rotating Spherical Cluster with the Exponential Distribution of the Density

The observations show that numerous classes of stellar systems, for instance, globular systems, are characterized by a considerable gradient of the stellar density from the center to the periphery. The numerical experiments (Agekjan and Baranov, 1969; Baranov, 1970) also confirm that in the central part of the spherical cluster the density is almost unchanged and then sharply falls towards the periphery. In our previous papers the distribution of the stellar density is traced as far as the boundary of the cluster, where approximately 0.9 of the whole mass of the system is concentrated inside the sphere with the radius 0.1 of the radius of the cluster. Therefore, in the motion of stars in such clusters the attraction of the central mass is of crucial significance.

Let us consider a spherical cluster uniformly rotating with a small angular velocity around a fixed axis and assume that the distribution of the stellar matter in the system is subject to the barometrical formula

$$D = D_0 \exp(-R/R_0), \tag{23}$$

where R is the distance of the test star from the center of the cluster, and D_0 and R_0 are the constants. The potential U corresponding to the barometrical formula (23) is easily determined from Poisson's equation:

$$U = -4\pi G\bar{m}D_0R_0^2\{\exp(-R/R_0) + (2R_0/R)\,[\exp(-R/R_0) - 1]\}. \tag{24}$$

In particular, it follows from the formula (24) that for large R the field of force becomes similar to the Newtonian one.

Since in the problem under consideration the main attraction force acting on the star moving in the periphery of the cluster is created by its central mass, one can conveniently investigate the motion by classical methods of celestial mechanics using Kepler's osculating elements a, e, Ω, i, ϖ, M_0. The differential equations for the osculating Kepler's elements are taken in Euler's form. The right-hand sides of these equations are expressed by means of the components of the disturbing acceleration S, T and W, which are the projections of the vector of the disturbing acceleration to the directions along the radius-vector, perpendicularly to the radius-vector in the orbital plane and along the normal to the orbital plane, respectively. Since these equations are well-known (see, for example, Subbotin, 1968), they are not listed here.

The star orbits lie in the plane passing through the center of the cluster. The orbits are not closed and cannot be expressed in the elementary functions.

Besides the central mass and the disturbing force of dynamical friction the attraction of the peripheral part of the cluster also influences the star motion. Since we assume the distribution of mass in the system to be spherical, the disturbing force due to the attraction of the peripheral part of the cluster is directed along the radius-vector. The attraction of the peripheral part of the cluster may be described as the difference between the attraction of the mass bounded by the sphere passing through the test star and the attraction of some reference sphere. In the following we assume the undisturbed semimajor axis a to be the radius of the reference sphere. The vector equation of the motion of the test star takes the form

$$\ddot{\mathbf{R}} = -\frac{G\mathfrak{M}(a)}{R^3}\mathbf{R} - \frac{G\Delta\mathfrak{M}}{R^3}\mathbf{R} + \frac{\mathbf{F}}{m},$$

where $\mathfrak{M}(a)$ is the mass of the sphere of the radius a, $\Delta\mathfrak{M}$ is the mass of the spherical layer between the spheres with radii R and a, $\mathbf{F}$ is the vector of the force of dynamical friction defined by formula (7).

The absolute value of the disturbing acceleration Q_R due to the attraction of the peripheral part of the cluster is determined by the formula

$$Q_R = G\Delta\mathfrak{M}/R^2, \tag{25}$$

where

$$\Delta\mathfrak{M} = -4\pi\bar{m}D_0R_0\{\exp(-R/R_0)\,[(R+R_0)^2+R_0^2] - \exp(-a/R_0)\,[(a+R_0)^2+R_0^2]\}. \tag{26}$$

The components S, T and W of the disturbing acceleration due to the force of dynamical friction take the form

$$\begin{aligned} S &= -\frac{F_0}{m}\frac{D}{v^3}\sqrt{\left(\frac{\mu}{p}\right)}\, e\sin f, \\ T &= -\frac{F_0}{m}\frac{D}{v^3}\left[\sqrt{\left(\frac{\mu}{p}\right)}(1+e\cos f) - \omega R\cos i\right], \\ W &= -\frac{F_0}{m}\frac{D}{v^3}\,\omega R\sin i\cos(f+\varpi), \end{aligned} \tag{27}$$

where

$$v = \left\{\frac{\mu}{p}(1+2e\cos f+e^2) - 2\sqrt{\left(\frac{\mu}{p}\right)}(1+e\cos f)\,\omega R\cos i + \omega^2R^2[\cos^2 i + \sin^2 i\cos^2(f+\varpi)]\right\}^{1/2}, \tag{28}$$

$p = a(1-e^2)$, f is the true anomaly, μ is the product of the constant of gravitation on the mass of the central part of the cluster. When deriving (27), the formulae by Fominov (1963) were used. In these formulae the absolute value of the disturbing resistance force was taken in the form (7).

Introducing now the eccentric anomaly E as the independent variable into the differential equations of the motion according to the formula

$$\frac{\mathrm{d}t}{\mathrm{d}E} = \frac{a^{3/2}}{\mu^{1/2}}(1-e\cos E),$$

taking into account the well-known relations

$$\begin{aligned} R\cos f &= a(\cos E - e), \\ R\sin f &= a\sqrt{1-e^2}\sin E, \end{aligned}$$

and the relations (25), (25)–(28) and integrating numerically the right-hand sides of the equations of motion for one revolution of the star around the center of the cluster, we obtain the estimates of the changes of the elements for one revolution. We accept $a=30$ pc, $e=0.2$ for the globular cluster and $a=3$ pc, $e=0.2$ for the open cluster. The angular velocity of the rotation is taken to be $\omega=0.1n$, where n is the mean motion of the test star. Some numerical data are given in Table I.

The results demonstrate the complicated character of dynamical friction effects on the evolution of the orbit. One notices that the dimensions of the orbit decrease, and the orbit tends to become circular. The changes of r_p and r_a in the rotating cluster surpass the corresponding changes of these elements in the nonrotating cluster. In the rotating cluster the greatest deformation is of the orbit of the test star moving in the equatorial plane ($i=0$).

The changes of the orbital elements increase noticeably when the ratio $m/\bar{m}$ increases. The changes of the elements in an open cluster are considerably greater than in a globular cluster.

An important peculiarity of the evolution of the orbit in the rotating cluster is the decrease of orientation with passage of time. Since the right-hand sides of the equations of motion for i and Ω contain the angular velocity of the rotation of the cluster ω as a factor, in the nonrotating cluster ($\omega=0$) the changes of i and Ω are equal to zero, that is, the test star moves in an invariable plane. The changes of the dimensions in this case are not dependent on i, Ω and ϖ. In the general case of the rotating cluster i, Ω and ϖ vary with time, and r_p and r_a depend on i, ϖ (the dependence of r_p and r_a on ϖ is quite weak), but all these parameters do not depend on Ω.

In the case under consideration (in quite the same way as in the case of the homogeneous cluster) the angles Ω and $\tilde{\omega}$ are connected by Equation (22). This follows from the properties of evenness (oddness) and periodicity of the right-hand sides of the equations of motion.

The change of inclination is greater than the changes of other elements defining the orientation of the orbit. The change of i is maximum for the polar orbits ($i=\frac{1}{2}\pi$), while for the equatorial orbits ($i=0$) it becomes zero. The changes of Ω and ϖ are comparable in the order of magnitude with the changes of inclination. The changes of i, Ω and ϖ depend essentially on the initial value of the parameter ϖ.

Thus the evolution of the orbits in the rotating cluster with the exponential distribution of the density tends to concentrate the massive stars both near the equatorial plane of the cluster and near the center of the cluster.

5. Conclusions

The estimation of dynamical friction effects on the orbit of the test star under the influence of the regular gravitational field and dynamical friction shows that in stellar systems with relatively short times of relaxation dynamical friction effects are not negligible. Both density distributions in the cluster are considered: the homogeneous and the exponential ones. The numerical examples show that the orbit of the test star in the cluster, in general, is subject to complicated variations: the originally elliptic orbit tends to become circular and its axis and inclination decrease with passage of time. Due to the influence of dynamical friction the massive stars tend to concentrate both in the vicinity of the center of the cluster and of the equatorial plane of the cluster.

References

Agekjan, T. A. and Baranov, A. S.: 1969, *Astrofiz.* **5**, 305.
Baranov, A. S.: 1970, *Astrofiz.* **6**, 261.
Chandrasekhar, S.: 1942, *Principles of Stellar Dynamics*, The University of Chicago Press, Chicago.
Chandrasekhar, S.: 1943a, *Rev. Mod. Phys.* **15**, 1.
Chandrasekhar, S.: 1943b, *Astrophys. J.* **97**, 255 (Parts I, II); **98**, 54 (Part III).
Fominov, A. M.: 1963, *Bull. Inst. Theor. Astron.* **9**, 185 (in Russian).
Rosenbluth, M. N., MacDonald, W. M., and Judd, D. L.: 1957, *Phys. Rev.* **107**, 1.
Subbotin, M. F.: 1968, *Vvedenie v Teoreticheskuyu Astronomiyu*, Moscow.

DISCUSSION

R. H. Miller: When inclined orbits tend to decrease their inclinations, approaching the equatorial plane, do they approach monotonically, or can they overshoot and approach in a damped oscillation mode?

A. S. Baranov: They approach monotonically.

ON THE DISAPPEARANCE OF ISOLATING INTEGRALS IN DYNAMICAL SYSTEMS WITH MORE THAN TWO DEGREES OF FREEDOM

C. FROESCHLÉ and J.-P. SCHEIDECKER
Observatoire de Nice, 06300 Nice, France

Abstract. We continue to study the number of isolating integrals in dynamical systems with three and four degrees of freedom, using as models the measure preserving mappings T already introduced in previous papers (Froeschlé, 1972; Froeschlé and Scheidecker, 1973a).

Thus, we use here a new numerical method which enables us to take as indicator of stochasticity the variation with n of the two (respectively three) largest eigenvalues – in absolute magnitude – of the linear tangential mapping T^{n*} of T^n. This variation appears to be a very good tool for studying the diffusion process which occurs during the disappearance of the isolating integrals, already shown in a previous paper (Froeschlé, 1971). In the case of systems with three degrees of freedom, we define and give an estimation of the diffusion time, and show that the gambler's ruin model is an approximation of this diffusion process.

1. Introduction

In a previous paper (Froeschlé, 1971), it has been found, using a four-dimensional mapping T as a model problem, that a dynamical system with three degrees of freedom has, in general, either two or zero isolating integrals (beside the usual energy integral).

Let T be a measure preserving mapping of the (x, y, z, t) space over itself defined by:

$$T \quad \begin{cases} x_1 = x_0 + a_1 \sin(x_0 + y_0) + b \sin(x_0 + y_0 + z_0 + t_0), \\ y_1 = x_0 + y_0, \\ z_1 = z_0 + a_2 \sin(z_0 + t_0) + b \sin(x_0 + y_0 + z_0 + t_0), \\ t_1 = z_0 + t_0. \end{cases} \pmod{2\pi} \qquad (1)$$

If $b = 0$, then this mapping T is the product of two area-preserving mappings T_1 of (x, y) on itself and T_2 of (z, t) on itself.

The initial conditions (x_0, y_0, z_0, t_0) are taken such that an invariant curve exists for T_1 (integrable case) and not for T_2 (wild or 'ergodic' case).

In this case Froeschlé (1971) has observed that as soon as $b \neq 0$, the value of the isolating integral of T_1 is subjected to a kind of random walk. This integral slowly disappears by some diffusion process due to the coupling term $b \sin(x_n + y_n + z_n + t_n)$, which produces a quasi-random perturbation of (x_n, y_n), as the points (z_n, t_n) behave in a quasi-random fashion.

In this paper, we study more precisely numerically this diffusion process and some characteristic parameters of an orbit during this diffusion. One of our tools is the variation of the eigenvalues of the linear tangential mapping T^{n*} of T^n, which is a good indicator of the stochasticity of an orbit. In particular, we study the character of C-system (Arnold and Avez, 1967) of T during the diffusion process.

Hence, we look for the number of eigenvalues which grow exponentially (Froeschlé and Scheidecker, 1973b).

Y. Kozai (ed.), The Stability of the Solar System and of Small Stellar Systems, 297–316. All Rights Reserved.

In Section 2 we study the link between linear tangential mappings and the diffusion process. In Section 3 we define and estimate the diffusion time. In Section 4 we study the variation of this diffusion time with the coupling term and with the initial conditions. In Section 5 we study the case of a dynamical system with four degrees of freedom, i.e. a six-dimensional mapping.

2. Linear Tangential Mappings and Diffusion Process

In order to study more precisely the dissolution of the isolating integrals of the discrete dynamical system T, we give the topology of the mapping T_1 in two characteristic cases. This mapping is defined by,

$$T_1 \quad \begin{cases} x_1 = x_0 + a_1 \sin(x_0 + y_0), \\ y_1 = x_0 + y_0. \end{cases} \quad (\text{mod}\, 2\pi) \tag{2}$$

Figures 1 and 2 display typical sets of points for the mapping T_1. The initial conditions

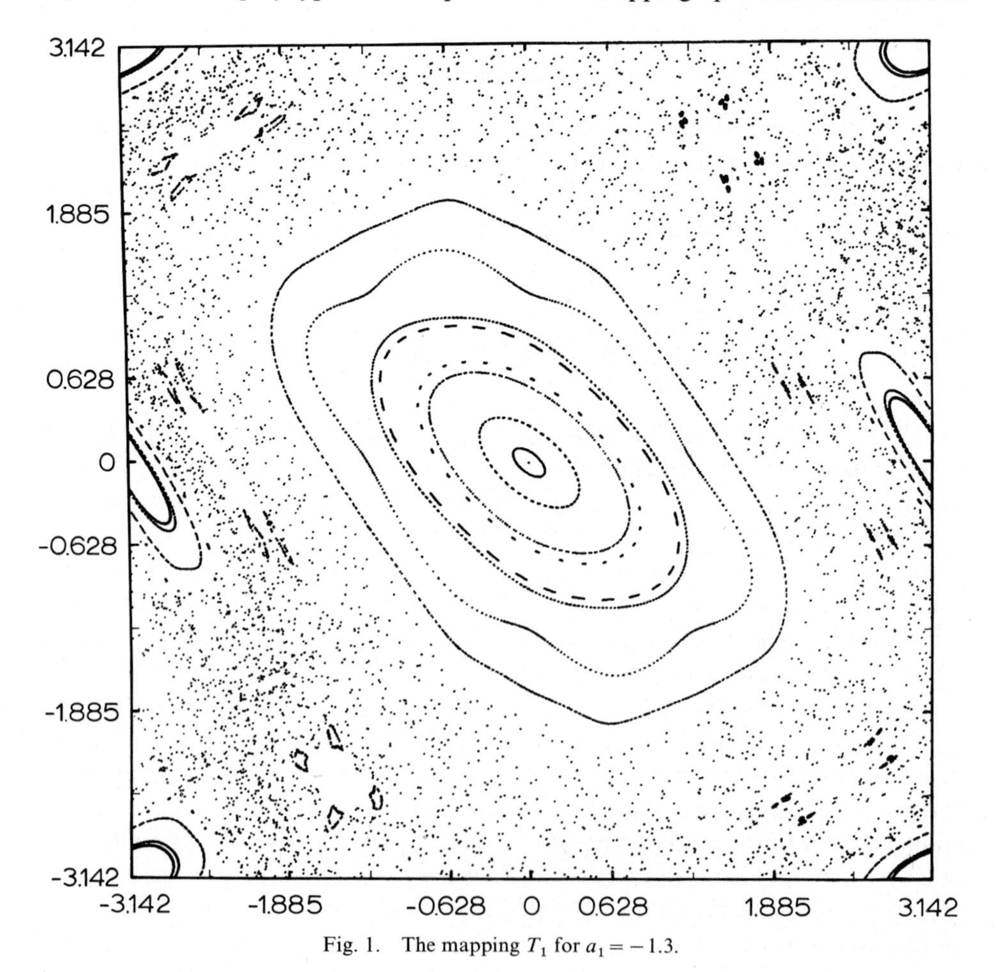

Fig. 1. The mapping T_1 for $a_1 = -1.3$.

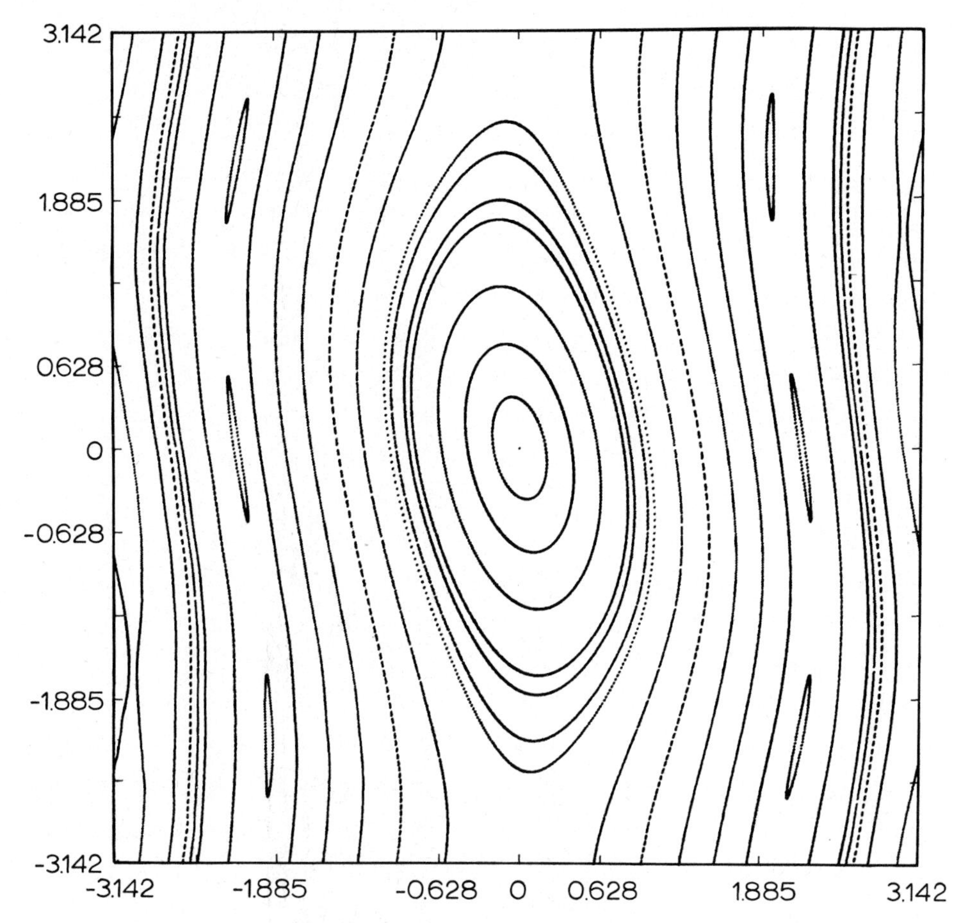

Fig. 2. The mapping T_1 for $a_1 = -0.3$.

and values of the parameter a_1 are presented in Table I. N is the total number of points plotted for each orbit.

Figure 1 exhibits all the characteristics and well-known features of problems with two degrees of freedom, i.e. invariant curves and islands, which correspond to the existence of isolating integrals; and also wild zones, sometimes called 'ergodic', where the points seem to fill a broad region in the plane, and which correspond to the non-existence of isolating integrals. On the other hand, on Figure 2 isolating integrals seem to exist everywhere: this is a case very close to an integral case. All the points are either on *libration* curves corresponding to the stable invariant point (0., 0.), or on *circulation* curves corresponding to the unstable invariant point (π, 0.).

Now, we study the behavior of the linear tangential mappings during the disappearance of the isolating integrals, i.e. during the diffusion process mentioned in Section 1. Thus, we use the variations with n of the two largest eigenvalues λ_1^n and λ_2^n – in absolute magnitude – of the linear tangential mapping T^{n*} of T^n (since the characteristic

TABLE I

Data for Figures 1 and 2

Figure	a_1	x_0	y_0	N
1	−1.3	2.8274	−3.1416	700
		2,6000	−3.1416	700
		2.5133	−3.1416	700
		3.0000	0	700
		2.9845	0	700
		2.8274	0	1000
		2.5133	0	900
		2.1991	0	800
		1.8860	0	800
		1.5708	0	700
		1.2566	0	500
		1.0000	0	400
		0.9425	0	400
		0.7000	0	200
		0.6283	0	300
		0.3142	0	200
		0.1000	0	10
		0.	0	3
		−2.0741	−1.7318	700
		−1.7880	−2.3114	700
		0.1969	−2.2867	700
2	−0.3	±3.1400	0	10
		±3.0000	0	700
		±2.8000	0	700
		±2.7000	0	1000
		±2.6500	0	400
		±2.6000	0	700
		±2.4000	0	700
		±2.2000	0	700
		±2.0000	0	700
		±1.8000	0	500
		±1.6000	0	500
		±1.4000	0	500
		±1.2000	0	500
		±1.0000	0	500
		±0.9500	0	500
		±0.8500	0	500
		±0.8000	0	500
		±0.6000	0	500
		±0.4000	0	500
		±0.2000	0	200
		0.	0	1

equation of the mapping T is reciprocal, the two other eigenvalues are the inverses of the previous ones; hence, they will not be plotted). λ_1^n and λ_2^n have been seen to be characteristics of the stochasticity of an orbit (Froeschlé and Scheidecker, 1973a, b). Numerical results are shown on Figures 3 and 4. Figure 3 displays the variations with

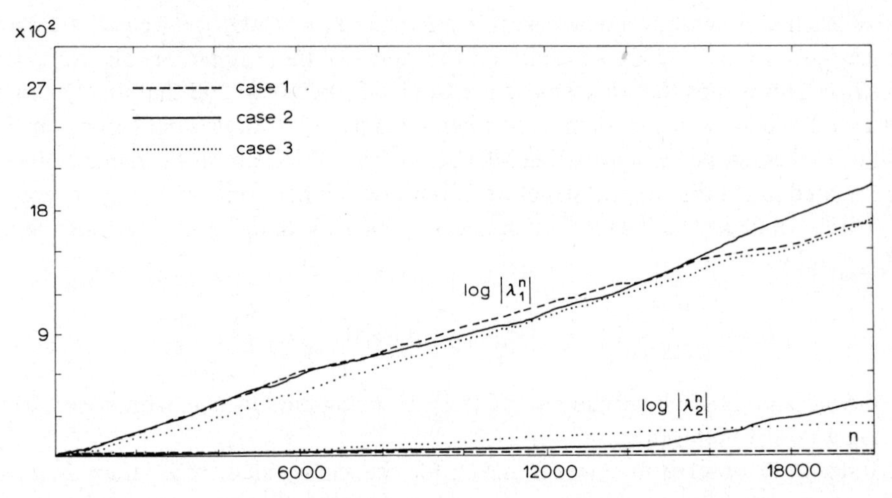

Fig. 3. Eigenvalues of the four dimensional mapping (1). Upper curves: variations of $\log_{10}|\lambda_1^n|$ with n. Lower curves: variations of $\log_{10}|\lambda_2^n|$ with n.

n of $\log_{10}|\lambda_1^n|$ and $\log_{10}|\lambda_2^n|$ in the three following cases:

Case 1: (uncoupled case) $a_1=-1.3$, $a_2=-1.3$, $b=0$, $x_0=0.2$, $y_0=0.2$, $z_0=0.5$, $t_0=3$.

Case 2: $a_1=-1.3$, $a_2=-1.3$, $b=0.05$, $x_0=0.2$, $y_0=0.2$, $z_0=0.5$, $t_0=3$.

Case 3: $a_1=-0.3$, $a_2=-1.3$, $b=0.05$, $x_0=0.2$, $y_0=0.2$, $z_0=0.5$, $t_0=3$.

Remark that in Case 2 (and Case 3), the initial conditions are such that in the corresponding uncoupled case (x_0, y_0) is taken in an integrable (or libration) zone for T_1 and (z_0, t_0) in an ergodic zone for T_2 (cf. Figures 1 and 2).

In the uncoupled case (Case 1), the lowest curve shows that one isolating integral does exist because its slope is equal to zero. In the coupled cases (Cases 2 and 3), although a sudden change in the slopes of $\log_{10}|\lambda_2^n|$ occurs, when the points (x_n, y_n)

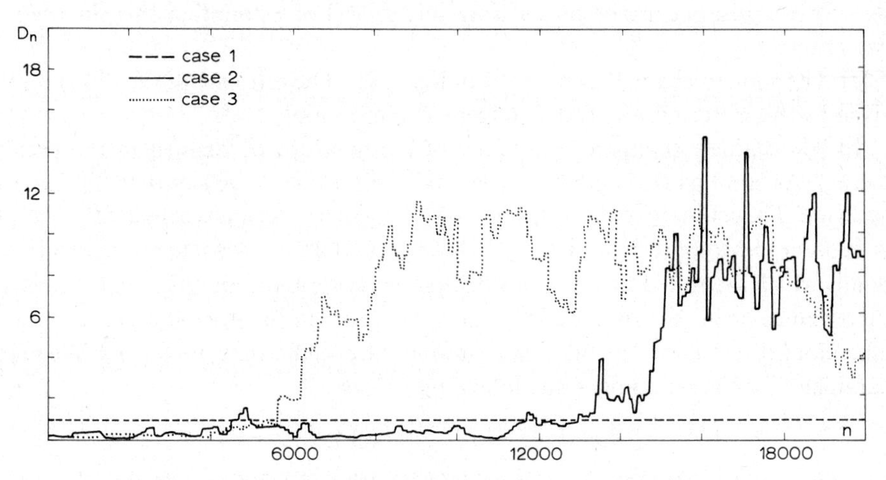

Fig. 4. D_n, a measure of the dimension of the curves, against n.

have reached either the ergodic zone (Case 2, for $n=n_2=14800$) or the circulatory zone (Case 3, for $n=n_3=6200$), we note that the values of these slopes are always *strictly positive*. This means that the orbits have an ergodic behavior and that the dynamical system T is close to a C-system, even when the diffusion process is still going on. The values of these slopes which are characteristics of the orbits and which change suddenly are related to the topological structure of both two-dimensional mappings T_1 and T_2.

This is confirmed by the results displayed on Figure 4, in the same three cases, where D_n given by

$$D_n = \left[\sum_{m=n-99}^{m=n} \left(x_m^2 - a_1 \left(y_m^2 + x_m y_m\right)\right) \right] / 100 \tag{3}$$

is the measure of the invariant curves of T_1, D_n being plotted vs n, with $n=k \times 100$, k being a positive integer.

Indeed, we observe in Cases 2 and 3, for the same values of n, (Case 2: $n=n_2$, Case 3: $n=n_3$), either a sudden increase of the value of D_n (Figure 4) or a sudden change in the slope of $\log_{10}|\lambda_2^n|$ (Figure 3). Furthermore, since, in these two cases, we start at the same initial point, the fact that $n_3 < n_2$ may be due to the topology of the curves in Figures 2 and 1. (In Case 3, the libration zone is narrower than in Cases 2. Hence the time necessary for escaping from this zone is smaller in Case 3.)

Moreover, the fact that, in the uncoupled case $\log_{10}|\lambda_2^n|$ remains constant, shows clearly that the effects of the rounding errors of the computer are negligible although they could have produced the same effects as the coupling.

3. Definition and Estimation of the Diffusion Time

We call 'diffusion time' the number of iterations of the mapping T which are necessary for the point (x_n, y_n), starting in the integrable zone, (or the libratory zone) of T_1, to reach the wild (ergodic) zone (or the circulatory zone) – in other words, the time necessary for the disappearance of isolating integrals. For estimating this time, we use two criteria:

(a) The sudden change in the slope of $\log_{10}|\lambda_2^n|$. The estimate of the diffusion time given by this criterion is called T_E (eigenvalue criterion).

(b) The sudden change of the value of the measure D_n defined in the previous section. As soon as D_n is greater or equal to the value d, we say that diffusion has occurred. The estimate of the *diffusion* time using this criterion is called T_D^d $(4 \leqslant d \leqslant 10)$.

In order to compare T_D^d and T_E we choose at random in the square $(-\pi, \pi) \times (-\pi, \pi)$ points which are tested by the orbits divergence criterion to be either in the invariant curves zone, or in the ergodic zone. The m first ones in the invariant curves zone are taken for (x_0, y_0) as well as the m first ones in the ergodic zone, for (z_0, t_0). We take as parameters of the mapping T the following values:

$$b=0.05, \qquad a_1=a_2=-1.3, \qquad m=14.$$

Figure 5 displays for two fixed values of d ($d=10$, $d=4$) and for each initial condition

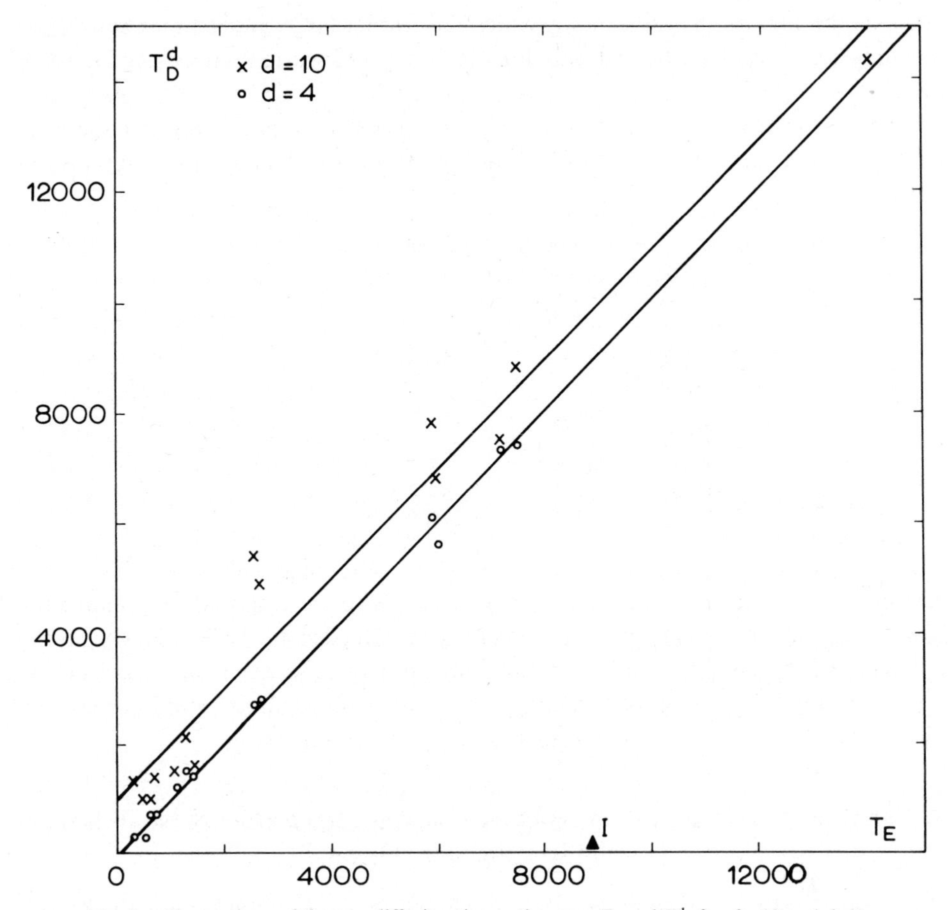

Fig. 5. Comparison of the two diffusion time estimators T_E and T_D^d, for $d=10$ and $d=4$.

(x_0, y_0, z_0, t_0) the different values of T_D^d plotted vs T_E. The correlation coefficients of the point clouds, and the coefficients a and b of the least square straight line $T_D^d = aT_E + b$ have been computed. The numerical results are given in Table II the straight lines are also shown on Figure 5.

It appears worthwhile to note that for the different values of the parameter d, the values of a are close to 1, with a good precision. This indicates that, for the different experiments, the values of T_D^d are either overestimated or underestimated by a same amount.

The definition of D_n (first-order Birkhoff's approximation) as well as the choice of the different values of d induces errors in the estimation of T_D^d. Moreover, Figure 5 shows also the point I corresponding to integrable initial conditions (x_0, y_0) taken in an island (cf. Figure 1), i.e. $(x_0 = -\pi, y_0 = 0)$, the point $(z_0 = 1.185494, t_0 = -2.520556)$ being taken in an ergodic zone. (Of course, this point has not been taken into account for the computation of the least square straight line.) For this point, T_D^d is equal to 100, which is the smallest possible value, while T_E is rather large. Indeed the use of D_n is meaningless, since it is allowed only for the family of invariant curves surrounding the origin.

These facts show us that T_E is a more general criterion of diffusion time than T_D^d. However, the computation cost of T_E is about 20 times as large as the computation

TABLE II

Values of the correlation coefficients and coefficients of the least square straight line $T_D^d = aT_E + b$ of the point clouds for different values of d

d	Correlation coefficient	a	b
4.	0.99905	1.0096	0.04989
6.	0.99751	0.9915	195.765
8.	0.99706	1.011	360.719
10.	0.9795	0.9912	1017.92

cost of T_D^d. This is due to the fact that it is necessary to use a special programme for computing the second eigenvalue (Froeschlé and Scheidecker, 1973b). Hence, in the following we shall use T_D^d as diffusion time estimator since we can avoid cases such as point I on Figure 5, using the topology of the curves on Figures 1 and 2. Of course, T_E has been used to check some results given by this last method.

4. Variation of the Diffusion Time with the Coupling Term b and with the Distance of the Initial Point to the Origin

We intend to estimate the diffusion time as a function of the coupling parameter b and as a function of the generalized distance of the initial point to the origin.

We take the *gambler's ruin model* as an approximation of the diffusion process, since the problem is reduced to the study of the jumps of (x_n, y_n) from one elliptic curve to another, up to the ergodic zone, considered as an absorbing barrier. Indeed, the family of invariant curves surrounding the origin can be taken, in a first-order approximation, as a continuous elliptic family (cf. Figures 1 and 2).

In order to handle the problem more easily, we take a discrete family of homofocal ellipses as a model. The iterated point is supposed to jump from one ellipse to the next one. In fact, this problem reduces itself to a *one-dimensional* gambler's ruin problem, since the equations of the family of homofocal ellipses are given by:

$$x^2 - a_1(y^2 + xy) = C_k, \quad k = 0, \ldots, k_1, \tag{4}$$

k_1 being the value corresponding to the largest ellipse, that is to say, to the absorbing barrier. The point (x_n, y_n) starts from the initial condition (x_0, y_0) belonging to the ellipse C_k, and moves a step at random at each iteration, backward or forward, to C_{k+1} or to C_{k-1}.

We put $X^2 = C_k$. X can be interpreted as one of the two values of the intersection of the ellipse C_k with the x-axis. We call ε the increase ΔX, taken as a constant, when we jump from one ellipse to the next one.

Let $N(X)$ be the expected number of iterations, which are necessary for reaching the absorbing barrier, when starting on the ellipse C_k, at the generalized distance X from the origin.

We use the well-known formula (Feller, 1971)

$$N(X) = \frac{N(X+\varepsilon) + N(X-\varepsilon)}{2} + 1. \tag{5}$$

The boundary conditions are:

$$N(+C_{k_1}^{1/2}) = 0, \qquad N(-C_{k_1}^{1/2}) = 0. \tag{6}$$

Taking a second-order Taylor development, we get:

$$\mathrm{d}^2 N(X)/\mathrm{d}X^2 \simeq -2/\varepsilon^2. \tag{7}$$

Hence, by integration:

$$N(X) \simeq -(2/\varepsilon^2)\, X^2 + AX + B. \tag{8}$$

Using (6), we get:

$$N(X) \simeq 2(C_{k_1} - X^2)/\varepsilon^2. \tag{9}$$

4.1. Variations with the Coupling Term b

We use Equation (9) with X equal to a fixed value $\bar{X}$, and take as initial conditions $x_0 = y_0 = 0.5$. Hence, by (4), we compute $\bar{X} = 0.9486$; from the measure of $C_{k_1}^{1/2}$ taken on Figure 1 we get the approximate value $C_{k_1} = 2.89$.

In a first approximation, we consider the length of the step to be ε, and to be proportional to the coupling term b. Therefore, we have plotted the variation of $T_D^9 b^2$ vs b on Figure 6 (b takes values from 0.01 to 0.2). The straight line which is displayed is the average of the values of $T_D^9 b^2$. (This value is equal to 7.77.)

The points displayed by Figure 6 are obtained by using the computed values T_D^9 of the diffusion time $N(X)$: each of these values is itself the average of 25 computed values, corresponding to initial conditions $(x_1^\theta, y_1^\theta, z_1^\theta, t_1^\theta)$ surrounding the point (x_0, y_0, z_0, t_0), at random:

$$\begin{aligned} x_1^\theta &= x_0, \\ y_1^\theta &= y_0, \\ z_1^\theta &= z_0 + r_0 \cos\theta, \\ t_1^\theta &= t_0 + r_0 \sin\theta, \end{aligned} \tag{10}$$

where $r_0 = 10^{-4}$ and θ is chosen at random between $-\pi$ and $+\pi$. Such a method tends to give us more precision in the result.

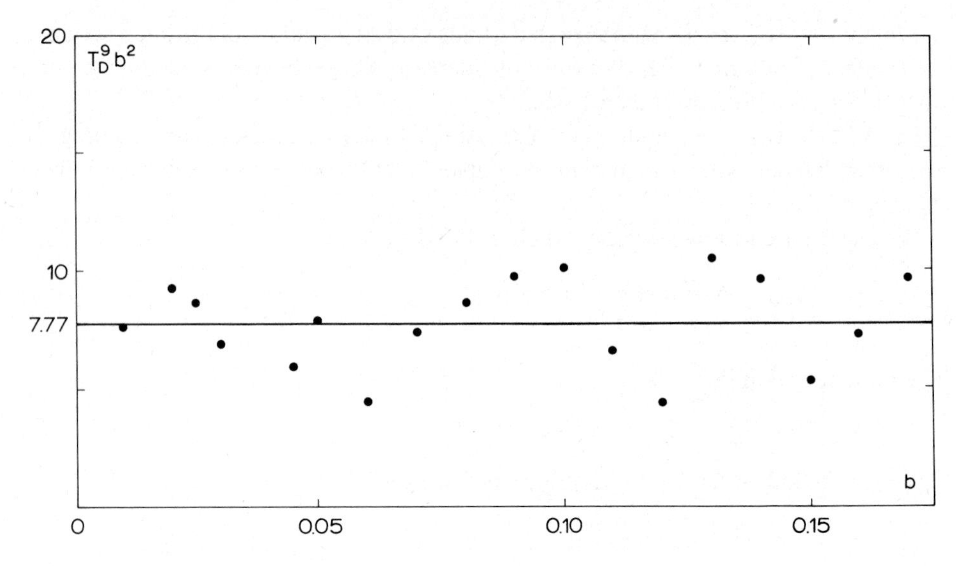

Fig. 6. Verification of the law $N(X)\,\varepsilon^2 \simeq$constant, for a given X.

Considering the crude approximations which have been made for obtaining Equation (9) the results are in rather good agreement with the gambler's ruin model.

From the preceding numerical results (mean value of $T_D^9 b^2$, values of $\bar{X}$ and $C_{k_1}^{1/2}$), using Equation (9) we get:

$$\varepsilon^2/b^2 \cong \tfrac{1}{2}. \tag{11}$$

4.2. Variations with the initial condition X

We take as fixed value of the coupling term: $b=0.05$. From (9), we get:

$$\varepsilon^2 N(X) + 2X^2 \simeq 2C_{k_1}, \tag{12}$$

where

$$X^2 = x_0^2 - a_1(y_0^2 + x_0 y_0), \qquad a_1 = -1.3. \tag{13}$$

And from (11):

$$\varepsilon^2 \simeq 12.5 \times 10^{-4}.$$

Thus, we plot the values of $T_D^9 b^2 + 4X^2$ vs X on Figure 7, since the T_D^9 are the experimental values of $N(X)$. Each value of T_D^9 is computed in the same way as previously (average of 25 random points). The straight line represents the average of the values of $T_D^9 b^2 + 4X^2$, which is found to be equal to 11.48, hence, to be very close to $4C_{k_1} = 11.56$.

Also, in this case, the gambler's ruin model seems to be a good approximation of the diffusion problem.

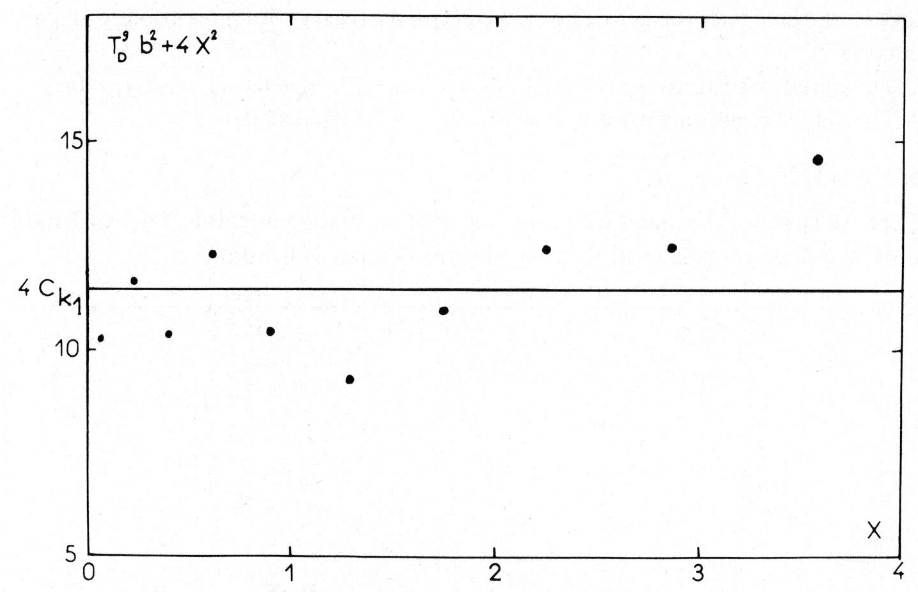

Fig. 7. Verification of the law $N(X)\,\varepsilon^2+2X^2\simeq$constant, for a given ε.

5. The Case of Dynamical Systems with Four Degrees of Freedom

The results given by Froeschlé (1971), and the previous sections in this paper suggest that, apart from particular cases such as $b=0$ for the present model, dynamical systems with three degrees of freedom have in general either two or zero isolating integrals, beside the usual energy integral. A similar effect probably exists in systems with more than three degrees of freedom.

We generalize the mapping T to the following six-dimensional mapping T_6, that is to say, to a dynamical system with four degrees of freedom:

$$T_6 \begin{cases} x_1=x_0+a_1\sin(x_0+y_0)+b\sin(x_0+y_0+z_0+t_0+u_0+v_0), \\ y_1=x_0+y_0, \\ z_1=z_0+a_2\sin(z_0+t_0)+b\sin(x_0+y_0+z_0+t_0+u_0+v_0), \\ t_1=z_0+t_0, \\ u_1=u_0+a_3\sin(u_0+v_0)+b\sin(x_0+y_0+z_0+t_0+u_0+v_0), \\ v_1=u_0+v_0. \end{cases} \pmod{2\pi} \qquad (14)$$

If $b=0$, this mapping T_6 is the product of three two-dimensional area preserving mappings: T_1 of (x, y) on itself, T_2 of (z, t) on itself, and T_3 of (u, v) on itself, and as in previous sections, we take initial conditions such that for $b=0$, we have either $N=0$, 1, 2 or 3 isolating integrals.

5.1. $N=0$

This corresponds to a purely ergodic case. Figure 8 displays the variation vs n of the logs of the three largest eigenvalues, λ_1^n, λ_2^n, λ_3^n(in decreasing order of absolute magnitude) of T^{n*}.

The initial conditions are: $x_0=2.0$, $y_0=0.0$, $z_0=2.1$, $t_0=0.0$, $u_0=2.2$, $v_0=0$.
The values of parameters are: $a_1=a_2=a_3=-1.3$, $b=0.050$.

5.2. $N=1, 2$

Here, we have, for the uncoupled case, one or two isolating integrals. Figures 9 and 10 show the same process as in the case of three degrees of freedom.

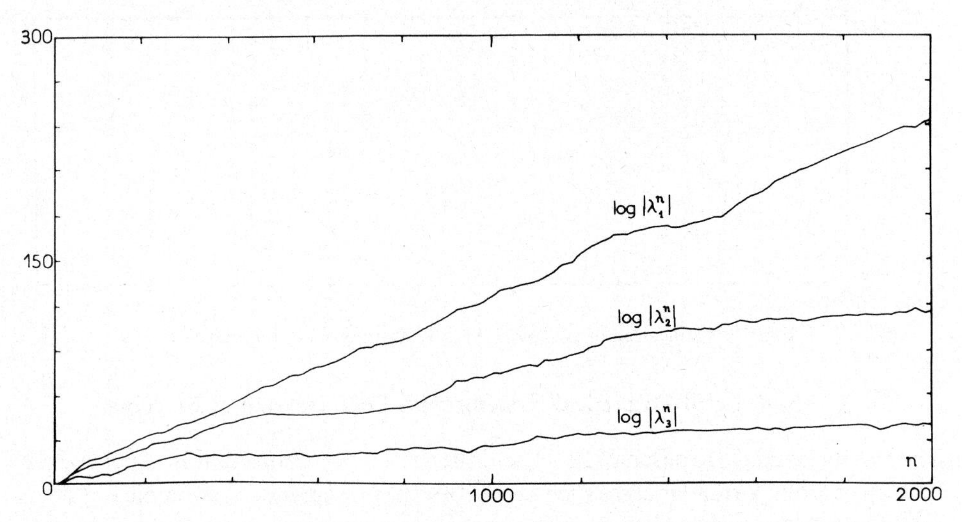

Fig. 8. Eigenvalues of the six-dimensional mapping (14): variations of $\log_{10}|\lambda_1^n|$, $\log_{10}|\lambda_2^n|$, $\log_{10}|\lambda_3^n|$ with n, when no isolating integrals exist.

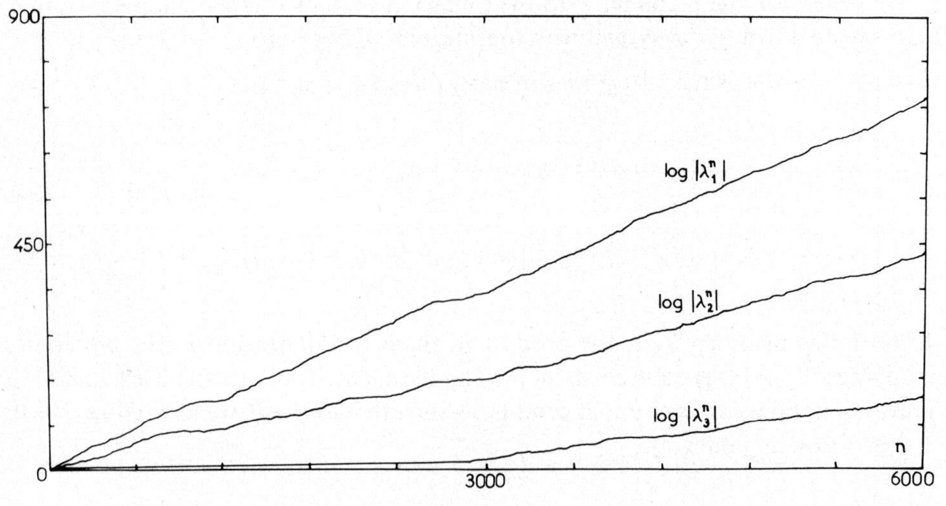

Fig. 9. Variations of $\log_{10}|\lambda_1^n|$, $\log_{10}|\lambda_2^n|$, $\log_{10}|\lambda_3^n|$ with n, when, for the corresponding uncoupled case, one isolating integral exists.

Initial conditions for Figure 9 are: $x_0=0.5, y_0=0.5, z_0=0.5, t_0=3.0, u_0=0.5, v_0=3.1$.

Initial conditions for Figure 10 are: $x_0=0.5$, $y_0=0.5$, $z_0=0.1$, $t_0=0.1$, $u_0=0.5$, $v_0=3.0$.

Parameters are: $a_1=a_2=a_3=-1.3$; $b=0.052$ for both cases. The value of b has been slightly increased in order to decrease the number of iterations necessary to reach the ergodic zone.

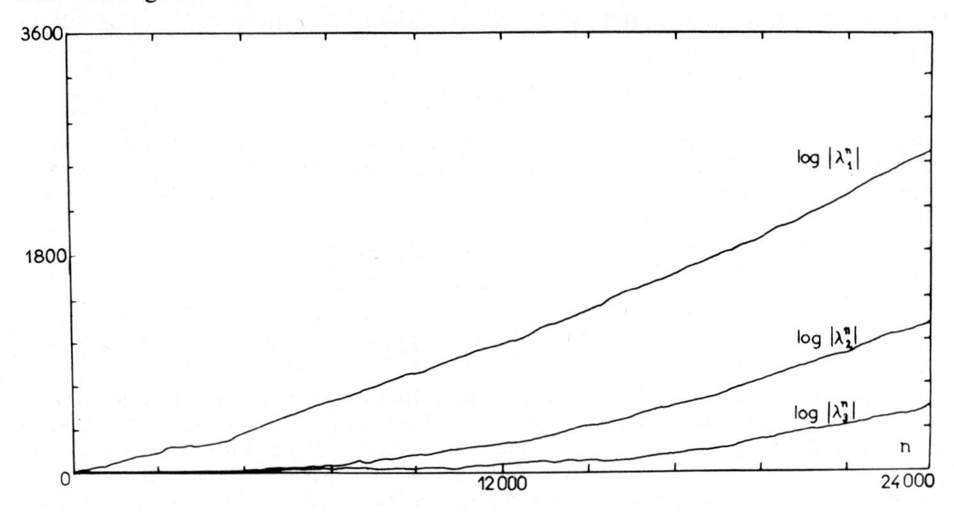

Fig. 10. Variations of $\log_{10}|\lambda_1^n|$, $\log_{10}|\lambda_2^n|$, $\log_{10}|\lambda_3^n|$, when, for the corresponding uncoupled case, two isolating integrals exist.

These figures show a sudden change in the slopes of the logs of the third eigenvalue (respectively 2nd and 3rd eigenvalues), and we note also that the values of these slopes are always strictly positive. Hence, the orbits have an ergodic behavior, and T_6 is close to a C-system.

5.3. $N=3$

It corresponds to a purely integrable case. In this case the slopes of the logs of the three eigenvalues remain equal to zero. Thus, these results confirm the phenomena already observed in the case of three degrees of freedom.

6. Conclusions

The results obtained in particular by means of the study of the variation with n of the logs of the largest eigenvalues of the linear tangential mapping T^{n*} of T^n confirm those given by Froeschlé (1971).

The character of C-system of the orbits has been shown to appear even during the diffusion process occuring when the isolating integrals disappear.

Furthermore, it has been shown that the gambler's ruin model is a rather good approximation of the diffusion process.

Finally, using a six–dimensional mapping, we have confirmed numerically the conjecture given by Froeschlé (1971) that a dynamical system with n degrees of freedom has in general either $n-1$ or 0 isolating integrals, beside the usual energy integral.

Acknowledgement

We have the pleasure of thanking Dr Michel Hénon for helpful discussions during this work.

References

Arnold, V. I. and Avez, A.: 1967, *Problèmes ergodiques de la mécanique classique*, Gauthier-Villars, Paris.
Feller, W.: 1971, *An Introduction to Probability Theory and Its Applications*, Vol. 1, Wiley, New York.
Froeschlé, C.: 1971, *Astrophys. Space Sci.* **14**, 110.
Froeschlé, C.: 1972, *Astron. Astrophys.* **16**, 172.
Froeschlé, C. and Scheidecker, J.-P.: 1973a, *Astron. Astrophys.* **22**, 431.
Froeschlé, C. and Scheidecker, J.-P.: 1973b, *J. Comput. Phys.* **11-3**, 423.

DISCUSSION

G. Contopoulos: You have taken the point (x_0, y_0) in the isolating region and (t_0, z_0) in the ergodic region. I would like to ask: (a) What happens if (t_0, z_0) is in the isolating region also? Did you find any indication of Arnold diffusion? (b) What happens at the transition region, as (t_0, z_0) goes from the isolating to the ergodic region?

C. Froeschlé: (a) If (z_0, t_0) is also in the isolating region we have found that if the coupling term is not large the points remain on a two-dimensional manifold of the four-dimensional space (x_0, y_0, z_0, t_0). We have not found any indication of Arnold diffusion. (b) This case has not been studied. We start either in the isolating region or in the ergodic region.

J. Moser: What is the order of time considered in your numerical experiments? I want to comment that recent calculation by a physicist at Brookhaven has indicated that for about 10^5 periods one had satisfactory bounds for the solution although after that to about 10^7 periods definitive deterioration was observed.

C. Froeschlé: The order of magnitude was about 2×10^5 periods.

G. Contopoulos: I would like to report on some recent work by a group of theoretical physicists in Milan, Italy, under Prof. Scotti. They studied the motion of N (non-linear) coupled oscillators ($N = 10$). At the same time they calculated analytically N formal integrals of motion, using a computer programme that I developed a few years ago. They found that as the energy (or the coupling constant) increases there is a threshold, above which the motion becomes ergodic. This threshold seems to remain finite (different from zero) as the number of degrees of freedom increases. Thus they conjecture that, in general, as N becomes large, the motion does not become ergodic if the coupling is below certain limit.

SUBJECT INDEX